S0-AVT-599

GENE LESSARD

Grandfather's Tale

The Story of Human Evolution and Dispersal

ISBN: 1-4776-1810-4
ISBN-13: 9781477618103

TABLE OF CONTENTS

LIST OF FIGURES

List of Tables

PREFACE

***

With all the great books out there that deal with human evolution and colonization, you could justifiably ask why another? I confess a frustration with all these books because they tend to deal with only a small part of the entire story — that portion that the authors are most familiar with. I found no books that illuminate the entire story. Often the authors spend an inordinate amount of verbiage in describing their travels to and from the site, the dingy hotels they had to stay in, the bars and restaurants with their often colorful cast of characters or the bureaucratic nightmares they had to go through to stick a shovel in the ground. You'll find none of that here. What you will find is a fairly detailed summary of the most important work of these authors. I have simply put the pieces together making it possible to tell the entire story. I hope I've done them justice!

My hope is that when you've finished this book you will pick up any number of the references listed and get a much more in depth discussion of the various authors' particular pieces to the story. Even more importantly, I hope you enjoy the entire story I've put together for you.

Before I launch into our journey, I have a few pieces of business intended to enhance your enjoyment of this book. Archeology is mostly about bones and stones. Archaeologists have developed a basic set of guidelines for excavating the past. Fagan (2003) outlined them fairly concisely:

"Clearly defined stratigraphy, and a clear understanding of the stratigraphic context of the finds, and the formation of the layers in which they were found.

Reliable and consistent radiocarbon dates, or dating established by some other widely accepted chronological method.

If possible, field and laboratory evidence from other disciplines to support the chronological and geological context. A good example of such evidence would be concordant pollen samples.

The presence of humanly made artifacts in a primary stratigraphic context, artifacts that are established as being of human manufacture according to strict scientific criteria."

These guidelines work well for bones and stones. But bones and stones are only one aspect of today's anthropological investigations. And, I certainly don't mean to denigrate the archaeologists work. Without their hard work and diligence, we would not know any more now than we did 150 years ago. However, genetics, linguistics, climatology, psychology, sociology, and a host of other disciplines, are providing the necessary holistic approach called for in these guidelines. "Operationally, holistic endeavors utilize two scientific principles of evidence evaluation — parsimony and concordance. The best single hypothesis will be the one that integrates the most information (holistic), with the fewest number of assumptions (parsimony), and where several lines of evidence lead to the same or similar inferences (concordance). In addition, the quality and types of evidence should meet reasonable standards of robusticity, redundancy, independence, and consensus. Reality, ultimately, is that which we agree it is (Turner 2002)." We'll see these guidelines at work throughout this book.

A brief note about taxonomy, the naming of species, is in order early on in our discussion. Taxonomy provides the greatest source of confusion in our evolutionary discussion. Species names are given to differentiate groups of organisms that are unable to mate successfully and reproduce, or, if they do produce offspring, the offspring are often sterile. Species names are given to separate one group from another closely related group of organisms. The convention for naming species is as follows: *Genus name / species name / varietal name.* Modern man would be named *Homo sapiens sapiens.* There is not always a variety within a species as in *Australopithecus africanus.* For a collection of species within the same genera you'll find *Genus name / spp.* as in *Homo spp.* When you first change a name or propose a name change you'll often find the original name in parenthesis following the change. Thus you might see *Homo (Australopithecus) africanus.* Sometimes you'll find the genus name abbreviated to save time and space such as *H. sapiens sapiens* or *A. africanus.* There will appear to be a contradiction when we speak of, say, *Homo spp.* sharing a common ancestor with *Australopithecus spp.* Obviously, by definition, *Australopithecus spp.* parents did not sire *Homo spp.* children. The boundary between two species is a fuzzy one designed strictly to note that a significant change has occurred.

Species naming is just a convenient way for scientists to put things into boxes but it is not without its flaws. Sometimes things don't neatly fit into convenient boxes. My dad used to say in jest "If it doesn't fit, get a bigger hammer." Taxonomic units "... are sometimes applied according to idiosyncratic proclivities of particular researchers, rather than according to testable hypotheses that can be independently validated or invalidated (Lynch and Granger 2008)." The "bigger hammer" approach to naming species and putting them into an evolutionary context has led to the development of a linear model for human evolution. "Think about the first specimen of *Homo habilis* to be born. Her parents were *Australopithecus.* She belonged to a different genus from her parents? That's just dopey! Yes it certainly is. But it is not reality that's at fault; it's our human insistence on shoving everything into a named category. In reality, there was no first specimen of any species or any genus or any order or any class or any phylum. Every creature that has ever been born would have been classified – had there been a zoologist around to do the classifying – as belonging to exactly the same species as its parents and its children (Dawkins 2009)." Evolution does not fit in boxes. Kingdon (1993) said it well. "It is wiser to regard fossils as guides for a known progression than to appoint them as actual ancestors."

Please remember the following when you read this book! I'll try to clarify this notion where appropriate. I have highlighted these two sentences so you just cannot miss them.

Our taxonomic system simply does not work for discussions of evolution. What happens in the continuum between boxes is much more significant than what happens in the boxes themselves!

One of the reasons I like Richard Dawkins is that he understands this and is upfront about it in his discussions on evolution. He coined the term "concestor" to help reduce the confusion when we speak of ancestors. He talks about a rendezvous in time as a point where we meet our various concestors. He is the only author I have found to date that provides a clear discussion of the taxonomic woes we face in moving from the Australopithecines to modern *Homo spp.*

Evolution is a continuum of change and is not always linear. This linear model for evolution from ape to humans has always left me with more questions than answers. It seems with each new discovery it is more and more difficult to "hammer" the various puzzle pieces together to maintain the continuity of the model. It appears to discount any possibility for co-evolution for something like bipedalism. The consensus seems to be if an animal walks upright it is automatically "hammered" into the puzzle come hell or high water. Instead, I'll follow the lead provided by Jonathan Kingdon (2003) in proposing that there were at least three lines of evolution of bipedalism that could have led to modern humans. And they all developed in different habitats. However, only one was successful! And, as it turns out, he/she wasn't the best walker of the bunch!

Carolus Linnaeus, a famed Swedish botanist, gave modern man his scientific name *Homo sapiens*. He then proceeded to divide the species into four separate subcategories: red Americans, yellow Asians, black Africans, and white Europeans. Linnaeus was probably not "racist" per se, but his categories later formed the basis for racism. He described *Homo sapiens americanus* as "ill-tempered ... obstinate, contented, free"; *Homo sapiens asiaticus* as "severe, haughty, desirous"; *Homo sapiens afer* as "crafty, slow, foolish"; and, *Homo sapiens europaeus* as "active, very smart, inventive." He obviously showed a distinct bias toward Europeans from whom he believed he was solely descended. It would have been more appropriate had Linnaeus named the species Homo curiositus — curious man. It is this very same curiosity or "need to know" that has proven there is no justification for racism. Human populations really are *tous parents, tous differents* — all related, all different. "The main genetic differences are between individuals and not between populations, or so-called 'races'. Differences of genetic origin among the latter are not only small but also superficial, attributable mostly to responses to the different climates in which we live. Moreover, there are serious difficulties in distinguishing between genetic and cultural differences, between nature and nurture (Cavalli-Sforza 2000)."

***

Often, books of this type are written by people, mostly journalists, immediately outside of the disciplines of anthropology or archeology, who use personal interviews to create the story line. I'll use the individual scientist's writings to put together my tale. You'll see extensive use of quotations. I prefer that the scientists themselves tell the story as much as possible in lieu of my interpretation of their work. You'll also note that I will have a preference for the work of some scientists over others. I take it as my prerogative to do so. Good examples are the works of Jonathan Kingdon, Richard Dawkins, Clive Finlayson and numerous others who take an ecological versus the bones and stones approach to the subject. In Finlayson's words "I have pursued my primates' highly specific habitats and places, not consigned them to generalized expanses." Being an ecologist myself, I find it easier to relate to their theories and proposals than those of the "bones and stones" folks.

***

I'll follow a somewhat different tact in the structure of the book. The book is divided into two separate parts - Part 1, On Human Evolution and Part 2, On Conquering the Planet. Each chapter will describe the state of science behind human development for various eras. Each chapter will be followed, as the title implies, by a discussion based on the oral tradition of my 10th great Grandfather. This discussion will be based on appropriate Native American stories and myths. At the end of Chapter I my Grandfather will introduce himself and tell the Algonquin Creation Myth.

Part 1 — In Chapter I, I'll provide an introduction to some basic evolutionary concepts with a much abbreviated discussion of early evolution from the beginning to about 5-7 million years ago. Chapters II through IV — will deal primarily with human evolution from about 6.0 million years ago until about 100 thousand years ago.

In Chapter II, Our First Ancestors — Ape-man or Man-ape, I'll explore our common ancestry with the apes. I'll also discuss the evolution of bipedalism and the evolution of our very first ancestors from about 6.0 to 2.5 million years ago. A proposed lineage will be developed based on the anatomical and ecological approach developed by Kingdon (2003).

In Chapter III, Transition to Humanity, I'll continue the story of our ancestry with the first truly Human ancestors (*Homo spp.*) from about 2.5 to 1.0 million years ago. My emphasis will be on the next big achievement of humanity - the development of a large brain and its social and cultural implications.

In Chapter IV, The Road to Modern Man covers the period from about 1.0 million years ago to 100 thousand years ago. I'll continue to explore our lineage from our beginning with *Homo habilis* through the emergence of *Homo sapiens* in southern Africa.

Part 2 — Chapters V through VII will cover man's conquering of the planet from their migrations out of Africa about 2 million years ago to the settling of the Americas and first contact by Native Americans with Europeans.

In Chapter V, Who Wants to Stay and Who Wants to Go? I'll discuss migrations out of Africa with an emphasis on the concordant information available through genetics and linguistics.

In Chapter VI, The Settling of the Americas, I'll continue the migration discussion with the various theories on the settling of the Americas beginning with the "Clovis First Theory" and continuing on with two alternative theories centered on Pacific and Atlantic coastal migrations.

And, finally in Chapter VII, The Great Extinctions and the Rise of Cultural Diversity in the Americas, I'll discuss three major events in the settling of the Americas: The extinction of "charismatic megafauna", the development of agriculture and the coming of Europeans with the intent to stay on.

Olson (2002) noted that "Modern humans seem to have a unique capacity and desire to learn new things. That doesn't mean the early moderns learned to speak - or learned anything else in a single moment of illumination. Their new behaviors must have emerged in stages. Maybe we are the first creatures on earth driven by a need to know." "As Richard Dawkins puts it, 'Intelligent life on a planet comes of age when it first works out the reason for its own existence' (in Kingdon 2003)." It is this same need to know that drove me to write this book. And, it is likely your need to know that drove you to read it.

PART 1

ON HUMAN EVOLUTION

Chapter I.
Early Beginnings

To-morrow, and to-morrow, and to-morrow,
Creeps in this petty pace from day to day,
To the last syllable of recorded time;
And all our yesterdays have lighted fools
The way to dusty death. Out, out, brief candle!
Life's but a walking shadow, a poor player
That struts and frets his hour upon the stage
And then is heard no more. It is a tale
Told by an idiot, full of sound and fury
Signifying nothing.

William Shakespeare, Macbeth (Act 5, Score 5)

Humans have been asking questions about their origins and their place in nature ever since true self-awareness began to flicker in the human mind, we can be sure of that. Every society in the modern and recent world has, or had, an explanation of how its people come to be, encapsulated in creation myths, stories that explain its people's place in the world.

Carl Swisher, Garniss Curtis and Roger Lewin, Java Man

Primitive man is known to us by the stages of development through which he has passed … through our knowledge of his art, his religion and his attitudes towards life … [and] through the medium of legends, myths and fairytales.

Sigmund Freud, Totem and Taboo

When you have eliminated the impossible, whatever remains, however improbable, must be the truth.

Sherlock Holmes, in Sir Arthur Conan Doyle's, The Sign of Four

This book is about the sum of many tomorrows that mark the passing of human events and is directed by the slow, grinding, petty pace of natural selection; of once lit candles, snuffed out after a brief moment on the stage of human evolution to be heard no more; of yesterdays that led to dusty death. Some would say it is a "tale told by an idiot, full of sound and fury signifying nothing." I think not. We can't really know who we are until we understand where we came from - until we understand our roots. It does seem at times that the arguments waged by the scientist in this tale are indeed full of sound and fury, signifying nothing. But you need to read between the lines, get beyond the petty arguments, and be your own judge of what seems reasonable and what does not! Sherlock Holmes would have balked at such a daunting task as to try to make heads or tails of our evolutionary past. Science and deductive reasoning continue to enlighten our heritage. We will probably never have all the answers to many of the currently unanswered questions. However, science works on approximations and probabilities and each new line of evidence improves our odds of understanding - often sending us back to the drawing board and starting anew.

The way of science is to propose a possible hypothesis for a particular theory, develop a strategy for collecting the data and testing it as rigorously as possible to either confirm or reject the hypothesis and, finally, publish the results. "Paradoxically, this scientific method works not because experiments prove facts, but because one by one, they rule out alternatives. The more we observe, the more we can dismiss some explanations that don't fit all the observations. What you're left with is the set of possible real facts. Sherlock Holmes said it well: 'When you exclude the impossible, then whatever remains, however improbable, must be the truth'. [This] web of observations constrains the possible facts ever more tightly, until almost all are ruled out, and the real explanation emerges (Lynch and Granger 2008)." Such is the case for gravity. No one can honestly believe that the theory of gravity is not fact. Shoot an arrow into the air and it will come down. I know this for sure because I have tested it myself. Armed with the calculus and knowing the angle the arrow is shot at and the initial velocity, etc. you can reliably predict where it will come down. Unfortunately for my early test as a youngster, I knew nothing of the calculus and my intended target remained without a mark on it. I did take particular pride in the next Sunday's chicken dinner even though grandmother was not necessarily pleased with my hunting prowess. The same is the case for evolution. Evolution is not to be believed or disbelieved. It is fact. "Evolution, although resisted by those who may have been offended by the suggestion that their ancestors swung from trees, at least had a comforting punch line: though we humans evolved from apes, we had evolved into something sublime, with powers unlike any other animal (Lynch and Granger 2008)."

"Mankind is a self-domesticated animal; an ape, a social ape; an ape in which the male takes the initiative in courtship and females usually leave the society of their birth; an ape in which men are predators, women herbivorous foragers; an ape in which males are relatively hierarchical, females relatively egalitarian; an ape in which males contribute unusually large amounts of investment in the upbringing of their offspring by provisioning their mates and their children with food, protection, and company; an ape in which monogamous pair bonds are the rule but many males have affairs and occasional males achieve polygamy; an ape in which females mated to low-ranking males often cuckold their husbands in order to gain access to the genes of higher-ranking males; an ape that has been subject to unusually intense mutual sexual selection so that many of the features of the female body (lips, breasts, waists) and the mind of both sexes (songs, competitive ambition, status seeking) are designed for use in competition for mates; an ape that has developed an extraordinary range of new instincts to learn by association, to communicate by speech, and to pass on traditions. But still an ape (Ridley 1993a)." For those who refuse to believe that all humanity shares a common ancestor with the apes - get over it.

***

Let's note at this point that Darwin's Theory of Evolution was just that 150 or so years ago. It has been tested over and over again since then. It is now accepted as fact. Ridley (1993a) states rather emphatically that we have indeed made the transition. "What was the secret that the serpent told Eve? That she could eat a certain fruit? Pah. That was a euphemism. The fruit was carnal knowledge, and everybody from Thomas Aquinas to Milton knew it. How did they know it? Nowhere in Genesis is there even the merest hint of the equation: Forbidden fruit equals sin equals sex. We know it to be true because there can only be one thing so central to mankind, sex." And evolution, again, is all about sex. It's about passing your genes on to the next generation. It is also about culture – nature vs. nurture - as we'll see in coming chapters. "Human nature is a product of culture, but culture is also a product of human nature, and both are the products of evolution (Ridley 1993a)." "There is no nature that exists devoid of nurture; there is no nurture that develops without nature. To say otherwise is like saying that the area of a field is determined by its length but not its width. Every behavior is the product of an instinct trained by experience (Ridley 1993a)."

In his book "Nature via Nurture" Ridley (2003b) is more definitive. He notes that "… [i]t is genes that allow the human mind to learn, to remember, to imitate, to imprint, to absorb culture, and to express instincts. Genes are not puppet masters or blueprints. Nor are they just the carriers of heredity. They are active during life, they switch each other on and off; they respond to the environment. They may direct the construction of the body and brain in the womb, but they set about dismantling and rebuilding what they have made almost at once – in response to experience. They are both cause and consequence of our actions. Somehow the adherents of the 'nurture' side of the argument have scared themselves silly at the power and inevitability of genes and missed the greatest lesson of all: the genes are on their side."

Physicists have an "...'anthropic' notion that the very laws of physics themselves, or the fundamental constants of the universe, are a carefully tuned put-up job, calculated to bring humanity eventually into existence. It is not necessarily founded on vanity. It doesn't have to mean that the universe was deliberately made in order that we should exist. It need mean only that we are here, and we could not be in a universe that lacked the capability of producing us. As physicists have pointed out, it is no accident that we see stars in our sky, for stars are a necessary part of any universe capable of generating us. Again, this does not imply that stars exist in order to make us. It is just that without stars there would be no atoms heavier than lithium in the periodic table, and a chemistry of only three elements is too impoverished to support life (Dawkins 2004)."

Did God create mankind directly? I don't believe so. I believe mankind is the product of about 3 billion years of evolution. Did God create the environment necessary for evolutionary events to occur? I believe so. I believe God created all the rules by which the universe should function - all the laws of physics, chemistry and biology. She then engineered the "Big Bang", or something like it, and allowed her rules to take effect. She stood back and admired her handy work and has not again interfered in the process. You may not share these same beliefs. And, that is perfectly fine. It could have happened in some other way. I just cannot imagine a better way. Evolution is not directed. It makes slight changes in organisms and not always for the better. It builds on itself and often produces functional but not necessarily aesthetic or optimum organisms. We are an excellent example of that. I am sure God would have created a better me. One that didn't end up with clogged arteries, and knees that needed replacing and an aching back that was never intended to support an upright organism. But here I am. I would never ask you to change your personal belief system, but, simply, to keep an open mind as you read through the book. Our belief system need not ensure or negate evolution as fact. Science does not have all the answers and probably never will. But that is what makes science so exciting. The act of discovery is what drives scientist to new levels of understanding.

Darwin in *On the Origin of Species by Means of Natural Selection* proposed the unthinkable at the time: "Maybe humans weren't created a few thousand years ago by a benevolent God but instead were descended from earlier forms of life, including primitive cavemen. The idea was horrifying to Victorian England as it is to many Christian fundamentalists today. As the Bishop of Worcester's wife famously exclaimed at the time 'Descended from apes, my dear? Let us hope that it is not true, but if it is, let us pray that it will not become generally known' (Olson 2002)." Darwin didn't do us any favors with his proposal and deniability in God. After all, if God created all the creatures he certainly created our ancestors! The Bishop's wife was wrong thinking in that to bury your head in the sand is the answer to difficult questions. Evolution need not conflict with religious beliefs. The Bible was never intended to be a treatise on science or history. If you take the Bible as a literal description of the history of mankind, you've totally missed the point of the Bible as a teaching device. Instead, using stories, it provides a guide for individual and social behavior. These "rules" for behavior will be shown to be important for the early development of kinship relationships and group interaction. In fact, the Bible tends to provide tantalizing information to energize man's curiosity. I'll give one example. In the Old Testament in Genesis we find two twin brothers born to Isaac and Rebekah and grandsons of Abraham and Sarah. These sons were named Jacob and Esau. Esau is described as a hunter while Jacob was known as a shy and simple man. In Genesis 25:19-25 Esau emerges from the womb with Jacob grasping his heel. He is described as follows: "Now the first came forth, red all over like a hairy garment; they named him Esau." Jacob was described as "light and smooth." In Genesis 25:22-23 Rebekah is concerned about the activity of the two brothers in her womb. Genesis states: "And the children struggled together within her; and she said, if it be so, why am I thus? And she went to enquire of the Lord. And the Lord said unto her, two nations are in thy womb, and two manners of people shall be separated from thy bowels; and the one people shall be stronger than the other people; and the elder shall serve the younger." So you have two totally different people born of the same mother - one hairy all over and a hunter, the second light and smooth and more of a thinker. Are we being introduced to the evolution of mankind? How can these twin boys be so totally different in appearance and personality? We'll see this same theme reappear in the Algonquin Creation Myth. It will unfold as the story of our evolution.

I can see God's hand in the entire universe. The universe is too complex and too beautiful not to see her hand in all aspects of existence from the beginning until the present. It has been said that nature is more complex than we think. In fact, it is more complex than we can think. However, it is not irreducibly complex. A quote from Albert Einstein taken from Dawkins (2006) best explains my personal belief system. "I don't try to imagine a personal God; it suffices to stand in awe at the structure of the world, insofar as it allows our inadequate senses to appreciate it."

***

Dawkins (1976) introduces us to the concept of the "replicator". "The fundamental unit, the prime mover of all life, is the replicator. A replicator is anything in the universe of which copies are made. Replicators come into existence, in the first place, by chance, by the random jostling of smaller particles. Once a replicator has come into existence it is capable of generating an indefinitely large set of copies of itself. No copying process is perfect, however, and the population of replicators comes to include varieties that differ from one another. Some of these varieties turn out to have lost the power of self-replication, and their kind ceases to exist when they themselves cease to exist. Others can still replicate, but less effectively. Yet other varieties happen to find themselves in possession of new tricks: they turn out to be even better self-replicators than their predecessors and contemporaries. It is their descendants that come to dominate the population. As time goes by, the world becomes filled with the most powerful and ingenious replicators."

These replicators formed in the primordial soup billions of years ago. Concentrations of organic molecules stimulated by some form of energy such as ultra-violate light from the sun provided the stimulus for the development of replicators. How and why this happened is a mystery of science! "We know a great deal about how evolution has worked ever since it got started, much more than Darwin knew. But we know little more than Darwin did about how it got started in the first place. It could have been an event of supreme rarity. It only had to happen once, and as far as we know it did happen only once." God's hand? This rare happening "… is true whether you think the original spontaneous generation was a natural or a supernatural event. The question of just how rare an event the origin of life was is an interesting one… Dawkins 2009)."

Dawkins (1976) notes that "In the beginning was simplicity. It is difficult enough explaining how even a simple universe began. I take it as agreed that it would be even harder to explain the sudden springing up, fully armed, of complex order – life, or a being capable of creating life. Darwin's theory of evolution by natural selection is satisfying because it shows us a way in which simplicity could change into complexity, how unordered atoms could group themselves into ever more complex patterns until they ended up manufacturing people. Darwin provides a solution, the only feasible one so far suggested, to the deep problem of our existence."

"As life emerged from the primeval soup several billion years ago, the molecules that caused themselves to be replicated at the expense of others became more numerous. Then some of those molecules discovered the virtues of cooperation and specialization, so they began to assemble in groups called chromosomes or run machines called cells that could replicate these chromosomes efficiently. In just the same way little groups of agriculturists joined with blacksmiths and carpenters to form cooperative units called villages. The chromosomes then discovered that several kinds of cells could merge to form a supercell, just as villages began to group together as tribes. This was the invention of the modern cell from a team of different bacteria. The cells then grouped together to make animals and plants and fungi, great big conglomerates of conglomerates of genes, just as tribes merged into countries and countries into empires (Ridley 1993a)."

"The 'Catch 22' of the origin of life is this. DNA can replicate, but it needs enzymes in order to catalyze the process. Proteins can catalyze DNA information, but they need DNA to specify the correct sequence of amino acids. How could the molecules of the early Earth break out of this bind and allow natural selection to get started. Enter RNA.

"RNA belongs to the same family of chain molecules as DNA, the polynucleotides. It is capable of carrying what amounts to the same four code 'letters' as DNA, and it indeed does so in living cells, carrying genetic information from DNA to where it can be used. DNA acts as the template for RNA code sequences to build up. And then protein sequences build up using RNA, not DNA, as their template. RNA is their genetic molecule, solely responsible for carrying genetic information from generation to generation.

"Now for the key point of the 'RNA World theory' of the origin of life. In addition to stretching out in a form suitable for passing on sequence information, RNA is also capable of self-assembling ... into three-dimensional shapes, which have enzymatic activity. RNA enzymes do exist. They are not as efficient as protein enzymes, but they do work. The RNA World theory suggests that RNA was a good enough enzyme to hold the fort until proteins evolved to take over the enzyme role, and that RNA was also a good enough replicator to muddle along in the role until DNA evolved (Dawkins 2009)."

There are two different sources of DNA in each of us. The first is nuclear DNA. This is the DNA that we pass down to our children - half from the mother and half from the father. We have 23 pairs of genes - each pair is numbered 1-22 and a 23rd pair that is labeled XY for the male and XX for the female. DNA is a very long filament containing various arrangements of four nucleotides: adenine, guanine, cytosine, and thymine referred to simply as A, G, C, and T. Any combination of three of these four nucleotides forms a codon. A single codon produces an amino acid and combinations of amino acids form proteins. There are only 20 amino acids that can be produced from these 64 possible codons. Therefore several different codons may specify the same amino acid - built in redundancy. Codons are hooked together like words in a sentence. These sentences or strings of codons are called genes. Therefore, you could have a chain that looked something like GCAATGGCC. A gene is a segment of DNA that has a specified, recognizable biological function. Each gene on a chromosome shares a similar gene on its opposing pair. These are referred to as alleles. As we'll see, alleles are rivals of each other for being passed on to the next generation.

The second source of DNA is found in an organelle in each of our cells called the mitochondria. The mitochondria produce the energy needed by the cell to perform its functions. Mitochondria are the descendants of bacteria that took up life inside another single celled organism more than a billion years ago. They've lived happily in this symbiotic relationship within their host providing energy to the host plants and animals in return for a place to "hang their hats" - so to speak. There is substantially less DNA in the mitochondria than you'll find in the 23 nuclear pairs of DNA. Mitochondria have about 16,500 base pairs while nuclear DNA has about 3 billion base pairs.

The spiral helix of DNA is composed of chromosomes and, it is from these chromosomes that we get genes. Dawkins (1976) defines a gene as "… any portion of chromosomal material that potentially lasts for enough generations to serve as a unit of natural selection." And a unit of natural selection must have the following properties: "… longevity, fecundity, and copying fidelity." He further explains that "Another aspect of the particulateness of the gene is that it does not grow senile; it is no more likely to die when it is a million years old than when it is only a hundred. It leaps from body to body down the generations, manipulating body after body in its own way and for its own ends, abandoning a succession of mortal bodies before they sink in senility and death."

"The genes are immortals, or rather, they are defined as genetic entities that come close to deserving the title. We, the individual survival machines in the world, can expect to live a few more decades. But the genes in the world have an expectation of life that must be measured not in decades but in thousands and millions of years (Dawkins 1976)."

Another property essential to Dawkins' (1976) discussion is that "... at the gene level, altruism must be bad and selfishness must be good. This follows inexorably from our definitions of altruism and selfishness. Genes are competing directly with their alleles for survival, since their alleles in the gene pool are rivals for their slot on the chromosomes of future generations. Any gene that behaves in such a way as to increase its own survival chances in the genepool at the expense of its alleles will, by definition, tautologously, tend to survive. The gene is the basic unit of selfishness." A gene gets itself into the genepool "... by helping to program the bodies in which it finds itself to survive and to reproduce. But now we are emphasizing that 'it' is a distributed agency, existing in many different individuals at once. The key ... is that a gene might be able to assist *replicas* of itself that are sitting in other bodies. If so, this would appear as individual altruism but it would be brought about by gene selfishness."

***

Dawkins (1976) recognizes that there is a "... tension between individual organisms and genes as rival candidates for the central role in natural selection." He goes on to note that "One way of sorting this whole matter out is to use the terms 'replicator' and 'vehicle'. The fundamental units of natural selection, the basic things that survive or fail to survive, that form lineages of identical copies with occasional random mutations, are called replicators. DNA molecules are replicators. They generally,... gang together into large communal survival machines or 'vehicles'. The vehicles that we know best are individual bodies like our own. A body, then, is not a replicator; it is a vehicle. I must emphasize this, since the point has been misunderstood. Vehicles don't replicate themselves; they work to propagate their replicators. Replicators don't behave, don't perceive the world, don't catch or run away from predators; they make vehicles that do all those things. For many purposes it is convenient for biologists to focus their attention at the level of vehicle. For other purposes it is convenient for them to focus their attention at the level of the replicator. Genes and individual organisms are not rivals for the same starting role in the Darwinian drama. They are cast in different, complimentary, and in many respects equally important roles, the role of replicator and the role of vehicle."

Evolution is all about sex; about reproduction; about passing your genes on to the next generation. It takes two to tango. Ridley (2003a) notes that "... there is no other way for evolution to work except by competitive reproduction. Those strains that reproduce persist; those that do not die out." He goes on to note that "... Richard Dawkins of Oxford University effectively invented the notion that because bodies do not replicate themselves but are grown, whereas genes do replicate themselves, it inevitably follows that the body is merely an evolutionary vehicle for the gene, rather than vice versa. If genes make their bodies do things that perpetuate the genes (such as eat, survive, have sex, and help rear children), then the genes themselves will be perpetuated. So other kinds of bodies will disappear. Only bodies that suit the survival and perpetuation of genes will remain."

"Within each species some individuals leave more surviving offspring than others, so that the inheritable traits (genes) of the reproductively successful become more numerous in the next generation. This is natural selection: the non-random differential reproduction of genes. Natural selection has built us, and it is natural selection we must understand if we are to comprehend our own identities (Robert Trivers in the forward to Dawkins 1976)."

"All along the tortuous path that leads back from us to ever earlier ancestors were people who had to fill their stomachs and with exquisite spasms of sexual chemistry pass on their genes. Time and again they faltered, as drought, poisoned ash and a magnitude of hazards conspired to destroy their frail substance. Fossil bones and footsteps and ruined homes are the solid facts of history, but the surest hints, the most enduring signs, lie in those miniscule genes. For a moment we protect them with our lives, then like relay runners with a baton, we pass them on to be carried by our descendants. There is a poetry in genetics which is more difficult to discern in broken bones, and genes are the only unbroken living thread that weaves back and forth through all those boneyards (Kingdon 1993)."

"But the uniqueness of the individual is only the first of the implications of sex [gender] for human nature. Another is that there are, in fact, two human natures: male and female. The basic symmetry of gender leads inevitably to different natures for the two genders, natures that suit the particular role of each gender. A third implication of sex for human natures is that every other human being alive is a potential source of genes for your children. And we are descended from only those people who sought the best genes, a habit we inherited from them. Therefore, if you spot somebody with good genes, it is your inherited habit to seek to buy some of those genes; or, put more prosaically, people are attracted to people of high reproductive and genetic potential – the healthy, the fit, and the powerful. The consequences of this fact, which goes under the name of sexual selection, are bizarre in the extreme… (Ridley 1993a)."

If we constrained ourselves to feeding and fornicating, we would not be where we are today as a species. As we evolved we came to a point where it was no longer an advantage to behave simply as an individual. Altruism became beneficial in the game of survival. We developed kinship bonds to hold families together. We formed groups with mutual interests often beyond the kinship group. We developed larger brains opening up new avenues to technological development, language, art, etc. – to culture.

Societies were formed from all of these adaptive attributes. However, the key to adaptation is still the replicator – the gene – and the vehicle – the individual. "Economists and sociologists who believe that individuals will usually act in their collective rather than their particular interests ('from each according to his ability, to each according to his needs' [The creed of socialism] versus 'Devil take the hindmost') are soon confounded. Society is composed of competing individuals as surely as markets are composed of competing merchants; the focus of economic and social theory is, and must be, the individual. Just as genes are the only things that replicate, so individuals, not societies, are the vehicles for genes. And the most formidable threats to reproductive destiny that a human individual faces come from other human individuals (Ridley 1993a)."

"The evolution of technology in prehistory may have proceeded by fits and starts but it has a recognizable theme and direction. The adaptation of technology mimics adaptation in evolution, in that every improvement or refinement of a tool enables it to proceed faster or take in a wider arc than its predecessors. It is commonly said that early humans broadened their niche; instead it might be truer to represent the invention of each new technique as the acquisition of a new niche and its addition to exclude and eventually replace a succession of other species (Kingdon 1993)."

"The more skills or 'niches' that humans have appropriated, the more versatility and flexibility has to be learnt. This is in marked contrast to any other animal. All the lower animals are rigidly programmed to perform their genetically predetermined roles. More developed species augment this with behavior learnt by imitation of the adults. In apes, the youngster's main incentive for imitating its mother seems to be sharing her food and keeping close to her. By contrast, a prime incentive for a human display of imitative skill seems to be parental praise (something that is rewarding to both child and parent).

"One of the peculiar features of history is that time always erodes advantage. Every invention sooner or later leads to a counterinvention.

Every success contains the seeds of its own overthrow. Every hegemony comes to an end. Evolutionary history is no different. Progress and success are always relative. This concept that all progress is relative, has come to be known in biology by the name of the Red Queen, after a chess piece that Alice meets in *Through the Looking-glass*, who perpetually runs without getting very far because the landscape moves with her. It is an increasingly influential idea in evolutionary theory ... The faster you run, the more the world moves with you and the less you make progress (Ridley 1993a)."

"Because so many technical skills have to be learnt, children are constantly competing among themselves or for their elders' attention. When this social dimension of learning is projected back to the earliest forms of *Homo* the need to teach simple techniques to children stands out as one of the most significant forces that might have shaped human evolution. Indeed, it may be in the biological origins of our earliest education techniques that humans can most truly be said to be self-made (Kingdon 1993)."

"Skills in the teaching and learning of tool-use imply an increase in the capacity (and therefore the volume) of the brain because Stone Age education, no less than any other, must have been a process and a program. Programs allow both the execution and learning of more complex techniques to proceed from a baseline of simpler ones. It is not necessary to posit language for the earliest human polytechnics; nonetheless the charades that would have been needed to communicate must have had the germs of a gestural language in them.

"In an extraordinary mimicry of natural adaptations, these primates began to manipulate elements and use materials in a way that rapidly multiplied the number of ecological niches they could invade. Each new tool opened possibilities that were formerly the prerogative of very specialized animals. Where diggers had needed heavy nails, now there were stone picks, cats no longer had the monopoly of sharp claws, spears mimicked horns, porcupine quills or canine teeth and so on. Here, for the first time, was an animal that was learning a multiplicity of roles via the invention of technology. An increasing number of animals now had a new competitor that would encroach on at least a part of their former niche. In some cases (perhaps some of the scavengers) the overlap may have been so great that the hominids took over (Kingdon 1993)."

***

"Food and food gathering are central to the study of ecology and are certainly no less relevant to human prehistory and understanding what humans are. What sort of eaters humans have been has been depended very largely on what sort of catchers they have been. A great part of human prehistory concerns change, innovation and diversification in food getting techniques, but first came the challenges that wild animals and plants presented to the ingenuity of early humans (Kingdon 1993)."

"There are four distinct but interrelated influences that would have made feeding ecologies differ very significantly from region to region. Firstly, humans have inbuilt nutritional needs and physiological limitations that must be met. These influence both the choice of foods and the situations in which they can be collected. Some foods and some places need adapting to and, once achieved, such specialization can become very specific and self-perpetuating. Secondly, the particular choice of food that would have been available in different parts of their range would have influenced ways of life and types of technology and here, too, established habits could have encouraged conservatism in diet and economy. Thirdly, there would have been constraints in terms of defenses put up by the plants and animals and limitations imposed by the places in which they were found. Diseases, predators, nuisances and direct competition from other animals would also have helped to define and maintain what foods could or could not be eaten in any one region. Fourthly, technology was the key giving access to otherwise closed food supplies. Technology also helped to overcome some of the constraints (Kingdon 1993)." The distribution of these resources and how our ancestors utilized them would be a key ingredient of our evolutionary development.

In ecology there is a term called ecotone. An ecotone is the overlap of two or more different ecological systems - often referred to as edge. It is in this edge or ecotone between systems that we are likely to find the maximum diversity of plants and animals. Finlayson (2009a) refers to the "innovators" that live in the edge as the most likely to adapt to change and evolve, though gradually, to the next level of development. Life is generally tougher in the edge because familiar resources are scarce and, therefore, requires more innovation just to survive, than in the core habitat where the resources you are familiar with and dependent on are more abundant. Those that live within the relative safety of the core habitat, the "conservatives", are less likely to adapt to change, particularly if the change comes quickly, and are most likely to become extinct.

Innovation in and of itself may not be enough to ensure survival of the group. Stringer (2012) notes that "... the best strategy for the average person in a variable environment might be to look around and rely on imitation, rather than individual learning. Some people might discover ways of coping with the changes, and, by imitation, this would provide a shortcut to success for the many. Moreover, the more copiers there were, the more chance that one of them might not copy accurately and by chance produce an accidental improvement on the original. If individual innovation was rare, progress would be slow – hence, perhaps, the 2 million years of boredom of the Lower Paleolithic. But if there were more innovators, the process could produce rapid cultural adaptations very quickly compared with individuals innovating in isolation, or with biological evolution. Thus through imitation and peer-group feedback, populations could adapt well beyond the abilities of an isolated genius, whose ideas might never get beyond his or her cave or might be lost through a sudden death."

Finlayson (2009b) notes that "... these edge people perceived their environment in a different way from those in the stable core. For them the critical resources that they needed for life, be it food, water, or shelter, were less evenly distributed across the landscape than in the center so it was harder for them to know where the good bits were. Not only were these good patches unevenly spaced out, they were also unpredictable in where they turned up. So the environment of the innovator was patchier in space and time than it was for the conservative. Any change that allowed the innovators to pick up cues that made them better at pining down the elusive and ephemeral resources would have been immediately favored. The less predictable resources became in space and time, the greater the pressure to find ways of improving flexible systems of detection that could allow for quick reaction." He adds "We should not forget that we are the product of marginal people who had to do a lot of improvising to get by."

I'll borrow these terms - innovator and conservative - as they are very descriptive of the actions taken by our ancestors - those that succeed and those that do not.

In some respects the history of mankind is more about how we went about feeding ourselves than about some conjured up taxonomic break down of various species. As Kingdon (2003) put it "Diets and the elaboration or invention on novel techniques to obtain food are central to understanding how a forest ape of restricted distribution expanded both its range and its foraging repertoire to become *Homo*, living in a variety of habitats." A major component of learning what is and is not edible may have developed from a technique coined by Kingdon (2003) as "niche stealing". Niche stealing involves watching what other animals are eating, trying their food out for taste, and if you like it, utilizing this new source of food for yourself - even to the point of chasing the original feeders away. Our evolutionary food line might look something like the figure below.

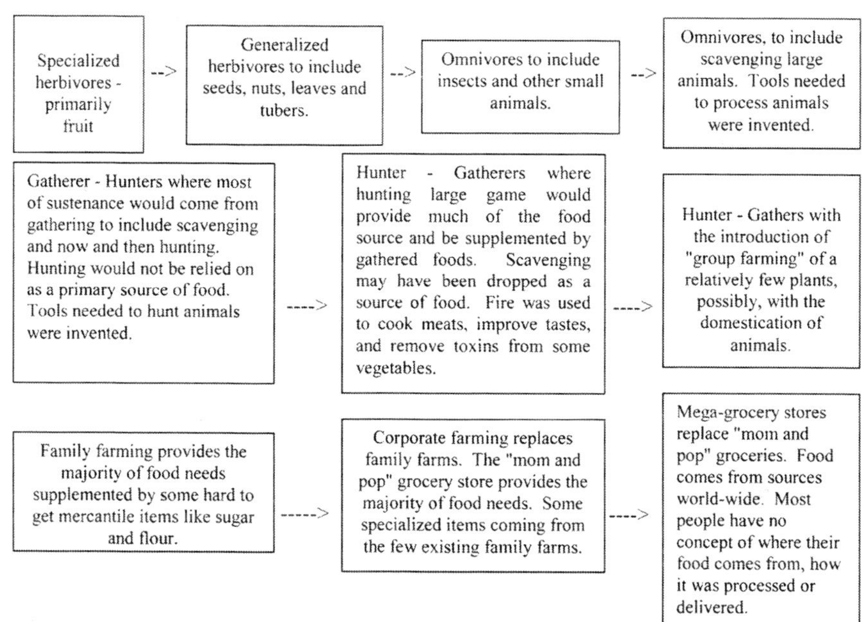

Figure1-1. Evolutionary food line

The first line in this scenario begins with the "Gatherers" and likely covered the first 3-4 million years of man's evolution. The second line would cover the remaining period up to about 10,000 years ago with the introduction of agriculture. We have a long history of "living off the land". The last line provides a look at the evolution of food production and our near total dependence on someone else to feed us. I'm sure there's an often thought but seldom asked question, "Does hamburger come from ham?" Or, "Does chocolate milk come from chocolate cows?" My favorite is "How do they get the pits out of maraschino cherries?"

***

Dawkins (2004) avoids the self-centered humanistic approach to the evolutionary history of mankind by working backwards in time to reveal each of our shared ancestors which he calls "Concestors". Each point in time is called a rendezvous. The first rendezvous of concestor

"0" is all humankind. All life shares a common ancestor at concestor 39. Some of these concestors were found on land and others in the sea.

He could have chosen any plant or animal as concestor 0 and achieved the same endpoint. This backward approach reduces us to the same level as any other species. The route back, however, would differ until we met with our common concestor. "Biological evolution has no privileged line of descent and no designated end. Evolution has many millions of interim ends (the number of surviving species at the time of observation), and there is no reason other than vanity – human vanity as it happens, since we are doing the talking – to designate any one a more privileged or climactic place than any other (Dawkins 2004)."

I'll provide a few high points in this pilgrimage – as Dawkins (2004) calls it. However, I'll work forward in time. At rendezvous 39 (concestor 39) are the eubacteria – all life's common ancestor. Moving forward in time we reach rendezvous 37 with the eukaryotes. Dawkins (2004) notes that "...one of the most momentous events in the history of life was the formation of the eukaryotic cell. Eukaryotic cells are the large complex cells, with walled nuclei and mitochondria that make up the bodies of all animals, plants and, indeed, all pilgrims who have so far joined us." Fast forward at this point to Concestor 17, the amphibians. Concestor 17 is the common ancestor of all tetrapods – animals with four feet. This was the first of ancestors that marks the "... emergence of fish onto the land [and] was a major transition in our evolutionary history (Dawkins 2004)." At concestor 15 we meet the monotremes – the first mammal-like reptiles. At rendezvous 9 we meet the culugos and the tree shrews. Dawkins (2004) notes that this is "... still in the time of the dinosaurs and before the flowering of mammalian diversity properly began. Actually, the flowering of flowers themselves had only just begun. Flowering plants, while diverse, had been previously restricted to disturbed habitats such as those uprooted by elephantine dinosaurs or ravaged by fire, but by now had gradually evolved to include a range of forest-canopy trees and understory bushes." Passing from rendezvous 9 to rendezvous 8, lemurs, bushbabies and their kin, we cross the K/T boundary "...which separates the Age of the Mammals from the much longer Age of the Dinosaurs that preceded it. The K/T was a watershed in the fortunes of the mammals. They had been small, shrew-like creatures, nocturnal insectivores, their evolutionary exuberance held down under the weight of reptilian hegemony for more than 100 million years. Suddenly the pressure was released and, in a geologically very short time, the descendants of those shrews expanded to fill the ecological spaces left by the dinosaurs (Dawkins 2004)." At rendezvous 7 we meet the tarsiers, at 6 the new world monkeys and at 5 the old world monkeys. It's at this point that the geneticists are able to shed more light on our ancestry.

Geneticists were able to measure differences in the amino acid sequences from apes and old world monkeys that were believed to have diverged about 30 million years ago. They used this information to establish a mutation rate of one amino acid per 1.25 million years in the protein albumin. They then compared changes in the amino acid sequence of albumin in humans and African apes, and found that humans and African apes diverged 5 to 7 million years ago. This time period corresponds with the timeline for the evolution of Kingdon's ground-ape. They also found that African apes (gorillas and chimpanzees) are more closely related to humans than to the Asian apes (orangutans) which broke off the line of apes about 10 to 13 million years ago. All human populations are descended from a single source population in Africa. We'll develop this theme throughout the book. For those who refuse to believe that we are all descendants of a common ancestor from Africa - get over it. It is from this juncture that we will begin our journey.

Grandfather's Tale

I am the Manitou. I live in all ages through many ancestors and descendants. I am telling my story through my grandson of ten generations in the future. My knowledge is his knowledge and I, in return, am informed through him. We are one and the same as in all past generations. I am the keeper of the past, the foreseer of the future and the interpreter of the present.

Traditionally, a Manitou can be either a spirit or a human with spirit powers. Humans with Manitou powers can often take the form of a rock, tree, or a star. The spirit Manitou usually appears as an animal, a rock, a tree, a star, or a sun, but can also take human shape. I am a human Manitou. Manitou powers can be used for evil but, should be used for the good of the community. Both spirit and human Manitous can have power over each other. The one that stands for good is often the more powerful. A mother who lives isolated from other humans and is impregnated by the sun will produce offspring that, inevitably, pose a threat to human and Manitou communities alike.

Time for Manitous is different than it is for humans. A Manitou day is equivalent to a human year - daytime is summer and night time is winter. Manitous should only be spoken of in the winter while they are sleeping so that they won't hear the discussion and become offended.

I am called Roch Manitouabeouich the son of Ancient Manitouabeouich. My father was born an Algonquin of the Western Abenaki in a place known today as Vermont. My father was, as was his father, a Manitou and passed that right to me. He was captured and adopted by the Huron, a band of the Iroquois and taken to Huronia. I was born in the late 1500's in what would be later called La Conception Mission, Ossossane, Huronia, Canada. My people are the Attignawantan or people of the bear. I speak both Iroquoian, my adopted tongue and Algonquian, my inherited tongue. I married Outchbahabanouk Oueou around 1636 and we had two children together - a son and a daughter.

My story is the story of my people. It will describe the journey of my people over millions of years and how that journey was affected by the land we lived in and our adaptations to our environment. But let me start at the beginning of creation with the story called Moskim (the rabbit).

Moskim

Above the clouds they lived. The chief had his lodge in the middle of the village. He had a wife, and he had a child, a little girl. One day he began to brood. He declared there was something unclean. He was jealous. He desired something. Everyone tried to enlighten him. No one could guess what this chief might desire. After a while, a man spoke up. He announced, "Now I really believe this might be what he wants. Perhaps he's thinking, 'I ought to pull up this tree of mine, then I ought to set it over there.' "

Then everybody came. Then these people pulled up this big tree. Then the ground fell out through the opening. All the people saw how the ground opened up, right while everyone was looking at it. And the chief too. He told his wife - he said "come on, let's go over to this hole, where everybody's looking!" Then the woman picked up her child and put the little girl on her back. Then they set off at once. They got there, and right away the chief looked into the hole. Then he said to his wife, "Come on! You look, too!" The woman said, "Alas! I'm frightened." "Come on now, take a look!" So the woman stepped forward, and she looked. The chief said, "Look closely, get down there!" She put her shawl between her teeth. Then she took hold of the ground there, on the far side and on the near side. She took hold on both sides - closed her fists on some huckleberry bushes with the soil attached. Then she bent her head far down. Right away the chief grabbed his wife and pushed her all the way in. Then the woman was really flying.

While they were traveling through the clouds, she met the fire serpent. He said to her, "You poor thing! I'm the one he's jealous of. That's actually why he wanted to kill you." He put his hand in his side and took out some corn, an ear of corn - just one. And a bone, a beaver bone. And a little kettle. And a mortar. And a little corn pounder. That might be all.

Then the manitous and spirit powers held a council and said, "Who will watch out for this poor one?" They said, "Maybe the sunfish?" And he spoke up: "I can take care of that person!" Then the others declared, "The sunfish couldn't take care of anything. He's an ugly one." They said, "The pike fish. He could be the one." Then the pike fish spoke up: "I'm a good man. I could take care of her forever." The manitous answered, Not you. You couldn't do it. You're too silly, too ill-tempered. You couldn't do it at all." Then next the turtle - the good one - spoke up: "I can take care of that person forever." "You're the one who can do it, all right." Then in fact he came over and raised his back, and the woman landed there.

Then, as she wept, she said to her child, "My little one, truly we are poor." Then she saw that she had some dirt in her hands, and she let it fall to the ground. What she spread out there grew larger. The more she spread it, the more it grew, and eventually it got much bigger. Then it kept on growing bigger and bigger. And before long this woman saw everything growing. Eventually trees grew, and potatoes, and grass. And whatever this woman might say, that's what would happen. Before long she thought, "The sun! There it is!" And immediately it came about. So there it was, the light of day. It walks the road each morning, that light of day. Then she spoke again. "Well," she said, "There are stars, too." Then it really happened. Then she used her hand and pointed to where the stars are. And she declared, "Now these are the Bear Followers - the Chasers. And these are the Bunched Ones - in wintertime, thanks to them, people will know how much of the winter has gone by."

Well then, the little girl was grown. Now, this girl, as she went out to play, found some woods where she could swing. When she grew tired, she would lie down. It felt very good, because of the wind. Now really, the way the wind was blowing she enjoyed it, it made her feel as if it might be inside her. One day the old woman took a good look at her and wondered. She thought, "This daughter of mine looks pregnant." Then she questioned her. She asked, "Didn't you ever see anyone? No one at all?" asked the old woman. But then it must have been so. Indeed, later on, twins grew in the belly of this girl. It was not long before these two - these twins - began to quarrel, arguing over which one would be born first. The one who was in position to be first born spoke up and asked his younger brother, "Which way would you come out?" "Right here at my mother's navel," he answered. Moskim said, "You'll kill our mother," and then he said to him,

"From now on you'll be called Flint." "So be it," said Flint. "As for you, you'll be called Moskim, because the mooshkiingw *likes to jump right out and creep into its burrow. From now on you'll be Moskim."*

When Moskim was born, the old woman was very glad. And then another one. And so Flint was born. In fact, when he was born, he went straight out. He was born through his mother's navel. He cut through it. When Flint was born, the young woman died. Then the old woman cried. When she had quite finished crying, the old woman dug a grave. As she laid her daughter's head toward the west, she spoke to her encouragingly. She made her daughter hear her. She told her, "You are the first to go. You prepare the road to the cold winds above the clouds. Everyone will go when they're finished here. That's where they'll go. Now you go first. That's where all will go. It will be that way forever here."

Chapter II.
Our First Ancestors - Ape-man or Man-ape
(6 to 2.5 Million years ago)

Even if it be granted that the difference between man and his nearest allies is as great in corporeal structure as some naturalists maintain, and although we must grant that the difference between them is immense in mental power, yet the facts ... appear to declare, in the plainest manner, that man is descended from some lower form, notwithstanding that connecting-links have not hitherto been discovered.

Charles Darwin, The Decent of Man

Our story begins in Africa about 6 million years ago (*Mya*) during the geological epoch known as the Miocene which lasted from 23.0 to 5.3 *Mya* and continued into the Pliocene (5.3 to 2.6 *Mya*). The Miocene was preceded by the warmer Oligocene (34.4 to 23.0 *Mya*). The Miocene epoch was a very active period in geological history. The continental plates were moving and creating mountains in the western part of North America, the Andes in South America, and throughout Europe and Africa. The Indian subcontinent was on the move and it's collision with Asia began the creation of the Himalayas. Uplift in East Africa produced the Great Rift Valley. The Great Rift Valley runs through Ethiopia, Kenya, Tanzania, and ending finally in Mozambique. Fissures of magma created domes of mountains in Ethiopia and Kenya. "New lakes and river systems were formed, while mountain ranges - pushed up on either side of the Rift Valley - dramatically altered climate continuity to the west, the east became drier, an effect that was probably intensified by the development of a seasonal monsoon-driven climate in the Indian Ocean, a meteorological effect produced by the rise of the Himalayas (Stringer and McKie 1996)." These geological forces that had been slowly splitting East Africa from the rest of the continent began to make a telling impact on the landscape changing the rain forest east of the Rift to woodland and grassland. These changes provided a constant background of environmental changes that have played a crucial role in driving evolution.

The ancestors of today's African apes contentedly thrived in the forested center and west, but hominids became increasingly isolated in the less densely forested side of East Africa. The mighty rivers that ran north-south down the Rift Valley only increased the hominids isolation (Stringer and McKie. 1996). As Coppens (in Stringer and McKie 1996) noted we are "unquestionably the pure product of a certain aridity."

The Pliocene was a period of global cooling. However, the average temperature was 2-3 degrees centigrade (4-5 degrees F) warmer and sea levels were 25 meters (82 feet) higher than today. The change to a cooler, drier seasonal climate led to a decrease in tropical species world-wide. Deciduous forests proliferated, coniferous forests and tundra covered much of the north and grasslands and savanna spread on all continents except Antarctica. Tropical forests were limited to a narrow band around the equator and deserts appeared in Africa and Asia. The continents continued to drift during the Pliocene (as they do today) as much as 112 miles over this 2.7 million year period (approximately 0.4 miles per 1000 years). North America and South America were joined together by the Isthmus of Panama. This linkage had a profound influence on global climate. The warm equatorial currents that flowed between the Atlantic and Pacific were cut off by the Isthmus leading to a cooling of the Atlantic Ocean by cold waters from the Arctic and Antarctic Oceans. Africa's collision with the European continent created the Mediterranean Sea by cutting off the remnants of what has been labeled as the Tethys Ocean.

There was a really amazing series of events that occurred during this period. The Messinian Salinity Crisis was a geological event that caused the Mediterranean Sea to dry up. The area that was the precursor to the Strait of Gibraltar closed up preventing flow of ocean water from the Atlantic into the Mediterranean Sea from about 5.96 to 5.33 *Mya*. During this period the Mediterranean Sea nearly completely desiccated into a deep dry basin in some places about 2 to 3 miles below sea level. Only a series of high salinity lakes would remain. Temperatures at the bottom of this basin could have reached as high as 176 degrees F. By comparison, Death Valley is 286 feet below sea level and the maximum temperature on record is 134 degrees F. About 5.33 *Mya* a dramatic event occurred. "Just as the great rivers of the Sahara cut deeply into the ground as they drained the mega lakes into the salty basin of Lake Cyrenaica in the eastern Mediterranean, in the extreme west a river was also eroding the land. But this river was not draining a lake; instead it collected rainwater from the nearby wet Atlantic coastal areas. Slowly, the river cut back towards the largest body of water of the region - the Atlantic Ocean. One day it reached the level of the Atlantic, thousands of meters above the dry western Mediterranean Basin, and oceanic water began to trickle in. For 26 years it was a mere drip but as the new channel opened, it became a torrent and the world saw the birth of a super-waterfall that gushed water into the hot and dry abyss 3 kilometers below. In 10 years the entire western basin had reached the level of the Atlantic and it then pored over into Lake Cyrenaica to fill the eastern basin in a year. The new Mediterranean would alter the climate of Europe and northern Africa, making it more arid and modern types of desert, semi-desert, and arid grasslands started to spread (Finlayson 2009a)." For comparison Niagara Falls is divided into the Horseshoe Falls and the American Falls. The Horseshoe Falls drop about 173 feet, the height of the American Falls varies between 70–100 feet because of the presence of giant boulders at its base. The larger Horseshoe Falls are about 2,600 feet wide, while the American Falls are 1,060 feet wide. The super-waterfall would have expanded to about the width of the Strait of Gibraltar, many miles wide, and had a drop of more than 50 times that of Niagara Falls. The sounds and fury must have been incredible!

***

The widely accepted ecological scenario that was the root cause of our evolution is as follows. A change in global ecosystems and climate would have a profound effect on the plants and animals that inhabited East Africa creating a mosaic of environments. The tropical forests to the west were rich in fruits - the primary food supply of the apes. The apes could therefore afford to be specialists in their food gathering and consumption in this tropical environment. With the creation of a mosaic of woodland and grassland environments, a specialist would be at a disadvantage with a declining amount and a wider more sparse distribution of fruits. A generalist who could easily move to areas where fruit was concentrated and also was more adapted to eating other foods, such as nuts, seeds and tubers, would gain a distinct advantage over the specialist. Kingdon (2003) notes that "Step by step, the predetermined behavior of a species with a single niche must have given way to the new competence of a species that could acquire multiple niches through an ever-expanding armory of technology, techniques, and eventually systems of communication to back them up." It was this change from specialist to generalist that triggered a number of changes leading, ultimately, to the evolution of humankind. We are the ultimate generalist! Although this explanation seems simple and straightforward, it doesn't take into consideration the large number of species in varying habitats that the archeologists have discovered in the past 100 or so years.

Kingdon (2003) takes an ecological approach to the evolution of mankind. "Previous approaches to drawing our family tree have focused on anatomical differences among the various fossil hominins. With each new discovery, these differences get re-examined and the system either redrawn and/or controversy erupts (most of the time) leading to unsolved difficulties." Finlayson (2009a) expresses the process more dramatically: "When something doesn't fit a long-established scheme, the usual response is disbelief, which rapidly turns to ridicule." Kingdon's (2003) approach takes into consideration these anatomical differences and the "detailed geographic, ecological, temporal, and genetic contexts" to address the environment our ancestors lived in at the time. "It is therefore significant to know whether a species, living or extinct, is newly evolved and actively expanding its range; or if it represents a stable, longer established form (Kingdon 2003)." He describes his basic concept as follows: "Decisive functional adaptation to living in drier habitats was less likely to have taken place in coastal forests than along the banks of distant inland rivers. During favorable climatic phases, the natural routes of population expansion would have been inland up broad forested valleys. The most likely times for early expansions away from the coast were during humid periods; just such a phase has been tentatively dated at 6 *Mya*, and another set of moist conditions is thought to have climaxed about 4.8 *Mya*." In addition "... a prime limitation on actively evolving bipeds would have been a dependence on gallery forests, rivers, and streams. Part of this limitation might have lain in the physiology of a shade- and moisture-dependent animal, but the main restriction would have been the need for exceptionally rich and concentrated food sources." And, "Greater mobility in their descendants could have substantially enlarged lateral ranges without changing the essentially riverine character of their adaptive niche." In summary, our early ancestors that were just beginning to develop bipedalism were living along the East African coast. The ecological conditions would have been similar to those that our ancestors experienced in West Africa prior to the development of the Great Rift Valley. As climatic conditions became favorable they moved upstream to richer more varied environments. With increased ability to walk they moved laterally inland to expand their range of habitats. Finlayson (2009a) introduces us to "conservatives" and "innovators". The innovators are those who are "forced" into the edge environments and must come to terms with living in the more difficult environments than those occupied by the conservatives. When sudden change occurs, they are well adapted to living with limited resources where the conservatives are unable to quickly adapt and fend for themselves - often leading to their extinction.

I'll briefly describe each of these river basins (Figure 2-1) starting in the north and ending in the south.

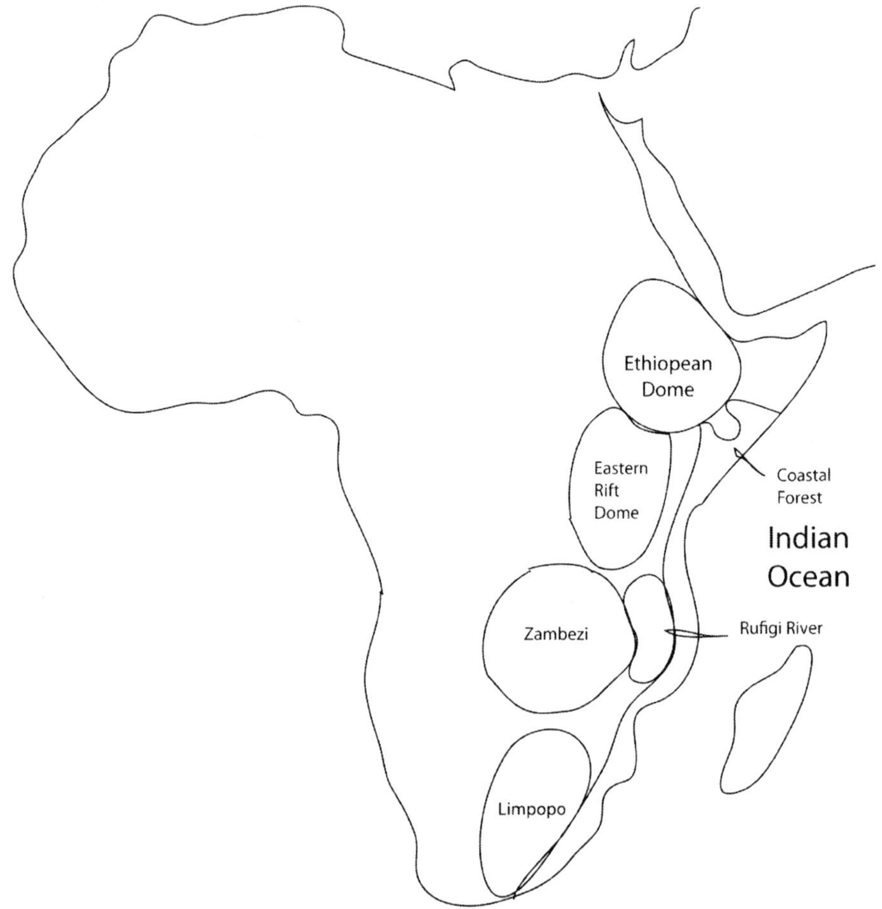

Figure 2-1 Map of Africa showing the Forested Coastal Zone and the 5 major basins draining into the Indian Ocean

1. The core of northeast Africa is the Ethiopian Dome. Two large rivers, the Web Shebelle and the Juba, drain the eastern face to the Indian Ocean just south of the equator. During cold, dry glacial periods the habitat would not be favorable for hominids; however, during warm, wet interglacial periods the habitat would have improved significantly allowing migration into the area from the coast. The Nile River would be a significant deterrent to immigration from the west.

2. The Tana, Galana, and Pangani rivers drain off a much smaller Eastern Rift Dome to the south of the Ethiopian Dome. Though the area is high and dry its temperatures are ameliorated by being on the equator. The main corridor between the Eastern Rift Dome and the Ethiopian Dome is along the shores of Lake Turkana. Kingdon notes that "The Eastern Rift was a close destination for upriver dispersals from the central and richest portion of the coastal forest strip (Opposite Pemba Island). Discovery of the 'Millennium Ancestor' (*Orrorin tugenensis*) near Lake Baringo confirms the central position of the Eastern Rift and 'Great Lakes' region in hominin evolutionary geography."

The first two river basins can be characterized as follows: "The Eastern Rift can be visualized as a north-south split that connects and runs through centers of two gigantic uplifted domes [Ethiopian Dome and Eastern Rift Dome] (those in turn, can be visualized as part of the buckling of Africa into a series of tectonic basins surrounded by plateaus and raised swells) ... The Northern Dome embraced the whole of the Ethiopian highlands and southwest Arabia, all of about 400 km [about 250 miles] across. The smaller but still impressive Eastern Rift Dome centers on southern Kenya and is outflanked by the giant volcanoes: Mt. Elgon and Mt. Kenya in the north, Mt. Kilimanjaro in the southeast, and further to the southwest, the now extinct crater Ngorongoro. Even more than a river valley, the Rift's floors and their surrounding slopes and mountains constitute an environment of unparalleled diversity and richness within a relatively small ambit (Kingston 2003)." Lake Turkana forms a corridor between these two major river basins. To date, the majority of the hominin fossils have been found in these two basins and corridor.

3. The Rufiji River is the largest of a tightly packed succession of tropical river basins. Notable among these basins are the Wami, Ruvu, and Rovuma River and some rivers in Mozambique. These rivers drain the mountains north and east of Lake Malawi. The whole area is close to the coast and would have approximated the ecology of the coast more than other areas. It was easily accessible. Evolution would have been slow and probably not much different than on the coast due

to the similarity of the habitats. No hominin fossils have been found in this basin to date. Since this basin is so similar to the coastal forest, evolutionary pressure would be similar to that in the coastal forest.

4. The Zambezi River, one of the largest in Africa, drains a vast area of flat inland plateau spanning the Tropic of Capricorn between 8 and 20 degrees south. The area is influenced by the climates from both the Indian and Atlantic Oceans. Temperatures range from about 10 C to 30 C degrees - about 58 to 86 degrees F. Rains occur from November through April (summer in the Southern Hemisphere) and drainage is slow with a substantial amount of ground water. Riverine habitats are common along the many slow moving streams, swamps and lakes. Waterlogged soils inhibit tree growth and favor grasslands. The headwaters of this river contain one of Africa's largest "edaphic [determined by soil conditions] grasslands". The area would have been easily accessible for early hominins however, once there, if climate became significantly drier, they would have been cut off from the coast by the long narrow neck of the river. The habitat would have facilitated evolution; however, no hominid fossils have been found to date in the area - primarily because the area has been overlooked by archeologists. Kingdon (2003) postulates that this basin was the original homeland of "Lucy", *Australopithecus afarensis*, and the robust Australopithecines, *Paranthropus spp.*

5. The Limpopo Valley marks the northern border of a radically different complex of South African rivers all draining the Drakensberg mountain chain. The area is in the temperate zone located between 21 and 34 degrees south latitude. The river systems in the numerous mini-river basins are short, steep, and hard to access - reaching elevations of 10,000 feet. The combination of high altitude and a more temperate climate provides for a wider range of temperatures and consequent seasonality. Southern winters are short (from May to August) with night time temperatures dropping below freezing in the inland portion of the basin. Summers are long and hot with rainfall ranging from 63 in. on the escarpments to14 in. or less in the west and northeast. With such a high degree of microclimates throughout the

basin, a large number of vegetation types have been described. The differences between the habitats in the interior and the coast are significant. However, because of the short distances traveled by early hominids and the ability to go up and down through various ecological zones with changes in seasons may have been a driving factor influencing evolutionary trends between migrants and those that stayed in the coastal area. There is an endemic fossil record within the basin. An adaptation to cold environments would have been one of the main advantages to those who entered and stayed in the upper reaches of the Limpopo Valley - a trait crucial to later migrations out of Africa. This basin would have provided far greater challenges to *Australopithecus africanus* and *Australopithecus sediba* than the Zambezi did to Lucy and the robust Australopithecines.

Where did Kingdon's hypothetical ground-apes come from? How did they get to the coastal zone of East Africa? How is it that they transitioned from primarily arboreal apes to primarily "ground apes"? The ground-ape would likely have come from west-central Africa. During cold dry periods this area was isolated from East Africa by desert conditions. There were two of these dry periods that peaked at 6.2 and 5.3 *Mya*. However, during warm moist periods the coastal forest would have expanded up the river drainages and the deserts would have been replaced by habitats more favorable to the apes that were the precursor to chimpanzees and humans alike. There were three of these warm moist periods when forests were very extensive with peaks at 6.9, 6.0 and 4.8 *Mya*. There then would be a corridor for the apes to move through from west-central Africa to East Africa. Creation of the Great Rift Valley provided a natural barrier to movement of ape populations between west and east Africa but, certainly not an impenetrable one. Natural selection provided the impetus which allowed the evolution of these separate populations - one line leading to the chimpanzees in West Africa and a second line, eventually, leading to modern humans in East Africa.

Kingdon (2003) starts with a primitive and unnamed "ground-ape" that lived in a 2500 mile long coastal zone in East Africa bordering the Indian Ocean (Figure 2-1). This zone stretches from just south of the "Horn of Africa" to nearly the "Cape of Good Hope". Throughout the length of this coastal ecosystem there are 12 major river drainages. Along this long, narrow coastal strip Kingdon has identified five major clusters of river basins that drain into the Indian Ocean and that are fairly distinct in their ecology. The recent discovery of *Sahelanthropus tchadensis* from Chad in Central Africa dated to 6 to 7 *Mya* may be a precursor to Kingdon's "ground-ape". *Sahelanthropus tchadensis* nick-named Toumai - a name given to children born at the start of a harsh, life threatening, dry-season - literally means "Hope of Life". Toumai lived along the shores of Lake Chad and may have walked upright. "The skull, holding a brain the size of a chimpanzee's, seems to combine features of an ancient ape with others that seem to anticipate later proto-humans (Finlayson (2009a)."

The entire area along the eastern coast of Africa would be available for colonization. As the climatic conditions changed back again to cool dry environments this new arrival (Kingdon's ground-apes) in East Africa would become isolated from the west-central population. In addition, populations along the coast would become isolated as their habitats were fragmented by the cool arid conditions. At this time "... there was a large ancestral population, which could have been around 50 - 70,000 individuals that slowly split into ... the coastal group and their west-central cousins (Finlayson 2009a)." Kingdon (2002) notes that "... periodic fluctuations in numbers and localized fragmentation within an already small gene pool could have speeded up both [natural] selection and genetic drift." Evolution within these isolated small coastal groups would act independently of each other and that of their west-central African cousins. In addition, over time, natural selection would act on populations differentially across the 2500 mile wide coastal forest. As climate conditions changed from warm-wet to cool-dry individual populations would become isolated across the range of environments from the relatively mild north, through the equatorial regions, to the southern temperate forests. Table 2-1 shows the origins of early hominins proposed by Kingdon (2003) and discussed in this chapter. I've added the recently discovered *Australopithecus sediba.*

Table 2-1. The proposed origin of early hominids by river basin from Kingdon (2003) and geographic distribution and age from Finlayson (2009a)

Location	Hominin species		Years *Mya*
Chad	*Sahelanthropus tchadensis*	Chad	6.0 - 7.0
	Australopithecus bahrelghazali	Chad	3.0 - 3.5
Ethiopian Dome	*Ardipithecus kadabba*	Ethiopia	5.54 - 5.77
	Ardipithecus ramidus	Ethiopia	4.32 - 4.51
	Australopithecus anamensis	Ethiopia, Kenya	3.9 - 4.2
	Australopithecus garhi	Ethiopia	2.5
Eastern Rift Dome	*Orrorin tugenensis*	Kenya	5.72 - 6.1
	Kenyanthropus platyops	Kenya	3.2 - 3.5
	Homo rudolfensis	Kenya	1.4 - 1.9
Rufiji River	None		
Zambezi River	*Australopithecus afarensis*	Ethiopia, Kenya, Tanzania	3.0 - 3.9
	Paranthropus aethiopicus	Ethiopia, Kenya	2.3 - 2.8
	Paranthropus boisei	Ethiopia, Kenya, Tanzania, Malawi	1.4 - 2.5
	Paranthropus robustus	South Africa	1.5 - 2.0
South Africa	*Australopithecus africanus*	South Africa	2.3 - 3.3
	Australopithecus sediba	South Africa	1.78 - 1.95
	Homo habilis	Ethiopia, Kenya, Tanzania, South Africa	1.8 - 2.33

Often we hear that we are descended from chimpanzees. This is not correct. We are, in fact, not descended from the chimpanzee. However, we do share a common ancestor. There is only a 1.6% difference in our genome and that of the chimps. "The often quoted figure of 98 percent for the shared genetic material of humans and chimps actually refers neither to numbers of chromosomes nor to numbers of whole genes, but to numbers of DNA 'letters' (technically, base pairs) that match each other within the respective human and chimp genes (Dawkins 2009)."

When we talk about a divergence in a lineage, I favor the concept developed by Richard Dawkins (2004) (working backward in time from modern humans) in which he discusses a rendezvous where "the ancestors of any set of species must eventually meet at a particular geologic moment". He calls this point a rendezvous and the "last common ancestor that they all share" as their "Concestor". Our rendezvous with the chimpanzees occurred about 6 *Mya*. It is at this point that the apes went one way and the humans went another. Our common Concestor is approximately our 250,000-greats-grandparent. Dawkins (2004) visualizes this moment of meeting "... the chimpanzee pilgrims meet us in some Pliocene forest clearing, and their dark brown eyes, like our less predictable ones [color wise], are fixed upon Concestor 1: their ancestor as well as ours." Is she more chimpanzee-like or more human-like? She was likely more chimpanzee-like since the chimpanzees are closer genetically to the apes than we are. As Dawkins puts it "Humans are the odd ones out among apes, both living and fossil. Which is only to say that more evolutionary change has occurred along the human line of descent from the common ancestor, than along the lines leading to the chimpanzees." The chimpanzees remained essentially chimpanzees until about 3 million years ago when the lineage split to form the chimpanzees on one branch and the bonobos on a second branch.

So what was our common ancestor like? She was:

"... likely hairy like a chimpanzee, and had a chimpanzee-sized brain;

Probably walked on [her] hands (knuckles) like a chimp, as well as [her] feet;

Probably spent time up in trees, but also lots of time on the ground;

Probably used and made tools, following local traditions as modern chimpanzees; and,

Probably was omnivorous, sometimes hunting but with a preference for fruit (Dawkins 2004)."

And, "...surely didn't have such advanced human features as language (Dawkins 2009)."

***

Before we launch into a discussion of each of our cousins, we need to take a look at the evolution of bipedalism. Kingdon (2003) noted that bipedalism was the first distinguishing feature leading from the apes to modern humans - "the central condition on which human evolution is predicated (Kingdon 2003)". Swisher et al. (2002) suggests that bipedalism is just one strategy for adapting to new environments. "A second strategy is to reduce the distance that a group has to travel each day, which is achieved by having smaller social groups. (A large group requires more total food resources each day than a small group, and therefore must travel farther to harvest it). It is a very flexible strategy, because when food is plentiful, social groups can be large, but when it is scarce, large groups can split up into smaller groups." This is the strategy used by chimps. However, gorillas use another strategy. That is, instead of forming smaller groups, gorillas change their diet by adding less nutritious foods like leaves. Richard Leaky (1994) noted that "... the origin of bipedal locomotion is so significant an adaptation that we are justified in calling all species of bipedal ape 'human'. This is not to say that the first bipedal apes species possessed a degree of technology, increased intellect, or any of the cultural attributes of humanity." And further "These humans were not like us, but without the bipedal adaptation they couldn't have become like us."

Bipedalism evolved through a series of "pre-adaptations". Kingdon (2003) gives four requisites to bipedalism: an erect back, balancing on two legs, walking, and running. It is standing, the precursor to these requirements, that is an evolutionary moment. He notes that each of these requisites developed piecemeal over a relatively long period of time. An erect back requires changes in the spine, pelvis, head-neck junction and feet. "It was during this phase that the feet changed from being claspers to becoming platforms. I [Kingdon] hold that it was foraging, mainly on the ground, in a squatting position that demanded these necessary modifications. The ability to stand without the expenditure of much energy requires good balance. Two legged balance, in the brief waddling of apes or the performing of poodles, is a precarious artifice because there is too much weight concentrated at the top of the column and too little stability at the bottom. In other words, these quadrupeds are top-heavy. Easy, non-energetic standing requires a downward displacement in the distribution of weight. If structures in the upper foreparts are no longer the largest and heaviest in the body, their smallest movements will no longer destabilize balance or threaten to topple the would-be walker. To achieve such shifts in the distribution of weight in living, functioning bodies, there must be substantial changes in the relative weight of muscles, bones, and organs. I [Kingdon] contend that this necessary slimming down of the upper part of the eastern ape's bodies came about through ecological and behavioral changes that rendered exceptionally heavy, powerful forelimbs redundant (Kingdon 2003)."

The orangutan and humans shared a trait not shared by the chimpanzees and the gorillas - "Tree-walking". Finlayson (2009a) noted that "All of them can stand upright but when chimpanzees and gorillas do so, the hind limbs are flexed. Orangutans and humans, on the other hand, stand on straight limbs. This way of walking on flimsy branches gives the orangutans great benefits. It can venture out onto the flimsy branches on the outside of the tree crown on its hind legs, transferring the center of gravity as necessary and hanging on, for safety, to other branches with the hands. It can then release one hand that can be stretched out to collect fruit that would not otherwise be accessible. The method of tree-walking is also used to move between trees without having to come to ground. It seems that this is an ancestral form of locomotion present in the ancestor of all great apes." He further explains that "What some proto-humans did was simply to carry the tree-walking way of life onto the ground. Deeper in the forests knuckle-walking did a better job as much of the time was spent in the trees, but once down that route the knuckle-walkers condemned themselves to a forest existence ..." It appears then, that tree-walking was the first pre-adaptation to bipedalism and gave our human ancestors a head start.

So what was the driving evolutionary force that initially led to bipedalism and its rapid progression over time? Of course no one knows for sure. There are, however, many theories. We know that bipedalism has many advantages. Are these advantages the cause or result of bipedalism? Each of these theories has some merit.

The first of these theories relates to the amount of energy consumed by a quadruped versus a biped. While it seems logical enough, studies by Taylor and Rowntree (1973) have shown that there is no significant difference between the amounts of energy required in locomotion by a quadruped versus a biped. Taylor and Rowntree concluded that "The relative energy cost of bipedal versus quadrupedal running should not be used in arguments about the evolution of bipedal locomotion in man." A later study by Ishida (in Kingdon 2003) in Japan, working with trained and untrained Japanese monkeys as models contradicts Taylor and Rowntree. They found a 30 percent saving in energy costs between fully upright and "bent" bipedalism.

Maxine Sheets-Johnstone proposed that bipedalism was a means for the male of the species to "show off his penis". Females, on the other hand were doing it for the opposite reason: concealing her genitals which are more prominent when on all fours. The biggest draw-back is that if there was a significant evolutionary benefit to these displays, why aren't all the apes bipedal creatures?

A whole different set of theories revolve around freeing your hands for other uses such as "bringing home the bacon" or making, carrying and using tools. Dawkins (2004) notes that "Many apes and monkeys feed on plant matter that is widely available but not particularly rich or concentrated, so you must eat as you go, more or less continuously like a cow. Other kinds of food such as meat or large ground tubers are harder to acquire but, when you do find them, they are valuable - worth carrying home in greater quantity than you can eat." Walking on two legs would then free your hands up to carry food home to a mate or to your children, or to trade favors with other members of your group. The best place to store food may have been in the stomach of a companion. For instance, a large amount of meat will surely go bad fairly quickly before you are able to eat it all. Therefore, you then give some to others with the understanding that when you need food they will share their catch with you. "The meat is no longer available, of course, but the goodwill it buys is safe in the long-term storage in a companion brain (Dawkins 2004)." Bipedal locomotion would not likely have evolved based on stone tool manufacture and use, since stone tools didn't come to the fore until about 2 million years passed.

An even more interesting theory was proposed by Owen Lovejoy (1981). Lovejoy "suggests that females would often have been hampered in their foraging by nursing infants, therefore unable to travel far and wide looking for food. The consequent poor nutrition and poor milk production would have delayed weaning. Suckling females are infertile. Any male who feeds a nursing female accelerates the weaning of her children and brings her into receptiveness earlier. When this happens, she might make her receptiveness especially available to the male whose provisioning accelerated it. So, a male who can bring lots of food home might gain a direct reproductive advantage over a rival male who just eats where he finds. Hence the evolution of bipedalism was to free the hands for carrying (Dawson 2004)."

Other theories involve the added height gained when standing on two legs versus all fours. It makes it easier to look over the tall grass for potential predators or to wade across deep streams and keep your head above water. Or bipedalism would limit your exposure to the sun to the top of your head which just happens to have a crop of hair to insulate it. This posture would also allow you to lose more heat from your body compared to a hunched over position where your body is closer to the heat radiating ground.

The squat-feeding posture proposed by Kingdon (2003), in my opinion, leads the pack. Kingdon suggests that squat-feeding, not bipedalism, induced changes in the upper body, backbone, and pelvis of ground apes. "The special legacy of squat-feeding was to disengage a heavy, cantilevered upper body from an equally oblique pelvis, and to rebalance a more lightly built head and thorax vertically over a compact, basin-like pelvis. That these changes could improve balance on two legs would have been an almost accidental bonus, an anatomical by-product." He provides a table of 40 factors that could have directed these anatomical changes over time. Kingdon contends that "only an ape population that was able to exploit an intensive and reliable (rather than an extensive and irregular) food supply could permit a radical shift in the priority functions of their forelimbs. Hands and arms could, most frequently, and for long periods, be devoted to turning over leaf litter, selecting, processing, and handling foods; also, but rather less frequently, to vertical climbing and to some diagonal 'propping' in the trees. As the incidence of bearing weight declined, there would come an identifiable point at which four-legged movement ceased to be as efficient as simple straightening of the legs. I [Kingdon] contend that this point existed when the species was balanced vertically."

This pretty well covers the first two requisites given by Kingdon (2003) for bipedalism: an erect back and balancing on two legs. Dawkins (2004) provides a very plausible theory on the mechanisms that drove the evolution of walking and then likely running. His theory is not only logical but incorporates the concept of sexual selection developed by Charles Darwin in his the *Descent of Man*. Dawkins (2004) notes that "Sexual selection produces quirky, whimsical evolution that runs away in apparently arbitrary directions, feeding on itself to produce wild flights of evolutionary fancy." He goes on to say "The essential point is that male appearance and female taste evolve together in a kind of explosive chain reaction. Innovations in the consensus of female taste within a species, and corresponding changes in male appearance, are amplified in a runaway process which drives both of them in lockstep, further and further in one direction." And, "What matters for our purpose is that sexual selection, according to a sound mathematical theory, is apt to drive evolution to take off in arbitrary directions and push things to non-utilitarian excess." It is the drive in the non-utilitarian direction that is the first component of his theory of the evolution of bipedalism. The next component is the tendency to imitate. The third component is "the widespread habit among apes generally of rising temporarily onto the hind legs, including during sexual and aggressive displays." With the pre-adaptation proposed by Kingdon (2003) of an erect back and ability to balance on two legs, the standing demanded in Dawkins (2004) theory could be quantified by a much more extended duration versus a temporary habit.

Dawkins (2004) puts these parts together in a scenario that could likely lead to bipedalism. "Our ancestors, like other apes, walked on all fours when not up in trees, but reared up on their hind legs from time to time, perhaps in something like a rain dance [performed by chimpanzees], or to pick fruits off low branches, or to move from one squat-feeding position to another, or to wade across rivers, or to show off their penises, or for any combination of reasons, just as modern apes and monkeys do. Then - this is the crucial additional suggestion I [Dawkins] am adding - something unusual happened in one of those ape species, the one from which we are descended. A *fashion* for walking bipedally arose, and it arose as suddenly and capriciously as fashions do. It was a gimmick." He continues, "... an admired or dominant ape ... gained sexual attractiveness and social status through his unusual virtuosity in maintaining the bipedal posture ... Others imitated his gimmicky habit and it became 'cool, 'it', 'the thing to do'..." He goes on to say that "...females in the local area of the fashion preferred to mate with males who adopted the new way of walking. They preferred them for the same reason as individuals wanted to join the fashion; because it was admired in their social group. And now the next step in the argument is crucial. Those who were especially good at the fashionable new walk would be most likely to attract mates and sire children. But this would be of evolutionary significance only if there was a genetic component to the variation in ability to do 'the walk'. And this is entirely plausible. We are talking, remember, about a quantitative shift in the amount of time spent doing an existing activity. It is unusual for a quantitative shift in an existing variable *not* to have a genetic component." This quantitative shift would be in the form of temporary standing to extended standing.

The final step in Dawkins (2004) theory follows standard sexual selection theory. "Those choosers whose taste conforms to the majority taste will tend to have children who inherit, from their mother's choice of mate, skill in walking according to the bipedal fashion. They will also have daughters who inherit their mothers taste in males. This dual selection - on males for possessing some quality and on females for admiring the self-same quality - is the ingredient for explosive, runaway selection... The key point is the precise direction of the runaway evolution is arbitrary and unpredictable. It could have been opposite. Indeed, in another local population perhaps it was in the opposite direction. An explosive evolutionary excursion, in an arbitrary and unpredictable direction, is just the kind of thing we need, if we are to explain why one group of apes (who became our ancestors) suddenly evolved in the direction of bipedality while another group of apes (the ancestors of the chimpanzees) did not."

If you think that Dawkins scenario is a bit far-fetched, just harken back to your high school days and think cheerleaders and football players! Or, consider the admiration shown by many females for the "bad boy" image of James Dean. Or, the wild behavior of many young women when they first saw Elvis "the pelvis" Presley sing and swivel his hips in mock sexual behavior. Ed Sullivan in his 1950's T.V. show refused to show Elvis from the waist down during his appearances.

***

Kingdon (2003) recognizes three distinct lineages where bipedalism was adapted as the primary development toward modern mankind. All three begin with his hypothetical East African Ground Ape. "So far, the only fossils from this period are the 9 million-year-old fragments of the 'Samburu Ape,' *Samburupithecus koptalami,* and the much younger 6-7 million-year-old 'Toumai,' *Sahelanthropus tchadensis,* a short-faced hominid with blunt, somewhat human-like teeth. In its cranial anatomy this fossil from north of Lake Chad provides a rough approximation for the common ancestor of chimps and humans (Kingdon 2003)."

Kingdon (2003) suggests that these lineages evolved bipedalism independently in different basins (Figure 2-2). For our discussion we will assume that all three had a common ancestor in Toumai, *Sahelanthropus tchadensis*. Toumai may have been the very first ancestor to "fix her gaze" on Concestor 1. She may have been the first of our ancestors to begin the transition to upright walking. In the words of Neil Armstrong as he stepped onto the moon "That's one small step for man, one giant leap for mankind." She may be the root of all humanity, that first creature that distinguished herself from the apes. The genesis of each of two of these three lineages may have been through *Orrorin tugenensis* (Millennium Man), or *Ardipithecus kadabba*. Finlayson (2009a) refers to Toumai, Millennium Man and Kadabba as the "Three Amigos". He notes that the paucity of information about the "Three Amigos" from the fossil record leaves the lineage open at this point - they "...may have all belonged to the same genus or even the same species. All three could have been ancestors of ours but maybe none of them were." He notes that "For the long period between 7 million and 5.54 million years ago, we have a handful of partial specimens from three tropical African sites. For some we have a head but no body, for others legs but no head, and for others mainly jaws and teeth. In all, they amount to fewer than twenty individual beings that lived their anonymous lives across this vast land at some point within a one-and-a-half-million-year period of the remote past.

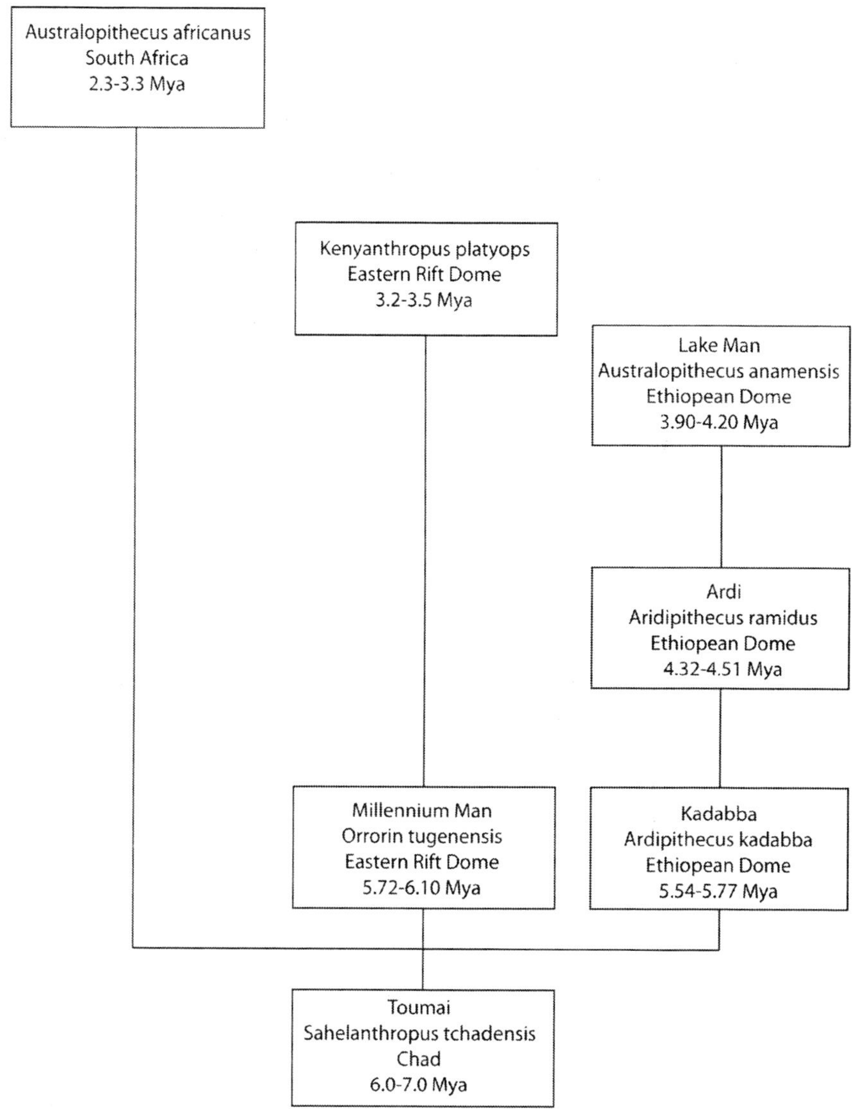

Figure 2-2. Kingdon's (2003) proposed lines of descent for the early hominids

You would think that we would treat these few and incomplete finds with some sense of humility and proportion, but no, despite the dearth of material, the three candidates are hotly contested as our ancestors." A third line of descent, favored by Kingdon (2003), may have been through *Australopithecus africanus.*

"The first Australopithecine to be discovered and the type specimen of the genus, was called the Taung Child. At the age of three and a half the Taung Child was eaten by an eagle. The evidence is that damage marks to the eye sockets of the fossil are identical to marks made by modern eagles on modern monkeys as they rip out their eyes. Poor little Taung Child, shrieking on the wind as you were borne aloft by the aquiline fury, you would have found no comfort in your destined fame, two and a half million years on, as the type specimen of *Australopithecus africanus*. Poor Taung mother, weeping in the Pliocene (Dawkins 2009)."

What would the "Three Amigos" behavior be like? Richard Leakey (1994) describes the following scenario: "Rather than living as aggregations of families in nomadic bands, as modern hunter-gathers do, the first humans probably lived like savanna baboons. Troops of thirty or so individuals would forage in a coordinated way over a large territory, returning to favored sleeping places at night, such as cliffs or clumps of trees. Mature females and their offspring would make up most of the troops numbers, with just a few mature males present. The males would be continually looking for mating opportunities, with the dominant individuals achieving the most success. Immature and low-ranking males would be very much on the periphery of the troop, often foraging by themselves. The individuals in the troop would have the human aspect of walking bipedally but would be behaving like savanna primates." However, Lovejoy (2009) believes that a "major shift in life-history strategy transformed the social structure of [these] early hominids. That shift probably reduced male-to-male conflict and combined three previously unseen behaviors associated with their ability to exploit both trees and the land surface: (i) regular food-carrying, (ii) pair-bonding, and (iii) reproductive crypsis (in which females did not advertise ovulation, unlike the case of chimpanzees). Together, these behaviors would have substantially intensified male parental investment - a breakthrough adaptation with anatomical, behavioral, and physiological consequences for early hominids and for all of their descendants, including ourselves (Lovejoy 2009)."

The oldest of these three Kingdon's lineages begins with *Orrorin tugenensis*, a small individual discovered in 6.0 million year old deposits in the Tugen Hills, above Lake Baringo in Kenya. *Orrorin tugenensis* replaced an earlier hominin *Ardipithecus spp.* discovered in 1994 in the Ethiopian Dome as the oldest of our ancestors - who lived between 3.9 and 5.8 *Mya*. There were fragments of five individuals unearthed: *Orrorin tugenensis* was named "Original Man" and given the nickname "Millennial Man". The location of this discovery was in the Eastern Rift Dome, which is located just south of the Ethiopian Dome. Kingdon (2003) describes *Orrorin tugenensis* as follows: "The hands and arms were essentially those of climbers, so in this respect were still apelike. The three femurs [from the original fragments of 5 individuals], instead, were quite unlike those of apes, and a well preserved femoral pelvis that was more similar to that of a biped than that of a quadrupedal ape. The vertebral column has been deduced to have been vertical. The relatively small, thickly enameled teeth are more human than apelike in both shape and alignment."

Orrorin tugenensis is the likely ancestor for *Kenyanthropus platyops* who was discovered on the shores of Lake Turkana in the Eastern Rift Dome, in Kenya, in 2001. She lived about 3.5 *Mya*, and may be the immediate ancestor of all *Homo spp. Kenyanthropus platyops* was a small-toothed creature with a brain the size of a chimpanzee. She "has so many facial and dental resemblances with the highly specialized but big brained [575 cc] *'Homo' rudolfensis* that the latter is almost certainly its descendant and so might even end up being called *Kenyanthropus rudolfensis*."

A second line of descent may have been through *Ardipithecus spp* to *Australopithecus anamensis* to *Australopithecus afarensis*. *Ardipithecus kadabba* began life about 5.8 *Mya*. We know very little about

Kadabba from the fossil record, just a few teeth and pieces of bone have been found to date. Kadabba is the likely immediate ancestor of *Ar. ramidus*, often referred to as Ardi and the only other member of the genus. Ardi lived in the Middle Awash River valley of Ethiopia somewhere around 3.9 to 4.5 *Mya*. The name stems from the Afar language, in which ardi means *ground or floor* and ramid means *root*. *Pithecus* is Greek for *ape*. "*Ar. ramidus* was a denizen of woodland with small patches of forest. We have learned that it was more omnivorous than chimpanzees (ripe fruit specialists) and likely fed both in trees and on the ground. (White et. al. 2009a)." In addition, "This woodland graded into nearby habitats that were open and are devoid of fossils of *Ardipithecus* and other forest-to-woodland community mammals (White et al. 2009 b)."

Ardi had a very small brain (300-350 cc), about 20% of the size of modern humans. As noted she was able to walk upright but likely could not walk far or run in an upright position. The rest of her anatomy would need significant evolutionary change to allow development of full upright walking and all its benefits. Though her feet were made for walking, she was still quite adept at life in the trees and likely retreated to the safety of the trees when threatened. Her anatomy points to a number of significant differences between her and her immediate ancestors. "... some of *ramidus's* features are apelike: relatively small back (molar) teeth with only a thin enamel covering on their crowns, features found today in gorillas and chimpanzees, who generally eat soft foods (mainly fruits) which require crushing but not a lot of chewing. This was probably the diet of *ramidus*, in this case. In contrast, the large surface area of the back teeth and the thicker layer of resistant enamel of later hominids suggest we changed to food which was either much more abrasive, or needed a lot more chewing, or both, such as nuts, seeds, and tubers (Dawkins 2004)." Their hands "... show that *Ardipithecus* did not knuckle-walk like African apes and that it lacked virtually all of the specializations that protect great apes hands from injury while they climb and feed in trees. *Ardipithecus* also shows that our ability to use and make tools did not require us to greatly modify our hands. Rather, human grasp and dexterity were long ago inherited almost directly from our last common ancestor with chimpanzees. We now know that our earliest ancestors only had to slightly enlarge their thumbs and shorten their fingers to greatly improve their dexterity for tool-using (Lovejoy et. al. 2009b)."

"The picture emerging from *Ar. ramidus* is that this last common ancestor had limb proportions more like those of monkeys than apes. Its feet functioned only partly like those of apes and much more like those of living monkeys and early apes such as *Proconsul* (which lived more than 15 million years ago). Its lower back was mobile and probably had six lumbar vertebrae rather than the three to four seen in the stiff backs of African apes. Its hand was unpredictably unique: not only was its thumb musculature robust, unlike that of an ape, but its midcarpal joint (in the wrist) allowed the wrist to bend backward to a great degree, enhancing its ability to move along tree branches on its palms. None of these changes that apes have evolved to stiffen their hands for suspension and vertical climbing were present, so its locomotion did not resemble that of a living ape (Lovejoy et. al. 2009a; Lovejoy et. al. 2009c)." The Ardipithecines disappeared from the fossil record about 4.4 *Mya*. They either went extinct or evolved into another species. Some contend that they are the immediate ancestor of *Australopithecus anamensis* - known as "Lake Man". For our purposes here, we'll go along with this contention.

***

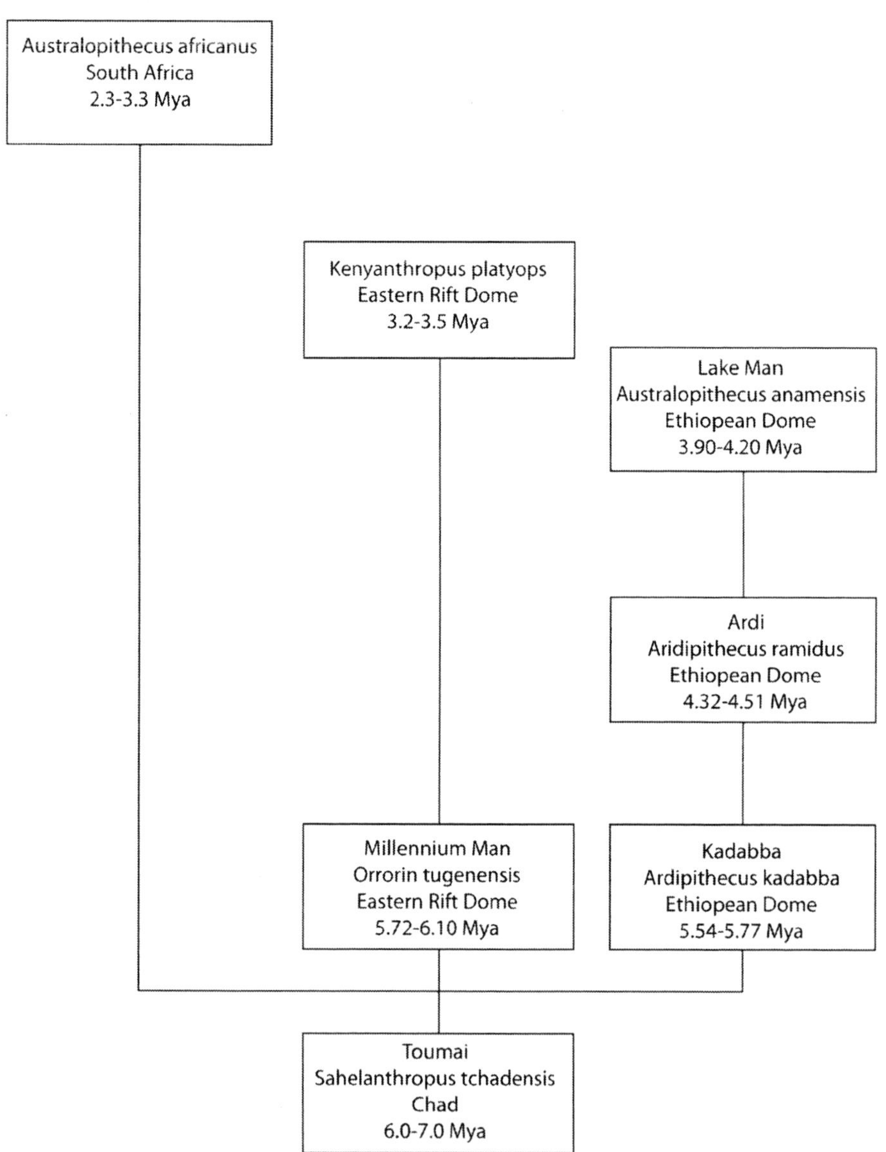

Figure 2-3 shows a chart of our proposed ancestral lineage based on the findings of Kingdon (2003). I've taken the liberty to add two species to his chart - *Australopithecus sediba* discovered in 2008, and *Ardipithecus kadabba,* which he likely treats as a single species *Ardipithecus ramidus*. As you can see we had a lot of cousins most of who died out at a relatively early date. There are two possible lines of descent through the Australopithecines - one through *A. anamensis* (Lake Man) and the other through *A. africanus*. The evolution of the Australopithecines, gracile apes, over about 2.4 million years (4.2 to 1.8 *Mya*) is characterized by a number of anatomical changes leading to changes in physical appearance and behavior as well as social behavior. The basic scenario goes as follows: The canine teeth of the males and females decreased in size and body size dimorphism between males and females decreased. These changes imply that there was less male-to-male conflict within and between groups of males and less need to fight for dominance of a "harem". This in turn led to an increase in pair bonding with an increase in parental investment in the new born offspring. Changes in the morphology of their molars allowed them to increase the diversity of foods they ate to include seeds, nuts and tubers which require more "chewing" than fruits. In fact, *Paranthropus* (*Australopithecus*) *boisei* was nicknamed the "Nutcracker". Anthropologists often refer to the entire assemblage of *Paranthropus spp.* as the "Nutcrackers". A gradual increase in brain size allowed for changes in social behavior among members of the troop and, likely, between troops. These were small incremental changes. Small incremental changes over a long period of time can have significant outcomes as we've seen with the development of bipedalism.

"Lake Man's wide habitat tolerance would have been favored among ancestral populations, probably Ramidus, on the margins of the wooded Ramidus heartland. It would have been a solution to living in the suboptimal rim - not being too fussy about where they ventured and taking what they could from each habitat. As climate changed and these edge habitats spread at the expense of the old woodland, many core Ramidus, specialized populations would have gone extinct but those on the margins found that they were kitted up to do very well in this new world (Finlayson 2009a)." And, "The innovative Lake Man emerged in the borders of the woodland in which conservative proto-humans lived (Finlayson (2009a)." Finlayson describes his innovators as those living on the edge and most likely to find new ways of living and thus most likely to survive during down times. His conservatives live in the middle ground where they are best adapted.

They become very specific in their habitat needs and when bad times come are unable to adapt quick enough and likely become extinct. Lake Man is considered to be the immediate ancestor of Lucy, *Australopithecus afarensis.* "Lucy and her kind seem to have been even more adventurous than all their predecessors and we find them in environments more open than any occupied by those that came before them (Finlayson 2009a)." Kingdon (2003) has their home of origin (Table 2-1) in these very same habitats as those in the Zambezi River basin described earlier. For this to be the case Lake Man had to venture into the Zambezi River Basin before about 3.0 *Mya.*

Conventional thinking has the line of descent going from *Australopithecus anamensis* to *A. afarensis* to *A. africanus* and finally to *Homo spp.* As you can see in Figure 2-3 Kingdon (2003) has *A. africanus* as a descendent of his hypothetical ground ape. A later discovery of *A. sediba* at the site of Malapa in South Africa in 2008 showed that she lived about 1.78 to 1.95 *Mya* (Berger, et. al. 2010). "The age and overall morphology of *A. sediba* imply that it is most likely descended from *A. africanus*, and appears more derived toward *Homo* than do *A. afarensis, A. garhi,* and *A. africanus."* Kingdon then shows *A. anamensis* as the ancestor of *A. afarensis. A. afarensis* is then the immediate ancestor of *A. garhi, A. bahrelghazali* and the robust Australopithecines now renamed *Paranthropus spp. P. aethiopicus* is the immediate descendent of *A. afarensis* and ancestor of *P. robustus* and *P. boisei.* This line then dies out with *P. boisei* at about 1.4 *Mya.* The robust Australopithecines became more specialized and conservative in their food resources. Specialization can be a problem when conditions change and negatively impact your special resource. This, alone, may have led to their ultimate fate.

Kingdon (2003) provides a good rationale for his lineage by separating *A. africanus* from *A. afarensis* based on anatomy and ecology. First, and foremost, *A. africanus* has never been found in the Zambezi, the proposed original home of *A. afarensis* nor has *A. afarensis* been found in South Africa. In addition, Kingdon (2003) notes that "They [*A. africanus*] differ from the Lucies [*A. afarensis*] and Robusts [*Paranthropus spp.*] in showing less sexual dimorphism implying significant differences in social structure and no sign of the anatomical or dental adaptation to specialized diets that typify the Robusts. These differences could be interpreted in terms of technological and behavioral solutions to obtaining food. The complexity of ecotypes in southeast Africa [as opposed to those in which the Lucies and Robusts lived] could have invited an adaptive shift that diversified food and ways of acquiring it. Equally important, an initially tropically adapted ground ape's year-round preference for stable home ranges in coastal forests would have had to change. Slowly, as seasonal shifts up and down the mountains became a routine annual movement, the narrow strip of coastal lowland forests would have become a winter retreat. Eventually, as the hominins' overall range enlarged, the narrow coastal shelf might have become a marginal habitat for a much larger population of now upland- or veldt-adapted animals." He gives a series of anatomical features that not only differentiate *A. africanus* from *A. afarensis* but lead more in the direction of *Homo spp.* "Distinguishing features of the South African Man-ape [*A. africanus*], when compared to Lucy, include a relatively larger brain; a smaller 'muzzle' or face; much less difference between the sexes [dimorphism]; and molars of similar type, arrangement, and rooting pattern to *Homo*. There are also clear hints of *Homo*-like ancestry in the bones of the hands, which are decisively less like those of the Lucies than those of humans. Likewise, the thumb is very unlike those of apes (and the Lucies) in being well developed and constructed to be opposed to the fingers in various subtle permutations of grip and clasp. As for the wrist, a humanlike flexibility contrasts dramatically with the 'tied up,' well-locked bones of Lucy. While the finger bones of the 3.3 million-year-old "Little Foot [*A. africanus*] skeleton are apishly curved, they are of similar proportions to those of later humans. By contrast, the limb proportions and feet, instead, are less like those of modern humans than those of apes, with the big toe still capable of deflection from the other toes, as in apes." In conclusion, He notes that the "linear" schools of thought that would have *A. africanus* a direct descendent of *A. afarensis* is "irreconcilable" with the facts.

The most important consequence of the habitat occupied by *A. africanus* is the seasonal requirements imposed on him. The seasonal migrations up and down the mountain may have led to individual groups establishing themselves in the numerous small river valleys available to them throughout this habitat. This would have been a subtle departure from the *A. afarensis'* behavior in the more uniform habitat of the Zambezi. "In many instances fellow valley dwellers, numbering 100 or more, might have been kin, but the environment alone would have encouraged social accommodation between larger numbers of more loosely connected individuals. Under these circumstances, narrowly territorial behavior could have declined, and some sort of (clan?) system, embracing a variety of associations within a larger area, could have developed. Tolerance between weakly acquainted individuals belonging to loosely connected, 'valley-by-valley' clans would have been encouraged by resources that were consistently well dispersed or by foods that were occasionally superabundant - the former promoting a safe dispersion, and the later promoting temporary, non-competitive convergences. In such circumstances, effective rallying of dispersed 'clan' members and the displacement of competitors could have become essential for 'lateral expansion' (away from river courses) and, quite possibly, for survival."

***

The "Tree Amigos" may have invented the art of bipedalism but the Australopithecines improved on that art. We see the first signs of social behavior in the possible development of clans in *A. africanus* and later in *A. sediba*. However, the small increase in brain size was not likely sufficient to lead to complex behavior changes or the invention of tools. Leakey's notion that all bipedal species should be classified as human seems to me as pushing the envelope a bit too far. Certainly, *A. sediba* was well on its way from ape to man. The question still remains, Are the Australopithecines ape-man or man-ape? We probably can't answer the question at this moment. However, as we get into the next line of evolution, that of *Homo spp,* things should start to clear up.

Lumping all of the Australopithecines together, Swisher et al. (2002) notes that "we can get an average weight for pre-erectus Australopithecine species of about 84 pounds and a stature of 4.2 feet. Although the average hides a small amount of variation among different species, much more important is the variation within species - that is, the difference between males and females. The average male Australopithecine weighed in at around 94 pounds and stood 4.6 feet tall; females where a diminutive 70 pounds and 3.75 feet. This degree of so-called sexual dimorphism in body size is very characteristic in primate species with a certain kind of social structure: namely, where there is keen competition among the males for the sexual access to the females. It's a blunt truth of the primate social world that the big guys usually win, so that some of the males will have the opportunity to produce many offspring, while others will be much less fortunate. This degree of sexual dimorphism also implies that females, with their young, range over smaller territories while foraging than do their males. In other words, the earliest members of the human family were apelike not only in much of their anatomy but also in their social behavior."

Early theory on human evolution required developing a large brain prior to bipedalism. We now know that this is just the reverse - bipedalism came first followed by large brain development many millennia later. Kingdon (2003) emphatically states that "Such naivety became totally unsustainable once it was clear that the first bipeds were cranial apes, creatures with ape heads mounted on human-like bodies." The real change in brain size comes with the onset of *Homo spp.* which we'll discuss later. Table 2-2 presents a view of how brain size changed among and within hominin species over time. We'll return to this table in the next chapter.

Table 2-2. Estimated average cranial capacity (brain size) among hominids

Species in ascending order of age	Years *Mya*	Estimated average cranial capacity (cubic centimeters)
Chimpanzee		400
Late Miocene (11.6 to 5.33 *Mya*)		
Sahelanthropus tchadensis	6.0 - 7.0	350
Orrorin tugenensis	5.72 - 6.1	
Ardipithecus kadabba	5.54 - 5.77	
Pliocene (5.33 to 1.8 *Mya*)		
Ardipithecus ramidus	4.32 - 4.51	325
Australopithecus anamensis	3.9 - 4.2	
Australopithecus afarensis	3.0 - 3.9	415
Kenyanthropus platyops	3.2 - 3.5	
Australopithecus bahrelghazali	3.0 - 3.5	
Australopithecus garhi	2.5	450
Australopithecus africanus	2.3 - 3.3	442
Paranthropus aethiopicus	2.3 - 2.8	419
Australopithecus sediba	1.78 - 1.95	420
Early Pleistocene (1.8 to 0.78 *Mya*)		
Paranthropus robustus	1.5 - 2.0	515
Paranthropus boisei	1.4 - 2.5	530
Homo georgicus	1.77	
Homo habilis	1.44 - 2.33	631
Homo rudolfensis	1.4 - 1.9	575
Homo floresiensis	.017 - 2.0?	417

Grandfather's Tale

During this time the people did not have language. Therefore, there is no oral tradition about these ancient people. As noted earlier a human or spirit Manitou may express himself in the form of a rock. His spirit would live forever in the rock. If you can read the stories the rocks have to tell, you can gain an understanding of the people encased in the rocks. To this day there is a rock that the Huron had to pass when they were on their way to Québec to trade. The Indians named this rock tsanhohi, the home of the tsanhohi, a species of bird of prey. This rock looks very much like the head of a man with upraised arms. In the belly is a cavern that is difficult to reach. The Indians believe that this rock is a man who, while lifting up his arms and hands, was transformed into a rock. This is the main thesis of the story to follow - The Origin of the Spirit Rock. If the Indians were in doubt about their journey to Québec, they would offer tobacco to the rock when passing in their canoes and, tsanhohi would protect them on their journey (Tooker 1991).

The Manitou in The Origin of the Spirit Rock will tell you the story of the origins of mankind. This next story is about the origin of the Spirit Rock.

The Origin of the Spirit Rock

There was a young man and he was fasting. And as he slept the Manitou Maeqnapos called him and said to him: "Well, then, come Uncle! I'm the one who invites you to come here to my house. As for me, I'll give you what I have. And with you will go along nine young men who are clean, who have never been with girls; that's the kind of young man you will bring here to my house."

Now, when dawn came, and the young man carefully thought it over, "And I wonder how I might reach the place where Maeqnapos invited me; maybe I can't get there." Now, when night came again, when he was sleeping it was then that he heard him again; Maeqnapos spoke to that young man. "Yes Uncle, you will arrive there at Maeqnapos' house; it's really me who invites you!" Now, it was then that he woke up. "Well then, I'll try to do it. That's what I'll do. If I can convince the young men, I'll take them with me to Maeqnapos' house."

When dawn came, it was then that he sought out the other nine young men. When he had found them he told them about it. "Well, then, my friends, I'm going to go to Maeqnapos' house. You should come with me; we'll go to Maeqnapos' house." And each, in turn, agreed! "Well right, in ten days, it's then that we'll set out," he said to his young friends.

And so, it was then that they told their fathers and mothers. They explained that Maeqnapos asked them to come to his house. The old men marveled at the news. After a long time had passed, it was then that the old men spoke. Each said "well then, Son, be on your guard! Whatever you may have come to know, it must be done properly." That is the warning they gave their sons.

So it turns out that ten days had passed. It was then that they got into their elm-bark canoes to cross this small sea. They traveled and traveled; then it was that they came to land. From there where they had set out, they had traveled nine days. And on the tenth day of traveling by canoe, when it was noon, it was then that they saw land.

A river - it was there that it flowed by; about as far as a man can jump was the distance of the other side. It was then that the land moved; it was here that the ground moved from side to side. "Don't look down!" the leader said to his young friends. "Like I move, that's how you should move!" He said to them.

"All right!" the young men said. One man - it was then that he jumped; he stood on the far side. Soon another one - it was then that the young man did so. Just as the last young man was going to jump; it was then that he looked down below at the water. "Oh, no! Don't look there!" the leader said to him. Then again, as he was going to jump, it was then that he looked at the water; it was then that he almost fell. "Oh, no, our friend!"

All the while, the ground was shaking, moving side to side. Now, all at once the young man made up his mind; it was then that he jumped. It was then that falling short, he fell; it was then that he died, falling there where the ground moved from side to side. "Oh, no, our friend! Well, then he'll fare so" the leader said.

After they had started again, when it was evening, just as the sun had set, they came in sight of a wigwam. Arriving at the door, it was there that they stopped, it was then, the story goes, that someone spoke from inside. "And why, Uncles, don't you come in? Come in, Uncles! Thank you for coming to me!" Maeqnapos said to his nephews. When they had entered, a man sat there. "Now, sit down, sit down, sit down, Uncles!"

And so, when the young men were all seated, a kettle hung over his hearth. Maeqnapos started to take down his kettle -a large wooden bowl was Maeqnapos' bowl and a large wooden spoon was his spoon. That kettle was completely full of corn, sweet corn; half of it was venison, with which he had cooked it. It was then that he placed it in the bowl, and he fed them.

"Well, then, Uncles, eat!" he said to them. It was then that the young men ate, oh my! They ate everything in that large bowl, the young men all going at it together. After they had finished eating, they moved the bowl.

"Well, then, Uncles, thank you for coming to me! It was I who invited you, here. And it's good that you've come to me. After four days pass you will go home," Maeqnapos said to them. "You'll stay here four days. It's then that I'll be satisfied. Uncles, I'll cooperate with you. I will give you whatever you ask for, Uncles," Maeqnapos, their nephew, told those young men.

Then while they were there, it was then, the story goes, that a kettle came into Maeqnapos' house. Just imagine! Someone must have boiled meat, splendid! Truly - that was good meat. Tobacco also came along into the house. "Now, Uncles, eat what's in the kettle. We're being given food we can eat; and we "smoke the tobacco that's here." And so it is that a person gives a feast here, making an offering of tobacco; it's then that the person's feast offering along with the tobacco come into Maeqnapos' house.

And four winters -that's how long the young men were there; but Maeqnapos calls four years four days. And truly, those young men only slept four times during those four winters. "Now, Uncles, it is time for you to go home. When morning comes, you will start off," he said to them.

It was then that he asked the young man who had taken along his young friends: "Now, Uncles, and what do you want that I can give you?" And so, the young men thought it over. And each in his turn asked Maeqnapos for something they wanted. Some wanted to be skillful hunters, others to be brave fighters and war leaders, another a medicine man, another wanted to run fast to improve his hunting and fighting ability, another wanted all the young women to love him.

"And now, Uncle, what for you?" "Well, then, Uncle, for my part, I want to live forever; that as long as the earth exists, I will live. That's what, as for me, I want."

"Oh, all right, I'll give it to you. As long as the earth exists, you will live," Maeqnapos said to his uncle. He stood up, getting hold of him by the shoulders, it was then that he lifted him up; and by the door it was there that he sat him down. When he had set him down, a Spirit Rock was there. "Now, Uncle, it's here that you will always be; as long as the earth exists, you will live."

And it was so. A Spirit Rock was there by the door. And still, it's there that the young man sits. Well, then, he asked that as long as the earth exists, he would live.

Chapter III.

Transition to Humanity

(2.5 million years ago to 1.0 million years ago)

...as the reasoning powers and foresight of members become involved, each man would soon learn that if he aided his fellow-men, he would commonly receive aid in return. From this low motive he might acquire the habit of aiding his fellows; and the habit of performing benevolent actions certainly strengthens the feeling of sympathy which gives the first impulse to benevolent actions.

But another and much more powerful stimulus to development of social virtues is afforded by the praise and blame of our fellow-men. To the instinct of sympathy, as we have already seen, it is primarily due, that we habitually bestow both praises and blame on others, whilst we love the former and dread the latter when applied to ourselves; and this instinct no doubt was originally acquired, like all other social instincts, through natural selection.

Charles Darwin, The Decent of Man

We pick up our story in South Africa at the end of the Pliocene and the beginning of the Pleistocene. The Pleistocene occurred from about 2.5 *Mya* to 12 thousand years ago (*Kya*). The end of the Pleistocene is marked by the retreat of the last continental glaciers and the beginning of the Holocene - the era that we currently live in. The continents continued to move about 62 miles or about 0.002 miles per 1000 years during this period. Glaciations were a global event. Each of these glaciations is often referred to as an "Ice Age". Glaciers ebbed and flowed throughout this period. Global temperatures were significantly cooler and sea levels fluctuated as much as 300 or more feet during glacial periods.

So what is the basic cause of glaciations? They are defined by a series of cycles related to the position of the earth in respect to the sun called Milankovich Cycles. As determinates of climate they are, of necessity, determinants of glaciations and glacial and interglacial periods. The first of these cycles has a period of 105,000 years. It is due to a change in the earth's orbit from one that is minimally elliptical to one that is more elongated. The more elliptical the orbit, the further the earth is from the sun, the greater the contrast among seasons. The second cycle has a period of 41,000 years and is due to a shift in the tilt of the earths' axis relative to an axis perpendicular to the orbital plane. The tilt of the earth's axis is felt to a greater degree at high latitudes. When the tilt is large, 24.4 degrees, versus 21.8 degrees the contrast between seasons is greatest. The third cycle has a period of 21,000 years and is called the precession of the equinoxes. Over this period the moment in time that the earth is closest to the sun as it traverses the elliptical orbit shifts gradually from January through February then March ... around the calendar until it once again returns to January.

Shorter cycles within these Milankovich Cycles occur due to changes in solar activity. There are two cycles, one at about 2,500 years and the second at about 200 years, related to changes in solar output. In addition, there is an 11 year sun spot cycle. All of these combined make it very difficult to predict the direction the earth's climate might take at any point in time, but makes prediction over long time periods more reliable.

Continental drift also contributed to climatic change by creating changes in the physiography of the earth's surface. As noted previously, mountain building altered patterns of rainfall such as the creation of the Himalayas. Creation of the Isthmus of Panama had a major affect by cutting off the circulation of warm waters between the Atlantic and Pacific Oceans. Because of this, cold waters then circulated from the Polar Regions and significantly cooled the Atlantic Ocean and surroundings. "These climatic anomalies reveal the unpredictable nature of climate changes as orbital, atmospheric, and tectonic factors came together, often in unexpected ways (Finlayson 2009a)."

When the contrast among seasons is slight, summer temperatures are not warm enough to melt glaciers and winter temperatures are relatively mild. Snow in the winter occurs more regularly and glaciers build, determining, over time, a glacial period. When the contrast among seasons is great the glaciers melt faster than snow can add to their mass. Summers are longer and warmer causing increased melting of the glaciers and winters are shorter with less snowfall which slows ice accumulation. The glaciers, therefore, decrease in size. This period of time is referred to as an interglacial period. During a glacial cycle, periodic cooling (stadial) and warming (inter-stadial) periods occur. During these interglacial periods, carbon dioxide, trapped in the ice for millennia, is released to the atmosphere in large amounts. Also, as the oceans warm carbon dioxide is released. Hence, as the globe warms atmospheric carbon dioxide increases significantly. We know that there is a lag in the increase of atmospheric carbon dioxide due to increasing global temperatures. Increased levels of atmospheric carbon dioxide are the result of – not the cause of – global warming events. We are currently at about the mid-point of an interglacial period where a shift in the climate is transitioning to cooler summers and warmer winters, and increased glaciations.

During glacial periods the mass of the glaciers can increase dramatically, tying up vast amounts of water and causing sea levels to drop hundreds of feet. Between 1.6 *Mya* and 900 *Kya* a complete and rather modest glacial cycle occurred about every 40,000 years. Between 900 *Kya* and 450 *Kya* the cycles grew in length and severity to 70,000 years. Since then the cycles have been about 100,000 years long with rather extreme continental glaciations. Glacial periods do not necessarily occur at all places on the globe simultaneously. However, they are more prominent in the Northern Hemisphere.

***

We left the last chapter with the unanswered question -Ape-man or Man-ape? My hope in this chapter is to answer this question and put a time and place to the beginnings of modern humanity. Fagan (1990) provides insight into what makes us human. He notes that "The most striking differences between ourselves and other living species lie in our ability to design and make tools, in our articulate speech and self-awareness, in our intellectual and creative skills, and in what one might loosely call our 'psyche'. Consciousness, cognition, self-awareness, foresight, and the ability to express oneself and one's emotions: these are direct consequences of fluent speech. They can be linked with another attribute of the fully-fledged human psyche: the capacity for symbolic and spiritual thought, concerned not only with substance and technology, but with defining the boundaries of existence, and the relationship between the individual, the group, and the universe." My task in this chapter is to show that the foundations for humanness described by Fagan were established during this period of time.

"Until the appearance of *Homo*, all bipedal apes had small brains, large cheek teeth, and protruding jaws and pursued an apelike subsistence strategy. They ate mainly plant foods, and their social milieu probably resembled that of the modern savanna baboon. These species - the australopithecines - were humanlike in the way that they walked but in nothing more. At some time prior to 2.5 million years ago - we still can't say exactly when - the first large-brained human species evolved (Leakey 1994)." These individuals are collectively known as H*omo spp.* (Figure 3-1). We demonstrated in the last chapter that the most likely immediate ancestor to *Homo spp.* was *Australopithecus sediba.* The next question is "Who is the most likely immediate descendent of *Australopithecus sediba?"*

The discussion is not straight forward. As noted in the previous chapter *Homo rudolfensis* is the likely descendent of *kenyanthropus platyops.* In fact, Kingdon (2003) suggests that the Rudolfs be given a

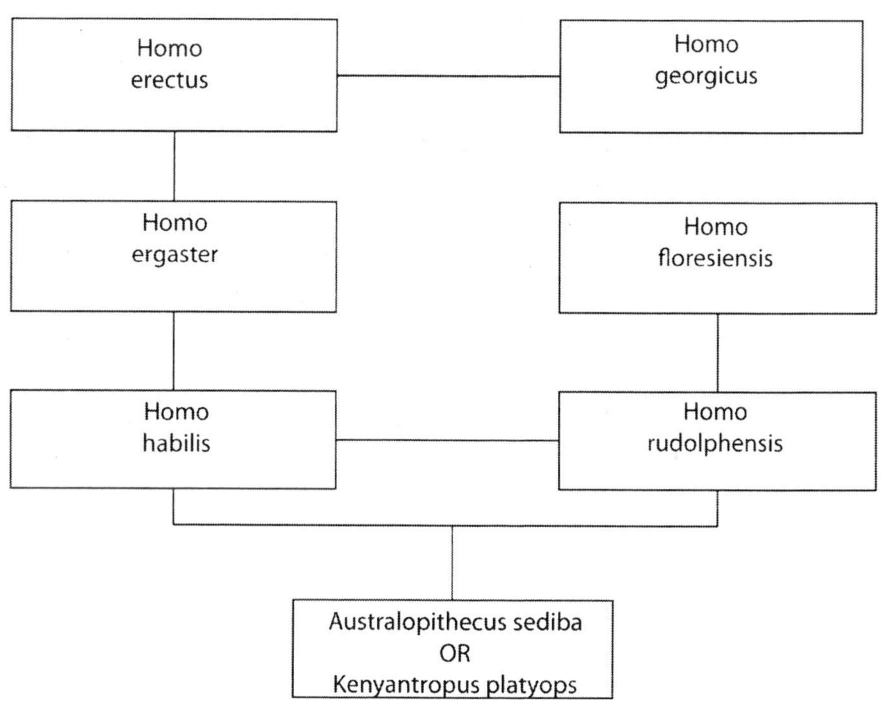

Figure 3-1. A proposed lineage of *Homo spp. (H.)*, and the relationships to *Ardipithecus (Ar.) spp., Orrorin spp. (O.), Kenyanthropus spp. (K.), Paranthropus spp. (P.)*, and *Australopithecus spp. (A.)* (From Kingdon 2003)

new name *Kenyanthropus rudolfensis.* In turn he notes that *Homo habilis* might best be called *Australopithecus habilis* (There's much agreement on this count). "If we assume, as almost every anthropologist today accepts, that all members of the genus *Homo* are descended from ancestors belonging to the genus we call *Australopithecus,* it necessarily follows that, somewhere along the chain of descent from one species to the other, there must have been at least one individual who sat exactly on the borderline. This is an important point, so let me stay with it a little longer (Dawkins 2009)." Recall from Chapter I, that I was emphatic about evolution being a continuum and taxonomy about placing species in boxes. And that, what are really important are not the boxes themselves, but what goes in between the boxes

– in this case between *Australopithecus* and *Homo*. Dawkins (2009) reviews the skulls of four "different species". In Figure 3-2 below, I've drawn the outlines of these four skulls and will summarize the important points made by Dawkins (2009). The four skulls are, in order, "Mrs. Ples", originally assigned the genus *Plesianthropus* and now in the box called *Australopithecus*, "Twiggy", KNM ER 1813, and KNM ER 1470. At the moment, KNM ER 1813 and KMN ER 1470 are classified as *Homo habilis*. However, there are folks who think that 1470 belongs to the species *Homo rudolfensis*. Dawkins notes that "In both respects, 1813 and 1470 seem more human, Mrs. Ples more 'ape-like'." Stay with me now! "Twiggy" is also classified as *Homo habilis*. "But her forward-pointing muzzle has more of a suggestion of Mrs. Ples about it than of 1470 or 1813. You will perhaps not be surprised to be told that Twiggy has been placed by some anthropologists in *Australopithecus* and by other anthropologists in *Homo*. In fact, each of these three fossils [excluding Mrs. Ples] has been, at various times, classified as *Homo habilis* and as *Australopithecus habilis* (Dawkins 2009)." The real point to be made here is not that there is so much confusion about the evolutionary placement of these various people, but that the confusion injected into the discussion by anthropologists fighting over naming priorities has distracted them from their true mission – placing modern man in the context of her ancestors! The taxonomic nightmare that has "evolved" to explain evolution is why some folks keep thinking there's a missing link! Boxes just don't work to describe the evolutionary continuum. What will the anthropologists do if there should be discovered a fifth skull that fits somewhere between these other four? What box will it go in? Maybe a whole new box? This then leaves the *Homo ergaster/Homo erectus* complex - many consider them to be a single species. I'll keep them separate just because I want more boxes on my diagram (You may laugh at this point and I won't be offended!). That leaves *Homo ergaster* by default. Most agree that *Homo ergaster/Homo erectus* are in the lineage to modern man. Note that I haven't closed off the lineage from *Kenyanthropus* [*Homo*] *rudolfensis*. I also show *Ardipithecus* ramidus as the immediate ancestor of *Australopithecus anamensis*. I've modified Kingdon's chart to reflect these issues. As I implied in Chapter I, the naming of species is often arbitrary and capricious.

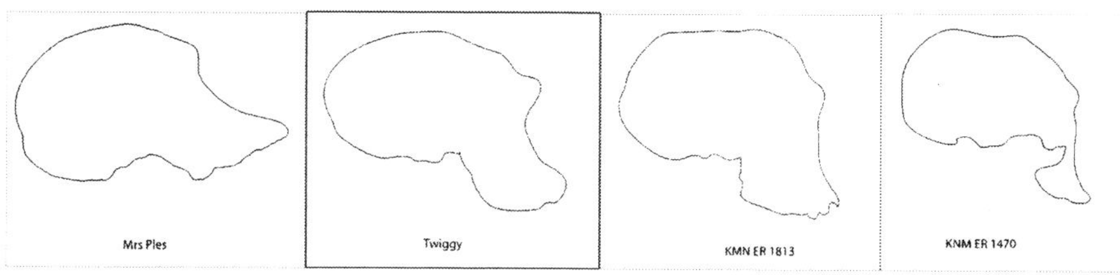

Figure 3-2. The continuum of cranial features in early human evolution

In an effort to sort all this out, we'll begin this new journey with *Homo habilis*. Our earliest records show this species emerging about 2.5 *Mya*. These early dates for *Homo habilis* come from Hadar, Ethiopia and Baringo, Kenya according to Kingdon (2003). *Homo habilis* probably had its origin in South Africa and would likely have migrated out to the east African highlands shortly after emerging from *Australopithecus sediba*. A global warming trend had set in sometime after 2.9 *Mya*. This warming would have negatively affected a cold adapted animal like *Homo habilis*. Kingdon gives a good sketch of the conditions that would have led to this migration. "Among the areas in which dispersing populations could have survived would have been any that resembled their cool, strongly seasonal homeland. On this count the uplands of eastern Africa were the obvious, if distant, zone for a secondary expansion. The main 'upland corridor' between southern and eastern Africa is the long south to north mountain range that marks today's frontier between Zimbabwe and Mozambique. This is but two valleys distant from the extreme northern end of the Drakensberg Mountains. Here, we can envision an early *Australopithecus africanus* [now likely *Australopithecus sediba*] population well positioned to initiate such a northward expansion when conditions were right. In spite of lowland discontinuities, this route has, nonetheless, been traversed in both directions by numerous species of upland-adapted animals and plants.

The ideal time for a successful traverse and colonization of uplands would have been during *cool* periods when entire temperature belts and their biotic communities drifted toward the tropical highlands and also slid downhill. The first Homo-like fossils, as well as the first worked stones, have been found in east Africa [dating from] between 2.4 and 2.6 *Mya*, a period that followed one of the sharpest, fastest, and most severe drops in global temperature during the Plio-Pleistocene [end of the Pliocene start of the Pleistocene]. Ecological conditions that had been restricted to southern latitudes became more widespread, especially in the uplands of eastern Africa. Considering its conservative anatomy and wide distribution in the east African highlands, *Homo habilis* is a good candidate for South African origins and colonization via an upland route at around 2.8 to 2.9 *Mya*. Once established in new but favorable regions and only weakly connected with the place of origin, a nucleus of colonists could rapidly proliferate and, less rapidly, differentiate from the parental stock (which, during climatic vicissitudes, may also have continued changing) (Kingdon 2003)."

H. habilis is known as the "Handy Man" for his ability to make fairly unsophisticated tools. "Handy Man" was a generalist and this is reflected in his tool chest. There is some discussion that the Australopithecines were making tools before *Homo spp.* If so, they were, quite likely, rudimentary. They were probably very similar to the tools made by chimps. "Chimpanzees are adept tool users, and use sticks to harvest termites, leaves as sponges, and stones to crack nuts. But - so far, at any rate - no chimpanzee in the wild has ever been seen to manufacture a stone tool (Leakey 1994)."

The oldest stone tools found to date are about 2.5 - 2.6 *Mya*. They are fairly simple and are created by banging two stones together - usually lava. This produced small very sharp flakes which could be used for a variety of jobs including skinning and butchering animals, cutting and shaping wood, or cutting various fruits, vegetables, or grasses. They were made on the spot as needed and discarded when the job was finished. This tool assembly is referred to as Oldowan Industry. The Oldowan Industry remained relatively unchanged for more than a million years. By 1.4 *Mya* a new much more sophisticated tool chest was built by *Homo erectus* called the Acheulean Industry. We'll talk more about this new "tool chest" later in the discussion.

The tool makers probably did not have a specific shape in mind but took what came to them. However, Toth (1985) notes "It seems clear that early tool making proto-humans had a good intuitive sense of the fundamentals of working stone. There's no question that the earliest tool-makers possessed a mental capacity beyond that of apes." Wilson (1980) noted that "tools are but extensions of the limbs and teeth, and in this sense, their advent is an extension of the generalization of hominid morphology." Kingdon (2003) notes that "Childhood, with its curiosity, delight in exploration, and playful expressions of discovery and invention, may have provided the essential grounding, the precondition for technological invention. Experience is essential for the maintenance and diffusion of traditions in many species, but it is less certain that mature individuals were significant sources of innovation in early '*Homo*.' Rather, inventiveness was more likely to have originated in less-experienced youngsters, in playful but ecologically relevant contexts." Tools allow man to become more generalized giving them access to more and newer resources. Whether they were hunters or scavengers (most likely scavengers), these new tools gave them access to a new source of energy - meat protein. Richer sources of food in the form of protein would lead to larger brains and larger brains to more innovative tools. As we'll see a little later in the discussion, Wrangham (2009) will add that not only this increase in protein but the cooking of food would provide even higher levels of energy.

"It was 2.5 million years ago that a climatic threshold was reached when the planet came under the grip of large-scale climatic cycles that were to become the hallmark of the Pleistocene. The latter part of the Pliocene was already signaling the shape of things to come. Between 3 and 2 million years ago grassland habitat became more abundant, especially after 2.5 million years ago, but important areas of woodland survived, maintaining a spectrum of habitats from closed to open. It was after 2.5 million years the robust proto-humans [*Paranthropus spp,*] capable of dealing with tough plant foods, appeared in the wooded, grassy savannahs. The great diversity of proto-humans soon came to an end though, as open habitats overwhelmed the landscape of 2 million years ago (Finlayson 2009a)." The edge habitat available to the innovators would have provided the ideal habitat for the emergence of *Homo spp.*

Our first evidence of *Homo ergaster* just happens to come from this same region. The earliest date for *Homo ergaster* in eastern Africa is about 2.0 *Mya.* He just appears in the archeological record as noted by Alan Walker (in Kingdon 2003) "without a past". He was given the name the Work-man. He was a tall slender individual not at all like his small ape-like ancestors. *Homo ergaster* was adapted to dry, open country. Kingdon (2003) notes that "To evolve adaptations to dry, open country, the Work-man would have had to suffer a period of prolonged isolation from other hominins, in an appropriate discrete habitat. Forced to evolve on the edge between woodland and grassland, *Homo ergaster* took a whole new approach to life in the grasslands and developed the anatomy needed for moving forward in time.

"If the Work-man emerged in an advanced condition and made up for a 'lost past' (by suddenly appearing, fossilized, in a wide scatter of localities), what was the secret of success?

It was not relative brain size because *kenyanthropus* [*Homo*] *rudolfensis* and *Australopithecus* [Homo] *habilis* had comparable brain-body ratios.

It was not walking per se: several Australopiths had long been accomplished walkers, although their shorter legs and somewhat gangly, long armed bodies would have been no match for the Work-man for pacing over long distances, still less for running.

It was not the use of stone tools alone, because crude stone tools of the 'Oldowan' type are recorded from various parts of eastern and southern Africa more than half a million years before the Work-man's appearance there.

It is possible, however, that another sort of tool provided the Work-man with an unassailable advantage: fire (Kingdon 2003)."

This key feature was likely invented by *H. ergaster* about 1.8 – 1.9 *Mya*. Fagan (1990) provides an overview of this historic event. "Like all hunters and gathers, *Homo erectus* had probably learned to live with natural fires and was not afraid of them. In time, the new hominid may have made a habit of conserving fire, taking advantage of smoldering tree stumps ignited by lightning strikes and other natural causes to light dry brush. Then came the biggest step of all: the *making* of fire. Perhaps as early as 1.5 million years ago, *Homo erectus* may have learnt to create fire in East Africa and at Swartkrans in South Africa ... Fire offers not only warmth, but protection against predators and an easy way of hunting game, even insects and rodents. The toxins from many common vegetable foods can be roasted or parched out in hot ashes, allowing people to use a wider range of foods in their diet." Kingdon notes "If fire became a general tool of *Homo ergaster* and their descendants, perhaps these new people were instrumental in making grasslands even more prevalent (rather than responding passively to expanding rangelands, as they are so commonly portrayed). Thus, the spread of fire-climax grasslands may not have preceded this burgeoning population of highly successful hominins but could have followed hard on their pyromaniac heels! Humans are undoubtedly niche stealers, but in this instance their activities may have vastly extended secondary grasslands, creating new niches for grassland and fire-climax species - above all for themselves."

"Of all the tools available to prehistoric peoples fire is quite the most significant and it is with the widespread use of fire that they began to shape the landscape. Originally fire would have been restricted to certain very limited regions, seasons, habitats and situations; most obviously close to volcanoes. Occasionally, oil, shale, bitumen or mineral oil seepages can be ignited and marsh fires are commonly triggered when methane gas is lit by lightning or by sun rays focused by dew drops above dry tinder. Lightning strikes are probably the most common starter of fires, but they only take hold and spread when both wind and fuel are right. Numerous plants are adapted to withstand regular firing – a capacity that was evolved long before humans came along – but the regions in which those plants evolved were originally quite limited: the Cape of Good Hope, parts of Central Africa, Australia and India for example.

"In some fibers and woods, embers can smolder for long periods without erupting into open flames. This is particularly true of underground burns where lightning ignites dead roots, peat or buried forest leaf-fall and then precedes much like a charcoal kiln. Prehistoric hominids living in regions where such fires were frequent would have noticed the tendency for long-lasting smoldering's to erupt into flames when exposed or fanned by a strong wind. This type of prolonged burning that would have offered them familiarity with and sustained observation of the behavior of fire. The situations in which they were of practical use would have been more urgent and immediate.

"In lightening-prone regions early hominids would have followed other scavengers to natural fires for the pickings they provided. Not only could fleeing animals be caught in more concentrated numbers, but the aftermath would have been richer still. In the embers could be found scorched pythons and adders, tortoises cooked in their shells and innumerable blackened lizards and insects. If a fire is sufficiently hot and widespread it will entrap much larger animals at vulnerable moments of their life cycle. Very old animals cannot move fast enough and newborn fawns are genetically programmed to lie tight.

"A host of scavengers, pirates, scroungers, call them what you will, have evolved to take advantage of these natural calamities of life. Before our ancestors out-competed them and appropriated all the major sources of free food there were a lot more various now extinct hyenas and jackals, bears, marsupial carrion eaters, giant lizards, condors and other carrion birds. Like the kills of large carnivores, natural fires have always created an important resource for such animals and the more omnivorous hominids became, the more they would have joined the throng of scavengers at any natural disaster, led by circling flocks of kites, storks, vultures and rollers. In fire-prone regions hominids would have been acquainted with and perhaps benefited from fire for a million years or more before they attempted to control it for their own ends. We can be reasonably sure that throughout this period they appreciated the way fire forces vulnerable prey to concentrate and that they relished the ease with which scorched skin and cooked flesh could be torn apart and consumed. But it would not have been 'cooking' that was the first use of fire.

"The most likely first invention by early hominids would have involved no more than furthering the natural utility of the fire to the individuals that were hanging about for the goodies it offered. Clusters of prey sheltered in clumps of bush that the fire had bypassed could be routed by no more than fetching a brand and renewing the blaze. An ability to keep upwind of a fire and maintain close observation of the progress would have had to precede the first act – but it would have been the easiest and most immediate initiative that would enhance a natural and well-used continuum. The act of renewing the fire would have first revealed that fire could be manipulated like a dangerous ally. Any further steps in its control would have depended upon this hands-on familiarity and possibly with its extension by observing that a fire could be held in one spot by being fed with fuel. However, that awareness is not necessary for other uses to be made of natural fires (and without immediate intervention).

"Early hominids were social, opportunistic, vulnerable to large carnivores and biting insects and relatively night-blind. Fire therefore had several attractions. It was warming. Flames and smoke were deterrents and gave some security from nocturnal nuisances. Fire gave light and, perhaps more important, held almost hypnotic attraction of living 'activity' in the long black inaction and emptiness of a moonless night.

"Of all these attractions, countering cold, particularly at higher altitudes and latitudes, would have been of the most practical use and would have provided the most immediate incentive to maintain a control of natural fires. If familiarity with fires was scattered and sporadic, it is likely that different hominid populations flirted with the various potentials of fire for many thousands of years before its control and creation became systematic and routine (Kingdon 1993)."

Wrangham (2009) indirectly agrees. "When our first ancestors first controlled fire, they would not have needed it for warmth, though fire would have saved them some energy in maintaining body temperature. But the opportunity to be warmed by fire created new options. Humans are exceptional runners, far better than any other primate at running long distances, and arguably better even than wolves and horses. The problem for most mammals is that they easily become overheated when they run. Most mammals cannot evolve a solution to this problem, because they need to retain an insulation system, such as a thick coat of hair. The insulation is needed to maintain body heat during rest or sleep, and of course it cannot be removed after exercise. At best it can be modified, such as by hair being erected to promote air flow."

However, humans did adapt to these hot, dry conditions in two ways. *Homo ergaster* was the "…first human species to have a forward projecting nose. In living humans, the external nose is usually cooler than the central body, and it thus tends to condense moisture that would otherwise be exhaled during periods of heightened activity (Klein and Edgar 2002)." But more important to this discussion is the fact that humans need to sweat to lose body heat during exertion. A covering of fur just wouldn't work. It would not allow them to sweat efficiently. The ability to sweat would be a great advantage over other animals that need to stay out of the sun during the hottest parts of the day and would soon be exhausted if pursued by an agile hunter such as man with the stamina to continue the pursuit long after the prey had given up the chase.

Once they had lost their hair, humans had an option available to them to keep warm at night. And it may have been the only alternative prior to the invention of clothing. "Fire offers that system. Once our ancestors controlled fire, they could keep warm even when they were inactive. The benefit would have been high: by losing their hair, humans would have been better able to travel long distances during hot periods, when most animals are inactive. They could then run for long distances in pursuit of prey or to reach carcasses quickly. By allowing body hair to be lost, the control of fire allowed extended periods of running to evolve, and made humans better able to hunt or steal meat form other predators (Wrangham 2009)."

"The hair loss that benefited adults would have been a problem for babies because babies spend a lot of time inactive and are therefore at risk to becoming cold unless cuddled or nested in warm surroundings. Perhaps at first babies retained their body hair even when their older siblings lost theirs. But an infant lying next to a fire would have risked burning his or her body hair. Nowadays, human babies are unique among primate infants in having an especially thick layer of fat close to the skin. Baby fat could well be partly a thermal adaptation to the loss of chimpanzee-like hair (Wrangham 2009)."

"The first firm evidence for a hearth comes from the cave of Escale in southeast France, is dated to about 1 million years ago and is presumed to have been made by Erect Man [*H. erectus*]. Effective control and setting of fire probably long predates this, but it is certain that cave-dwelling and human colonization of cold northern latitudes were totally dependent upon previous fire-controlling techniques that had been developed further south. The first signs of people (rather than cave bears and other carnivores) being present in European and Chinese caves coincides with layers of charcoal and burnt bones (Kingdon 1993)."

"Two biologists, Mark Pagel and Walter Bodmer, have recently invoked Darwin's idea of sexual selection as the driver of human hairlessness. They suggest that lack of hair was favored among early humans because it was a sure signal that no parasites were lurking in their fur

(Wade 2006)." The hair louse would have been a significant nuisance for early man. The loss of its fur over most of its body would certainly be of a benefit. However, it didn't take long for the body louse to evolve from the hair louse. While the hair louse requires hair to cling to, the body louse developed to cling to clothing which must have been invented right on the heels of hair loss. Genetics provides a way to put a date on this process. "The branch point at which the body louse first evolved from the head louse turned out to be around 72,000 years ago, give or take several thousand years either way. Assuming the body louse evolved almost immediately after its new niche was available to it, and then people first addressed their nakedness only in the most recent stage of their evolutionary history. It was about this time, or a few thousand years later, that people perfected language and broke out of Africa to colonize the rest of the world. It seems they had decided to get dressed for the occasion (Wade 2006)."

***

As with the jump from quadrapedalism to bipedalism the next big step in man-kinds evolution was an increase in brain size (Refer back to Table 2-1). Note that brain size in the early hominids did not change significantly from that of the chimp - 400 $\pm$ a few cc. There was a slight increase in the *Paranthropus spp.* to over 500 cc - about a 15 percent increase from *A. afarensis.* The real jump forward comes with *Homo spp.* It is this increase in brain size that leads immediately to tool development and later to the development of consciousness and therefore language, art, writing, kinship - in short, culture. Wilson (1980) sums it up nicely, "The struggle toward an efficient uprightness not only involves the brain in coordination and reorganization, then, but also requires the brain to solve more problems to make upright posture work." In addition, a new anatomical feature will be added to the mix beginning with *Homo spp.* In apes and the Australopithecines the rib cage is pyramid shaped. This configuration provides more room at the bottom to house a large gut needed to process the vast amounts of low quality, hard to digest plant materials. This new shape will be fully evident in the *Homo ergaster/Homo erectus* complex to follow.

With the availability of higher quality, higher energy, easy to digest foods, *Homo spp.* begins the process of turning the pyramid upside down. Not only these new foods but our ability to cook them has provided the major jump in energy. "Cooking gelatinizes starch, denatures protein, and softens everything. As a result of these and other processes, cooking substantially increases the amount of energy we obtain from our food (Wrangham 2009)."

Wrangham (2009) noted that "In humans, because we have adapted to cooked food, its spontaneous advantages are complemented by evolutionary benefits. The evolutionary benefits stem from the fact that digestion is a costly process that can account for a high proportion of an individual's energy budget – often as much as locomotion does. After our ancestors started eating cooked food every day, natural selection favored those with small guts, because they were able to digest their food well, but at a lower cost than before. The result was increased energetic efficiency."

Another anatomical feature of modern man is that the "Human chewing teeth, or molars, also are small – the smallest of any primate species in relation to body size. Again, the predictable physical changes in food that is associated with cooking account readily for our weak chewing and small teeth (Wrangham 2009)."

"A third change in anatomy can be found in our intestines. Wrangham (2009) notes that "... the human small intestine is only a little smaller than expected from the size of our bodies, reflecting that this organ is the main site of digestion and absorption, and humans have the same basal metabolic rate as other primates in relation to body weight. But the large intestine, or colon, is less than 60 percent of the mass that would be expected for a primate of our body weight. The colon is where our intestinal flora ferment plant fiber, producing fatty acids that are absorbed into the body and used for energy.

That the colon is relatively small in humans means we cannot retain as much fiber as the great apes can and therefore cannot utilize plant fiber as effectively for food. But that matters little. The high caloric density of cooked food means that normally we do not need the large fermenting potential that apes rely on.

"Finally, the volume of the entire human gut, comprising stomach, small intestine, and large intestine, is also relatively small, less than in any other primate measured so far. The weight of our guts is estimated at about 60 percent of what is expected for a primate of our size: the human digestive system as a whole is much smaller than would be predicted on the basis of size relations in primates.

"Our small mouths, teeth, and guts fit well with the softness, high caloric density, low fiber content, and high digestibility of cooked food. The reduction increases efficiency and saves us from wasting unnecessary metabolic costs on features whose only purpose would be to allow us to digest large amounts of high-fiber food. Mouths and teeth do not have to be large to chew soft, high-density food, and a reduction in the size of jaw muscles may help us produce the low forces appropriate to eating a cooked diet. The smaller scale may reduce tooth damage and subsequent disease. In the case of intestines ... compared to that of great apes, the reduction in the human gut size saves humans at least 10 percent of daily expenditure: the more gut tissue in the body, the more energy must be spent on its metabolism. Thanks to cooking, very high-fiber food of a type eaten by great apes is no longer a useful part of our diet. The suite of changes in the human digestive system makes sense (Wrangham 2009)."

So when did all these changes take place? When did humans learn to control fire and, in particular, when did they learn to cook their food. "Before two million years ago, there is no suggestion for the control of fire. Since then there have been only three periods when our ancestors' evolution was fast and strong enough to justify changes in the species names. They are the times that produced *Homo erectus* (1.8 million years ago), *Homo heidelbergensis* (800,000 years ago), and

Homo sapiens (200,000 years ago).These are therefore the only times when it is reasonable to infer that cooking could have been adopted (Wrangham 2009)."

Wrangham (2009) notes that "The timing of the *erectus-heidelbergensis* transition provides a reasonably comfortable fit with the archeological data on the control of fire... The main changes in anatomy from *Homo erectus* to *Homo heidelbergensis* were an increase in cranial capacity (brain volume) of about 30 percent, a higher forehead, and a flatter face." However, he notes that these "modifications show little correspondence to changes in diet. So this Pleistocene transition does not look favorable. It is a possibility for when cooking began, but not a promising one.

"The only other option is the original change, from habilines to *Homo erectus*. This shift happened between 1.9 million and 1.8 million years ago and involved much larger changes in anatomy than any subsequent transitions... The reduction in tooth size, the signs of increased energy availability in larger brains and bodies, the indication of smaller guts, and the ability to exploit new kinds of habitat all support the idea that cooking was responsible for the evolution of *Homo erectus*."

***

Corballis (2011) noted that there "... are two genes that seem to have resulted in increased brain size through negative rather than positive selection; that is, they were once active genes that were inactivated. Moreover, both of them seem to have been inactivated just prior to the time when brain size increased in our own genus, Homo, perhaps providing the initial thrust that propelled us toward humanity. One is a gene that encodes an enzyme that produces an acid that inhibits brain growth. This acid is absent in Neandertal fossils and in present-day humans, but present in other primates – although down-regulated in the chimpanzee. It has been estimated that the gene was inactivated around 2.8 million years ago. The other gene encodes the myosin-heavy chain (MYH 16) responsible for the strong chewing muscles in most primates, including chimpanzees and gorillas, as well as the early hominins. This gene was inactivated an estimated 2.4 million years ago, leading to speculation that the shrinkage of jaw muscles and their supporting bone structure removed a further constraint on brain growth – brawn gave way to brain. This change may have signaled a change of diet from tough vegetables to tender meat, or it may have had to do with the increasing use of the hands rather than the jaws to prepare food. Conclusions about roles of these last two genes in determining the size and shape of the human brain are of course speculative, and indeed controversial, but it is nevertheless interesting to reflect that we may owe our humanity in part to the loss of genetic information."

"Man's life history, physiology, and basic anatomy were beginning to make significant changes at this time that would lead to changes in social behavior, such as the development of kinship and culture. The first of these changes was of course an increase in brain size. We see about a 10% increase in the brain size from the first of our ancestors to *A. africanus* over a period of millions of years. The leap from *A.* africanus to *Homo habilis* represents about a 25 percent increase - and it occurred in the "blink of an evolutionary eye". Over about the next 2 million years average brain size would increase to about 800 cc in *Homo ergaster* and 900 cc in early *H. erectus* and 1100 cc in late *Homo erectus*, 1200 cc in *Homo heidelbergensis*, and 1350 cc in *Homo sapiens sapiens* - a nearly 70 percent increase.

To put brain size into context, a measure of braininess has been developed for all mammals called EQ (Encephalization Quotient). EQ gives a measure of actual brain size verses expected brain size for mammals of equivalent body size. You would expect an elephant to have a larger brain than a mouse. But does that mean the elephant is a lot smarter than a mouse? Not necessarily. The EQ for a chimpanzee is about 2. That is the chimps brain size is about twice as large compared to all other mammals of the same body size. So to start with, apes have brains that are 2 times larger than would be expected for all other mammals of similar body size. The chimp is, therefore, "more brainy" than the average mammal of equivalent size. The EQ for *A. Afarensis is about 3.1 and for A. africanus* is about 3.4 - as we would expect somewhat larger than that of the chimps. The EQ for *H. habilis is about 4.0* and for early *H. erectus* (*H. erectus* had a substantially larger body than *H. habilis*) is about 5.5 and for late *H. erectus* about 6.1. This is a remarkable change in so short a period of time. For interest at this point the EQ for *H. sapiens sapiens*, modern man, is over 7.6 (Table 3-1).

Table 3-1. Estimated average cranial capacity (brain size) among hominids - primarily *Homo spp.*

Species in ascending Order of age	Estimated average cranial capacity (cubic centimeters)	Encephalization Quotient
Australopithecus afarensis	413	3.1
Australopithecus africanus	442	3.4
Australopithecus sediba	420	
Homo habilis	631	4.0
Homo rudolfensis	575	
Homo georgicus	600	
Homo floresiensis	417	
Homo ergaster	800	
Homo erectus early	900	5.5
Homo erectus late	1100	6.1
Homo antecessor	1200	
Homo sapiens neanderthalensis	1450	
Homo sapiens sapiens	1350	7.6

But it is not just the volume of the brain that makes us smart. "The cerebral cortex of a mammal is a sheet of grey matter, wrapped around the outside of the brain. Getting brainier partly consists in increasing the area of the sheet. This could be done by increasing the total size of the brain, and of the skull that houses it. But there are downsides to having a big skull. It makes it harder to be born, for one thing. As a result, brainy mammals contrive to increase the area of the sheet while staying within limits set by the skull, and they do it by throwing the whole sheet into deep folds and fissures. This is why the human brain looks like a wrinkled walnut; and the brains of dolphins and whales are the only ones to rival those of us apes for wrinkliness (Dawkins 2009)."

Biologists have recently come to understand that brain size influences more than just intelligence. It correlates with a number of what are known as life-history factors, such as the age of weaning, the age at which sexual maturity is reached, gestation length, and longevity. In species with big brains, these factors tend to be stretched out: infants are weaned later than those in species with small brains, sexual maturity is reached later, gestation is longer, and individuals live longer. Humans today are born helpless. They have a long childhood ahead of them which involves a tremendous amount of learning based on the culture they are born into. The true gestation period for a human is about 21 months - for 9 of these they are carried internally by the mother and for 12 they are literally carried externally, again primarily by the mother. Leakey (1994) notes that "Culture can be said to be *the* human adaptation, and it is made possible by the unusual pattern of childhood and maturation." He also noted that "The helplessness of the newborn human infant is, however, less a cultural adaptation than a biological necessity. Human infants came into the world too early, a consequence of our large brain and the engineering constraints of the human pelvis. "The brain of a newborn ape is about 200 cc and doubles to that of the adult by maturity. The brain of a newborn human has to more than triple in size to reach maturity. With an average adult brain size of 1350 cc that means the human baby must be born relatively early before the brain size exceeds a physiological limit of the female pelvis of about 385 cc. If, like the ape, our brain size only needed to double it would reach about 675 cc prior to birth. What woman would ever give birth to a first child and if she did, and survived the ordeal, would she try for a second? It works out that an adult brain size of about 770 cc is the point at which humans departed from the apes. Now our friend the "Handy Man" is close to this figure - 631 cc and may be considered on the cusp of mankind. Alan Walker (in Swisher et. al. 2000) estimated the biological limit on head size at birth in *Homo erectus* at about 275 cc. *Homo erectus* would require an increase in brain size after birth by a factor of about 3 - the same as for *Homo sapiens. H. ergaster* and *H. erectus* definitely meet our criteria of man not ape. So we've answered the question posed at the end of the last chapter - Ape-man or Man-ape? The Australopithecines were Ape-men, *Australopithecus (Homo) habilis* was likely a transition to Man-ape and *Homo ergaster* and *Homo erectus* were both Man-apes.

***

I've brought up the topic of culture a number of times. "Culture is as much a mechanism for bonding as a mechanism for inter-cultural divergence, maintained through such institutions as religion, language, and custom, whether based on hunting and gathering, agriculture, industry, or trading (Corballis 2011)." "Culture provides a uniquely advantageous means for adapting to environmental change. Cultural innovations can accumulate far more rapidly than genetic mutation, and good ideas can spread horizontally across populations as well as vertically between generations. This strategy of cultural adaptation, more than anything else, has enabled our species to transform itself from a relatively large African mammal to the dominant life form on Earth. We have developed an unprecedented ability to adapt to a wide variety of environments and, sometimes, unfortunately, to alter them irrevocably. Having acquired this seminal cultural advantage, the earliest fully modern humans were able to disperse from Africa, northwards through the Near East to Europe and eastwards across Asia to China and beyond (Klein and Blake 2002)." This epic journey will be discussed in great detail in Chapter V.

"The premise of cultural evolution is that there is a single phenomenon – the march of progress, the ascent of man, apes to Armageddon – that Darwin explained only up to a point. My [Pinker] own view is that human brains evolved by one set of laws, those of natural selection and genetics, and now interact with one another according to other sets of laws, those of cognitive and social psychology, human ecology, and history (Pinker 2009)." This cognitive niche would affect many aspects of human behavior.

"Humans did not adapt to have the strength to compete directly against the large predators that would make a meal of them. Nor did they adapt the stealth to outrun them like many of the undulates on the African savannah. Instead, humans "...built on their primitive inheritance of intelligence and social structure rather than on physical attributes of strength or speed. This is what might be termed the third way, which was to evolve what has been termed the 'cognitive niche,' a mode of living built on social cohesion, cooperation, and efficient planning. It was a question of survival of the smartest (Corballis 2011)."

"Nevertheless it is unlikely that human intelligence can be explained simply in terms of the response to ecological challenges. Our Pleistocene forebears seemed to have conquered the threats posed by the inhospitable environment, but then discovered a further threat – themselves. Paraphrasing an earlier seminal paper by Nicholas K. Humphrey, Richard D. Alexander writes that 'the real challenge in the human environment throughout history that affected the evolution of the intellect was not climate, weather, food shortages, or parasites – or even predators. Rather, it was the necessity of dealing continually with our fellow humans in social circumstances that became ever more complex and unpredictable as the human line evolved.' Humans have proven as adept at killing each other as at killing nonhuman predators – in fact more so, if you consider the extraordinary range of lethal weapons we humans have invented. Nevertheless successful adaptation must have depended as much on cooperation as on competition, leading to what has been called 'runaway social selection.' Our lives depend on a subtle calculus of sharing and greed – of left-wing socialism and right-wing individualism, if you like (Corballis 2011)."

"The cognitive niche embraces many of the zoologically unusual features of our species. Tool manufacture and use is the application of knowledge about causes and effects among objects in the effort to bring about goals. Language is a means of exchanging knowledge. It multiplies the benefit of knowledge, which can not only be used but exchanged for other resources, and lowers its cost, because knowledge can be acquired from the hard-won wisdom, stroke of genius, and trial and error of others rather than only from risky exploration and experimentation. Information can be shared at a negligible cost: if I give you a fish, I no longer possess the fish, but if I give you information on how to fish, I still possess the information myself. So an information-exploiting lifestyle goes well with living in groups and pooling expertise – that is, with culture. Cultures differ from one another because they pool bodies of expertise fashioned in different times and places. A prolonged childhood is an apprenticeship for knowledge and skills. That shifts the balance of payoffs for males toward investing time and resources in their offspring and away from competing over sexual access to females. And that in turn makes kinship a concern of both sexes and all ages. Human lives are long to repay the investment of a long apprenticeship. New habitats can be colonized because even if their local conditions differ, they obey the laws of physics and biology that are already within humans' ken, and can be exploited and outsmarted in their turn (Pinker 2009)."

To answer the question of whether early *Homo spp.* developed culture we need to look at the implications of the physiological changes that have occurred and their effect on kinship. In other words, how did these changes effect adult social behavior. Leakey (1992) expresses it this way: "Humans became human through intense learning - not just of survival in the practical world, but customs and social mores; culture. The social and economic contract that is at the heart of the hunting-and-gathering life demands of individuals an understanding of their roles, their place in the community, the behavior that is expected of them. Among foraging people in the modern world, elaborate kinship systems define an individual's relations to his fellows, to members of other groups, to his ancestors, even to gods."

Dillehay (2000) describes the concept of "cultural landscape" as "the domain of mutual interaction between people and their natural world." He adds "...the environment is perceived and used by different societies and individuals in any number of ways. But it is in fact the complexity of the mutual interaction between society and environment that creates the cultural landscape. The concept most often used by archaeologists to explain this interaction is 'culture ecology' in which society, environment, and technology play roughly equal roles in human adaptation to a particular environment; that is, a society creates and perpetuates a sustaining technology that in turn extracts resources from the environment for human use. Of course technology itself is created as a response to the environment and its impact extends beyond the environment to the social and economic complexity of society itself. The notion of 'culture ecology' is an attempt to conceptualize the complexities of the interactions between these three facets of human existence."

Kingdon (2003) put culture in the context of evolution when he noted that "Unlike the Lucies and Robusts, *Homo* seems to have crossed a new threshold where biological adaptation was overtaken by cultural adaptation as the mechanism for ecological success. The metaphor of threshold may be misleading inasmuch as accretions of innovations must have been developed stage by stage. Nonetheless, the use of fire must represent quite the largest leap toward a 'cultural animal'. Alterations in anatomy continued to be meaningful manifestations of adaptive change, but direct adaptations to local environments were subordinate to - and, to an increasing degree, a consequence of - culture. The preeminence of culture became most marked in *H. sapiens* ." Culture had its genesis in the formation of kinship bonds.

Wrangham (2009) notes that "... cooking has made possible one of the most distinctive features of human society; the modern form of the sexual division of labor. The sexual division of labor refers to women and men making different and complementary contributions to the household economy. Though the specific activities of each sex vary by culture, the general division of labor is a human universal. It is therefore assumed to have appeared well before modern humans started spreading across the globe sixty thousand years ago. So discussion of the evolution of the sexual division of labor centers on hunter-gatherers."

"Hints of comparable sex differences in food procurement have been detected in primates. Female lemurs tend to eat more of the preferred foods than males. In various monkeys such as macaques, guenons, and mangabeys, females eat more insects and males eat more fruit. Among chimpanzees, females eat more termites and ants, and males eat more meat. But such differences are minor because in every nonhuman primate the overwhelming majority of the foods collected and eaten by females and males are the same.

"Even more distinctive of humans is that each sex eats not only from the food items they have collected themselves, but also from their partner's finds. Not even a hint of this complementarity is found among nonhuman primates. Plenty of primates, such as gibbons and gorillas, have family groups. Females and males in those species spend all day together, are nice to each other, and bring up their offspring together, but, unlike people, the adults never give each other food. Human couples, by contrast, are expected to do so.

"In foraging societies a woman always shares her food with her husband and children, and she gives little to anyone other than close kin. Men likewise share with their wives, whether they have received meat from other men or have brought it to camp themselves and shared part of it with other men. The exchanges between wife and husband permeate families in every society. The contributions might involve women digging roots and men hunting meat in one culture, or women shopping and men earning a salary in another. No matter the specific items each partner contributes, human families are unique compared to the social arrangements of other species because each household is a little economy (Wrangham 2009)."

"The division of labor by sex affects both household subsistence and society as a whole. Sociologist Emile Durkheim thought that it's most important result was to promote moral standards, by creating a bond within the family. Specialization of labor also increases productivity by allowing women and men to become more skilled at their particular tasks, which promotes efficient use of time and resources.

It is even thought to be associated with the evolution of some emotional and intellectual skills, because our reliance on sharing requires a cooperative temperament and exceptional intelligence. For such reasons anthropologists Jane and Chet Lancaster described the sexual division of labor as the 'fundamental platform of behavior for the genus *Homo*,' and the 'true watershed for differentiating ape from human lifeways.' Whether they were right in thinking the division began with the genus *Homo* is debated. Though I [Wrangham] agree with the Lancaster's, many think the division of labor by sex started much later. But there is no doubt of its importance in making us who we are (Wrangham 2009)."

"Relying on cooking food creates opportunities for cooperation, but just as important, it exposes cooks to being exploited. Cooking takes time, so lone cooks cannot easily guard their wares from determined thieves such as hungry males without their own food. Pair-bonds solve the problem. Having a husband insures that a woman's gathered foods will not be taken by others; having a wife ensures the man will have an evening meal. According to this idea, cooking created a simple marriage system; or perhaps it solidified a preexisting version of married life that could have been prompted by hunting or sexual competition. Either way, the result was a primitive protection racket in which husbands used their bonds with other men in the community to protect their wives from being robbed, and women returned the favor by preparing their husbands' meals. The beneficial aspects of the household, such as provisioning by males, increases in labor efficiency, and creation of a social network for child-rearing, were additions consequent to solving the more basic problem: females needed male protection, specifically because of cooking. A male used his social power to ensure that a female did not lose her food, and to guarantee his own meal by assigning the work of cooking to the female (Wrangham 2009)."

In summary, Wrangham (2009) notes that "The idea that cooking led to our pair-bonds suggests a worldwide irony. Cooking brought huge nutritional benefits. But for women, the adoption of cooking has also led to a major increase in the vulnerability to male authority. Men were the greater beneficiaries. Cooking freed women's time and fed their children, but it also trapped women into a newly subservient role enforced by male-dominated culture. Cooking created and perpetuated a novel system of male cultural superiority. It is not a pretty picture." I have to add at this point that with more and more women working outside the home, and with their domestic duties often unchanged, the picture is not improving!

Wilson (1980) provides an in-depth discussion on kinship. He cites three factors leading to kinship: the development of a "primary bond" between the mother and child, the lengthening of the pair bond due to the generalized social organization typical of the human species, and the development of a bond of fatherhood between the father and child. With the development of kinship, humans have "the most flexible, generalized and adaptable principle of group organization in the primate order." Wilson notes that "This principle provides the species with the means to organize groups of any (or almost any) size, from a minimal nuclear family to a maximal tribe, and according to any one of the number of possible specific forms. From the psychological point of view, the genesis of kinship engages the male in the entire process of reproduction and nurturance *directly*, not just indirectly, as one among many protectors of the females and infants as a group. The adult male is retained in the reproductive/nurturance cycle and the specialized primate adaptations, which depend on the division and specialization of the sexes and lead to the organization of groups according to separate principles, are overcome.

"The relationship between a mother and her child is fairly straightforward: Both are seeking roughly the same goal – the welfare of themselves and each other. The relationship between a man and his wife's lover or between a woman and her rival for a promotion is also fairly straightforward: Both want the worst for each other. One relationship is all about cooperation, the other all about conflict. But what is the relationship between a woman and her husband? It is cooperation in the sense that both want the best for each other. But why? In order to exploit each other. A man uses his wife to produce children for him. A woman uses her husband to make and help rear her children. Marriage teeters on the line between a cooperative venture and a form of mutual exploitation – ask any divorce lawyer. Successful marriages so submerge the costs under mutual benefits that the cooperation can predominate; unsuccessful ones do not."

Dawkins (1976) explains how this cooperation/exploitation venture could have evolved. "Natural selection favored the production of sex cells that were small and that actively sought out big ones to fuse with. So we can think of two divergent sexual 'strategies' evolving. There was the large-investment or 'honest' strategy. This automatically opened the way for a small-investment exploitive strategy. Once the divergence between the two strategies had started, it would have continued in runaway fashion. Medium-sized intermediates would have been penalized, because they did not enjoy the advantages of either of the two more extreme strategies. The exploiters would have evolved smaller and smaller size, and faster mobility. The honest ones would have evolved larger and larger size, to compensate for the ever-smaller investment contributed by the exploiters, and they became immobile because they would always be actively chased by the exploiters anyway. Each honest one would 'prefer' to fuse with another honest one. But the selection pressure to lock out exploiters would have been weaker than the pressure on exploiters to duck under the barrier: the exploiters had more to lose, and they therefore won the evolutionary battle. The honest ones became eggs, and the exploiters became sperms." He finishes his argument with "So, in mammals for example, it is the female who incubates the fetus in her own body, the female who makes the milk to suckle it when it is born, the female who bears the brunt of the load of bringing it up and protecting it. The female sex is exploited, and the fundamental evolutionary bias for the explanation is the fact that eggs are larger than sperm."

"This is one of the great recurring themes of human history, the balance between cooperation and conflict. It is the obsession of governments and families, of lovers and rivals. It is the key to economics.

It is ... one of the oldest themes in the history of life, for it is repeated right down to the level of the gene itself. And the principal cause of it is sex. Sex, like marriage, is a cooperative venture between two rival sets of genes. Your body is the scene of this uneasy coexistence (Ridley 1993a)."

Let's examine kinship factors more closely. Since the child is born helpless it relies on the female for feeding and nurturing. Wilson (1980) notes "Nursing entails close bodily contact and is a two-way affair that offers the infant not only sustenance but also tactile satisfaction, warmth, and security. For the mother, nursing provides relief and also a sense of emotional satisfaction and closeness to the infant. In addition, since the human infant is immobile, it must be carried, and this again affords close contact between mother and child." These conditions "result in a close tie between adult female and infant. With the relatively greater need for attachment, and we can assume from fossil evidence of early *Homo*, that it was a feature of life. It is this closeness of the adult female and the child over an extended period of development that leads to the formation of a primary bond between the two. This same bond is lacking for the adult male and can never be established between him and the child." Wilson (1980) continues "The relationship between an adult male and infant is possible only through their common connection with the adult female. The relation is premised not on the biological role of the male in the conception of the infant but only on the continuity of his relation to the female [lengthening of the pair bond], which must be sufficient to overlap with her involvement in the primary bond." It is the female who admits the male to the pair bond and then to the primary bond. The adult female, therefore, permits the adult male to be the father of her child.

The pair bond is a function of the generalized nature of social organization among humans expressed in their generalized pattern of sexual receptivity. Wilson (1980) notes that "A significant difference between the human and nonhuman primate is the loss of estrus among humans. Human females have a menstrual cycle that encompasses an ovulation period, at which time of impregnation is most likely to lead to conception. But women are sexually receptive at all times (though less sometimes) and can engage in intercourse at any time during the menstrual cycle and for most of the period of pregnancy. Human males, correspondingly, maintain a continuous interest in females and can engage with them in intercourse at any time during a span of sexual maturity lasting for fifty years or more. The human pattern of sexual receptivity is generalized; corresponding to the switch from specialized sexual changes, such as those found in nonhuman primates, to the permanent presence of elaborated secondary sexual characteristics, itself a generalized condition." It follows then that if all adults within a group are capable of having sex with each other at any time a set of rules governing their behavior must be established to maintain the pair bond. It is likely then that as a condition of survival these rules were the first of many to be established by society. And, in fact, "All known human societies have such a body of rules and that such provisions have a fundamental importance (Wilson 1980)." Rules against incest, polygamy, promiscuity, etc. may have developed from the praise-blame hypothesis suggested by Darwin. Fatherhood brings with it a whole new set of emotions and responsibilities. Recognition of the husband's and father's roll to care for the mother and child would have been understood and reinforced within the social group. Any behavior that deviates from the expected would bring "praise or blame" from society and provide reinforcement for positive behavior and castigation for negative behavior.

Wilson (1980) notes that "The transformation into kinship marks the point at which the human emerges, as the emphasis shifts from the primary and pair bonds to the kinship bond of father/child. At this point the child replaces the female as the new point for kinship relations. A series of transformations occurs: the male becomes 'father' (a different order of connection from any existing before); the offspring of the female acquires a kin identity (child, son/daughter); and the female now bears to the offspring the kinship relationship of mother." And, "...the relations of kinship must pass not through the female, but through the child, who connects the bonds."

Does the pair bond ensure fidelity? Write (1995) asks "Are human males and females born to form enduring bonds with one another? The answer is hardly an unqualified yes for either sex. Still, it is closer to a yes for both sexes than it is in the case of, say, chimpanzees. In every human culture on the anthropological record, marriage – whether monogamous or polygamous, permanent or temporary – is the norm, and the family is the atom of social organization. Fathers everywhere feel love for their children, and that's a lot more than you can say for chimp fathers and bonobo fathers, who don't seem to have much of a clue as to which youngsters are theirs. Their love leaves fathers to feed and defend their children, and teach them useful things."

"Socioecology studies indicate that whether a species is monogamous or polygamous can be determined by its ecology. "The logic behind it was that female mammals set out to distribute themselves without regard to sex, living alone or in small groups or in large groups according to the dictates of food and safety. Males then set out to monopolize as many females as possible either by guarding groups of females directly or by defending a territory in which females lived. Solitary, widely dispersed females gave a male only one option: to monopolize a single female's home range and be her faithful husband (for instance, the gibbon). Females that were solitary but less far apart gave him the chance to monopolize the home ranges of two or more separate females (for instance, the orangutan). Small groups of females gave him the chance to monopolize the whole group and call it his harem (for instance, the gorilla). He would have to share large groups with other males (for instance, the chimp) (Ridley 1993a)."

"With the invention of agriculture, the opportunity for some males to be polygamous arrived with a vengeance. Farming opened the way for one man to grow much more powerful than his peers by accumulating a surplus of food, whether grain or domestic animals, with which to buy the labor of other men. The labor of other men allowed him to increase his surplus still more. For the first time having wealth was the best way to get wealth. Luck does not determine why one farmer reaps more than his neighbor to the same degree that it determines the success of a hunter. Agriculture suddenly allowed the best farmer in the band to have not only the largest hoard of food but the most reliable supply. He had no need to share it freely, for he needed no favor in return.

"Pastoral societies are, almost without exception, traditionally polygamous. It is not hard to see why. A herd of cattle or sheep is almost as easy to tend if it contains fifty animals as twenty-five. Such scale economics allow a man to accumulate wealth at an ever-increasing rate. Positive feedback leads to inequities of wealth, which leads to inequities of sexual opportunity (Ridley 1993a)."

"Many human societies are indeed monogamous. In our society, parental investment by both parents is large and not obviously unbalanced. Mothers certainly do more direct work for children that fathers do, but fathers often work hard in a more indirect sense to provide the material resources that are poured into the children. [I'm not sure that the balance of care-giving hasn't shifted dramatically with the advent of both parents working – the females at work all day and at care-giving all evening.] On the other hand, some human societies are promiscuous, and many are harem-based. What this astonishing variety suggests is that man's way of life is largely determined by culture rather than by genes. However, it is still possible that human males in general have a tendency towards promiscuity, and females a tendency towards monogamy, as we would predict on evolutionary grounds. Which of these two tendencies wins in particular societies depends on details of cultural circumstance, just as in different animal species it depends on ecological details (Dawkins 1976)."

"The nature of the human male, then, is to take opportunities, if they are granted him, for polygamous mating and to use wealth, power, and violence as a means to sexual ends in the competition with other men – though usually not at the expense of sacrificing a secure monogamous relationship. It is not an especially flattering picture, and it depicts a nature that is very much at odds with modern ethical preferences – for monogamy, fidelity, equality, justice, and freedom from violence (Ridley 1993a)."

"It is our usual monogamy, not our occasional polygamy, that sets us apart from other mammals, including apes. Of the four other apes (gibbons, orangutans, gorillas, and chimpanzees), only the gibbon practices anything like marriage. Gibbons live in faithful pairs in the forests of Southeast-Asia, each pair living a solitary life within a territory (Ridley 1993a)."

"There has been no genetic change since we were hunter-gathers, but deep in the mind of the modern man is a simple male hunter-gather rule: Strive to acquire power and use it to lure women who will bear heirs; strive to acquire wealth and use it to buy other men's wives who will bear bastards. It began with a man who shared a piece of prized fish or honey with an attractive neighbor's wife in exchange for a brief affair and continues with a pop star ushering a model into his Mercedes. From fish to Mercedes, the history is unbroken: via skins and beads, plows and cattle, swords and castles. Wealth and power are means to women; women are means to genetic eternity.

"Likewise, deep in the mind of modern women is the same basic hunter-gatherer calculation, too recently evolved to have changed much: Strive to acquire a provider husband who will invest food and care in your children; strive to find a lover who can give those children first-class genes. Only if she is very lucky will they be the same man. It began with a woman who married the best unmarried hunter in the tribe and had an affair with the best married hunter, thus ensuring her children a rich supply of meat. It continues with a rich tycoon's wife bearing a baby that grows up to resemble her beefy bodyguard. Men are to be exploited as providers of parental care, wealth, and genes.

Cynical? Not half as cynical as most accounts of human history (Ridley 1993a)."

Through all of this we are primarily monogamous, raise our children in a family environment, and invest heavily in their futures. Write (1995) notes that "At some point, in other words, extensive *male parental investment* entered our evolutionary lineage. We are, as they say in zoology literature, high in MPI. We're not so high that male parental investment typically rivals female parental investment, but we're a lot higher than the average primate." He goes on to add that "... even with high MPI, and in some ways because of it, a basic underlying dynamic between men and women is mutual exploitation. They seem, at times, designed to make each other miserable." It is the uneasy coexistence, the constant pull between cooperation and exploitation, mentioned earlier by Ridley (1993a).

"That male commitment is in limited supply – that each man has only so much time and energy to invest in offspring – is one reason females in our species defy stereotypes prevalent elsewhere in the animal kingdom. Females in *low*-MPI species – that is, in most sexual species – have no great rivalry with one another. Even if dozens of them have their hearts set on a single, genetically optimal male, he can, and gladly will, fulfill their dreams; copulation doesn't take long. But in high-MPI species such as ours, where a female's ideal is to *monopolize* her dream mate – steer his social and material resources toward her offspring – competition with other females is inevitable. In other words, high male parental investment makes sexual selection work in two directions at once. Not only have males evolved to compete for scarce female eggs; females have evolved to compete for scarce male investment (Write 1995)." Recall that "... this dual – on males for possessing some quality and on females for admiring the self-same quality – is the ingredient for explosive, runaway selection (Dawkins 2004)."

This leads us to a second concept known as "kin selection." Wright (1995) notes that "... should a gene appear that *happens* to make its vehicle [read the kin A] behave in ways that help the survival or reproductive prospects of other vehicles [read the kin B] likely to contain a copy of that gene, then the gene may thrive, even if prospects for *its* vehicle [read kin A again] are lowered in the process.

This is kin selection. And the logic could apply to genes inclining human beings to sense early on who their siblings are and thereafter share food with them, give guidance to them, defend them, and so on – genes, in other words, leading to sympathy, empathy, compassion: genes for love."

Wright (1995) provides an example involving a drowning child. "A gene inclining you to jump in a river and save a drowning child, taking a one-in-ten chance of dying, could flourish so long as the child were your offspring or your brother or sister; the gene could even spread, at a slower rate, if that child were your first cousin, since first cousins share, on average, one eighth of your genes." He continues this thought that a gene for heroism could develop if the drowning event occurred "… in rather small populations where most of the children were fairly near relatives of the man who risked his life. In other words: indiscriminant heroism, reflecting the *average* degree of relatedness to people in the general vicinity, could evolve if that average were fairly high."

So how do we determine relatedness among individuals? It turns out that there is an index of relatedness based on genetics. The formula is given below:

$$RI = X(1/2)^g$$

Where RI = the relatedness index.

X = the number of common ancestors shared by the two people you're trying to relate.

g = the number of steps (generations) you need to count starting with individual A to the common ancestor and back again to individual B.

Let's take a simple example. What is the RI between you and your brother or sister? First you both share 2 common ancestors – a mother and a father – so X = 2. Now counting up from yourself to your next common ancestor you have 1 (generation) and back down to your sister you have 1 (generation) – therefore g = 2. Plug this info into the formula and you have:

$RI = 2(1/2)^2$ or $2(1/4) = 1/2$

Do the same for first cousins. *Normally*, you both share a grandmother and a grandfather – X = 2. And you have to count up 2 to get from you to your grandparents and down 2 to get to your first cousin - g = 4. Therefore:

$RI = 2(1/2)^4$ or $2(1/16) = 1/8$

Notice that above I said normally! My case is somewhat different. My father and my uncle were brothers and my mother and my aunt were sisters. Two Lessard brothers married two Rolewicz sisters. Therefore my first cousins and I share 4 grandparents. Our RI is then:

$RI = 4(1/2)^4$ or $4(1/16) = 1/4$. We are midway between brothers and sisters and first cousins!

The relatedness index gives a way to put a quantitative estimate on when altruism should pay off.

Dawkins (1976) explains that "For relationships as distant as third cousin ($2(1/2)^8 = 1/128$), we are getting down near the baseline probability that a particular gene possessed by A will be shared by a random individual taken from the population. A third cousin is not far from being equivalent to any old Tom, Dick, or Harry as far as an altruistic gene is concerned. A second cousin (relatedness = 1/32) is only a little bit special; a first cousin somewhat more so (1/8). Full brothers and sisters, and parents and children are very special (1/2), and identical twins (relatedness = 1) just as special as oneself. Aunts and uncles, nephews and nieces, grandparents and grandchildren, and half-brothers and half-sisters [and of course my special case of double-cousins], are intermediate with a relatedness of 1/4." I suspect that a relatedness index 1/4 is very special, since in my case, one of my double-cousins,

Mary Lee (affectionately known as Sis), risked her life to save me from drowning when we were younger.

"Now we are in a position to talk about genes for kin-altruism much more precisely. A gene for suicidal saving of five cousins would not become more numerous in the population, but a gene for saving five brothers or ten first cousins would. The minimum requirement for a suicidal altruistic gene to be successful is that it should save more than two siblings (or children or parents), or more than four half-brothers (or uncles, aunts, nephews nieces, grandparents, grandchildren [or in my special case double-cousins]), or more than eight first cousins, etc. Such a gene, on average, tends to live on in the bodies of enough individuals saved by the altruist to compensate for the death of the altruist itself.

"We can now see that parental care is just a special case of kin altruism. Generally speaking, an adult should devote just as much care and attention to its orphaned baby brother as it does to one of its own children. Its relatedness to both infants is exactly the same, ½. In gene selection terms, a gene for big sister altruistic behavior should have just as good a chance of spreading through the population as a gene for parental altruism. In practice, this is an over-simplification for various reasons ..., and brotherly or sisterly care is nothing like as common in nature as parental care. But the point I am making here is that there is nothing special *genetically* speaking about the parent/child relationship as against the brother/sister relationship. The fact that parents actually hand on genes to children, but sisters do not hand on genes to each other is irrelevant, since the sisters both receive identical replicas of the same genes from the same parents (Dawkins 1976)."

Dawkins (1976) summarizes his thinking by noting that "Kin selection accounts for within-family altruism; the closer the relationship, the stronger the selection. Within a small, closely related population, an indiscriminate altruism could indeed evolve. And that's true even though some of the altruism would get spent on people who weren't relatives. After all, even if you channel your altruism precisely toward siblings, some of it is wasted, in evolutionary terms, since siblings don't share all your genes, and any given sibling may not carry the gene responsible for the altruism. What matters, in both cases, is that the altruism gene *tends* to improve prospects for vehicles that will *tend* to carry copies of itself; what matters is that the gene does more good than harm, in the long run, to its own proliferation. Behavior always takes place amid uncertainty, and all natural selection can do is play the odds.

"Any genes that sharpen the precision with which altruism is channeled will thrive. A gene that leads a chimpanzee to give two ounces of meat to a sibling will eventually prevail over a gene that leads it to give an ounce to a sibling and an ounce to an unrelated chimp. So unless identifying kin is very hard, evolution should produce a strong and well-targeted strain of benevolence, not a weak and diffuse strain. And that is what has happened.

"Back before our species became high in male parental investment, there was no reason for males to be intensely altruistic toward offspring. That sort of affection was the exclusive province of females, in part because only they could be sure who their offspring were. But males could be pretty sure who their brothers and sisters were, so love crept into their psyches via kin selection. Had males not thus acquired the capacity for sibling love, they might not have been so readily steered toward high male parental investment, and the even deeper love it brings. Evolution can only work with the raw materials that happen to be lying around; if love for certain kinds of children – siblings – had not been part of males' minds several million years ago, the path to loving their own children – the path to high MPI – might have been too tortuous."

The bonds of kinship and relationships between other kinship groups are only possible if male, female, and child are able to see themselves and others as individuals. Wilson (1980) ties sexuality, the foundation of the initial attraction of male to female and female to male "as the primary source of individuation in the human species."

But, individuality goes beyond just recognizing the observable, external features of the individual and others. It allows for individuals to look within themselves; to refer to "my mind' or "my spirit" in addition to "my body" - the psyche. This looking within one's self requires self-consciousness. Wilson (1980) notes "In all probability, language and self-consciousness were, as they are now, bound up together, and maybe they evolved together." Language provides a "...means of *expressing* and *symbolizing* consciousness (Write 1995)."

It is possible to have "... mutual back-scratching, or 'reciprocal altruism', where there is a delay between the good deed and repayment (Dawkins 1976)." Dawkins (1976) goes on to say that "A long memory and a capacity for individual recognition are well developed in man. We might therefore expect reciprocal altruism to have played an important part in human evolution. Trivers goes so far as to suggest that many of our psychological characteristics – envy, guilt, gratitude, sympathy etc. – have been shaped by natural selection for improved ability to cheat, to detect cheats, and to avoid being thought of as a cheat. Of particular interest are 'subtle cheats' who appear to be reciprocating, but who consistently pay back slightly less than they receive. It is even possible that man's swollen brain, and his predisposition to reason mathematically, evolved as a mechanism of ever more devious cheating, and ever more penetrating detection of cheating in others. Money is a formal token of delayed reciprocal altruism."

Axelrod (2006) provides a treatise on the evolution of cooperation. The evolution of cooperation, as we've discussed in terms of kinship, requires a social structure within the community. Axelrod (2006) uses game theory, specifically the "Prisoners Dilemma" to develop his theory. "The way the game works is shown in Figure [3-3]. One player chooses a row, either cooperating or defecting. The other player simultaneously chooses a column, either cooperating or defecting. Together, these choices result in one of the four possible outcomes shown in the matrix. If both players cooperate, both do fairly well. Both get R, the *reward for cooperation*. In the concrete illustration Figure [3-3] the reward is 3 points. This number might, for example, be a payoff in dollars that each player gets for that outcome. If one player cooperates but the other defects, the defecting player gets the *temptation to defect*, while the cooperating player gets the *sucker's payoff*. In the example, these are 5 points and 0 points respectively. If both defect, both get 1 point, the punishment for *mutual defection* (Axelrod 2006)." Simple enough. You might think that always defecting would be the best strategy? As we'll find out, while this strategy has some merit, the best strategy is one called TIT FOR TAT in that it will ensure cooperation and therefore score the highest point total over a number of iterations of the game. In TIT FOR TAT the first player always cooperates on the first move. Axelrod (2006) calls it a "nice" strategy. After the first move she/he always retaliates by doing whatever the second player does. He notes that the Golden Rule: "Do unto others as you would have them do unto you" is a non-starter for promoting cooperation. TIT FOR TAT might be stated as a revised Golden Rule: "Do unto others as they would do unto you, and, if they don't cooperate, do as they do to you." Apparently, a little retaliation is a good thing if you're trying to elicit cooperation. Axelrod (2006) found that TIT FOR TAT always performs best to foster cooperation no matter what strategy is used against it.

	Cooperate	Defect
Cooperate	R=3 Reward for mutual cooperation	S=0, T=5 Sucker's payoff, and temptation to defect
Defect	T=5, S=0 Temptation to defect and sucker's payoff	P=1 Punishment for mutual defection

Figure 3-3. The Prisoner's Dilemma matrix

"The evolution of cooperation requires that individuals have a sufficiently large chance to meet again so that they have a stake in their future interaction. If this is true, cooperation can evolve in three stages.

1. The beginning of the story is that cooperation can get started even in a world of unconditional defection. The

development *cannot* take place if it is tried only by scattered individuals who have virtually no chance to interact with each other [hence the requirement of a social structure]. However, cooperation can evolve from small clusters of individuals who base their cooperation on reciprocity and have even a small proportion of their interactions with each other.

2. The middle of the story is that a strategy based on reciprocity can thrive in a world where many different kinds of strategies are being tried.
3. The end of the story is that cooperation, once established on the basis of reciprocity, can protect itself from invasion by less cooperative strategies. Thus, the gear wheels of social evolution have a ratchet (Axelrod 2006)."

"Turning to the development of the theory, the evolution of cooperation can be conceptualized in terms of three separate questions:

Robustness. What type of strategy can thrive in a variegated environment composed of others using a wide variety of more or less sophisticated strategies?

Stability. Under what conditions can such a strategy, once fully established, resist invasion by mutant *strategies*?

Initial viability. Even if a strategy is robust and stable, how can it ever get a foothold in an environment which is predominately noncooperative (Axelrod 2006)?"

The TIT FOR TAT strategy was the only strategy that could satisfactorily address all of these questions. However, the opposite of the Golden Rule – ALL D - "always defect" – or not cooperate – strategy under certain conditions also was robust, stable and initially viable. This strategy works well in a society where all the individuals always defect. It cannot be invaded by a single individual using TIT FOR TAT. However, if a cluster of people invade this society with a TIT FOR TAT strategy, TIT FOR TAT will eventually win out. As Axelrod (2006) notes, "Genetic kinship theory suggests a plausible escape from the equilibrium of ALL D. Close relatedness of players permits true altruism – sacrifice of fitness by one individual for the benefit of another. True altruism can evolve when the conditions of cost, benefit, and relatedness yield net gains for the altruism-causing genes that are resident in the related individuals."

"What accounts for TIT FOR TAT's robust success is its combination of being nice, retaliatory, forgiving, and clear. Its niceness prevents it from getting into unnecessary trouble. Its retaliation discourages the other side from persisting whenever defection is tried. Its forgiveness helps restore mutual cooperation. And, its clarity makes it intelligible to the other player, thereby eliciting long-term cooperation (Axelrod 2006)."

"The main results of Cooperation Theory are encouraging. They show that cooperation can get started by even a small cluster of individuals who are prepared to reciprocate cooperation, even in a world where no one else will cooperate. The analysis also shows that the two key requirements for cooperation to thrive are that the cooperation be based on reciprocity, and that the shadow of the future [the fact that the players will meet again] is important enough to make this reciprocity stable. But once cooperation based on reciprocity is established in a population, it can protect itself from invasion by uncooperative strategies.

"It is encouraging to see that cooperation can get started, can thrive in a variegated environment, and can protect itself once established. But what is most interesting is how little had to be assumed about individuals or the social setting to establish these results. The individuals do not have to be rational: the evolutionary process allows the successful strategies to thrive, even if the players do not know why or how. Nor do the players have to exchange messages or commitments: they do not need words, because their deeds speak for them. Likewise, there is no need to assume trust between the players: the use of reciprocity can be enough to make defection unproductive. Altruism is not needed: successful strategies can elicit cooperation even from an egoist. Finally, no central authority is needed; cooperation based on reciprocity can be self-policing.

"The emergence, growth, and maintenance of cooperation do require some assumptions about the individuals and the social setting. They require an individual to be able to recognize another player who has been dealt with before. They also require that one's prior history of interactions with this player can be remembered, so that a player can be responsive. Actually, these requirements for recognition and recall are not as strong as they might seem. Even bacteria can fulfill them by interacting with only one other organism and using a strategy (such as TIT FOR TAT) which responds only to the recent behavior of the other player. And if bacteria can play games, so can people and nations.

"For cooperation to prove stable, the future must have a sufficiently large shadow. This means that the importance of the next encounter between the same two individuals must be great enough to make defection an unprofitable strategy when the other player is provocable. It requires that the players have a large enough chance of meeting again and that they do not discount the significance of their next meeting too greatly.

"Finally, the evolution of cooperation requires that successful strategies can thrive and that there be a source of variation in the strategies which are being used. These mechanisms can be classical Darwinian survival of the fittest and the mutation, but they can also involve more deliberate processes such as imitation of successful patterns of behavior and intelligently designed new strategic ideas (Axelrod 2006)."

Should a parent show favoritism for one child over another? Dawkins (1976) points out a few cases where favoritism might be warranted, but notes "... there is no genetic reason for a mother to have favorites." There seems to be an uneasy "...conflict between parents and young, a battle of the generations. The battle is a subtle one, and no holds barred on either side. A child will not lose an opportunity of cheating. It will pretend to be hungrier than it is, perhaps younger than it is, more in danger than it really is. It is too small and weak to bully its parents physically, but it uses every psychological weapon at its disposal: lying, cheating, deceiving, exploiting, right up to the point where it starts to penalize its relatives more than its genetic relatedness to them should allow. Parents, on the other hand, must be alert to cheating and deceiving, and must try not to be fooled by it. This might seem an easy task. If the parent knows that its child is likely to lie about how hungry it is, it might employ the tactic of feeding it a fixed amount and no more, even though the child goes on screaming. One trouble with this is that the child may not have been lying, and if it dies as a result of not being fed the parent would have lost some of its precious genes (Dawkins 1976)." "There is, then, no general answer to the question of who is more likely to win the battle of the generations. What will finally emerge is a compromise between the ideal situation desired by the child and that desired by the parent (Dawkins 1976)."

Once all the children are out on their own, the nest is empty, there does come a time in life when parents should shift their investment from children to grandchildren. I have found no studies that indicate that parents should lock their teenagers in a closet until they are 21 years of age – although, I'm sure we've all considered it!

"This seems a good moment to mention the puzzling phenomenon known as the menopause, the rather abrupt termination of a human female's reproductive fertility in middle age. This may not have occurred too commonly in our wild ancestors, since not many women would have lived that long anyway. But still, the difference between the abrupt change of life in women and the gradual fading out of fertility in men suggests that there is something genetically 'deliberate' about the menopause – that it is an 'adaptation'. It is rather difficult to explain. At first sight we might expect that a woman should go on having children until she dropped, even if advancing years made it progressively less likely that any individual child would survive. Surely it would seem always worth trying? But we must remember that she is also related to her grandchildren, though half as closely.

"For various reasons, ... women in the natural state became gradually less efficient at bringing up children as they got older. Therefore the life expectancy of a child of an old mother was less than that of a child of a young mother. This means that, if a woman had a child and a grandchild born on the same day, the grandchild could expect to live longer than the child. When a woman reached the age where the average age of each child reaching adulthood was just less than half the chance of each grandchild of the same age reaching adulthood, any gene for investing in grandchildren in preference to children would tend to prosper. Such a gene is carried by only one in four grandchildren, whereas the rival gene is carried by one in two children, but the greater expectation of life of the grandchildren outweighs this, and the 'grandchild altruism' gene prevails in the gene pool. A woman could not invest fully in her grandchildren if she went on having children of her own. Therefore genes for becoming reproductively infertile in middle age become more numerous, since they were carried in the bodies of grandchildren whose survival was assisted by grandmotherly altruism.

This is a possible explanation of the evolution of the menopause in females. The reason why the fertility of males tails off gradually rather than abruptly is probably that males do not invest so much as females in each individual child anyway. Provided he can sire children by young women, it will always pay even very old men to invest in children rather than in grandchildren (Dawkins 1976)."

So, in respect to early *Homo spp.*, we need to ask: What was their social organization like? Were they in kinship groups? Did they have language or at least some means for expressing themselves, if not through words, then through symbols? What is the evidence we have available to us to answer these questions? Since we only have archeology to guide us in answering these questions, we have to examine the physical evidence. For the most part, this evidence is in the form of bones and tools, although, genetics offers some important insights. If we can show that *H. ergaster/H. erectus* complex had an ability to develop language, we can assume that they possessed a sense of self and, therefore must have entered into kinship relationships - in short, they developed culture.

***

Although there was an apparent need for language did early *Homo spp.* have the anatomical structure that actually allowed for the complex sounds required of modern language? There are three structural components available to us that are requisite to this discussion: A hyoid bone, the presence of a Broca's area, and the position of the larynx in the throat. The hyoid bone acts as an anchor point for the vocal tract in our throat. A hyoid bone has never been found in early *Homo spp.* There is only a single case of a hyoid bone being present and it was in association with Neandertal man - a later topic for discussion.

The second line of evidence is in the actual structure of the brain. There are two language centers in the brain - the Werinche's area and the Broca's area. The Werinche's area is associated with language comprehension and is located in the upper back part of the left temporal lobe. This area is not available to us in the fossil record. The Broca's area is a small lump in the brain found on the left side, toward the front of modern human brains and is associated with speech production. Leakey (1992) notes that "the presence of Broca's area cannot be taken as certain proof of language abilities, because in modern humans the linguistic machinery is buried beneath this neurological lump, and not in it. At best Broca's area is indirectly indicative of language abilities." Within the Broca's area is a structure known as the Brodmann's area. We share this area with the chimpanzees and gorillas and, therefore with the Ardipithecines and Australopithecines. Why is this significant? We know apes (other than ourselves) don't speak. It only makes sense once you "...think of linguistic ability as having a heterogeneous structure. If this ability has developed piecemeal over time, then ape brains should share some of the same structures we use for language. The ape asymmetry also means ...that the neuroanatomical substrates for left-hemisphere dominance for language were evident at least five million years ago and are not unique to human evolution (Kenneally 2010)." This will give us a connection between gesturing and speech which we'll discuss at more length in the next chapter. There is evidence that *H. habilis* and *H. ergaster/H. erectus* both demonstrated a distinct Broca's area and were, therefore, likely capable of producing speech. So, if they had the capability to produce speech, did they actually produce language? We have to return once again to anatomy to provide clues to this question.

The third factor in our discussion of language revolves around the location of the larynx in the throat. In mammals the larynx is located high in the throat. This allows them to breath and drink at the same time. This is true of the chimps, probably of the Australopithecines, and to varying degrees in *Homo spp.* as they evolved. The range of sounds in mammals is limited because of this anatomical construct. However, in modern humans the larynx is located down in the throat. "The lower position of the larynx creates a much larger pharyngeal space above the vocal cords, allowing for a far greater range of sound modification. The expanded pharynx is the key to our ability to produce fully articulate speech (Leakey 1992)." The trade off in this structure is that modern humans are prone to chocking. This is a very large, and somewhat dangerous, tradeoff but, the result, development of language, is a huge benefit. In newborn humans the larynx is high in the throat allowing it to drink and breathe at the same time - a useful adaptation for nursing. This position of the larynx hampers language and is why they seem to only have two forms of vocalization - cooing and crying. As the baby is weaned, the larynx begins to move down the throat and at a young age is low in the throat allowing them to speak.

Once again we are faced with a problem. The larynx and pharynx are made of cartilage and do not preserve well. However, Laitman (in Leaky 1992) found that there was a relationship between the location of the larynx in the throat and the curve of the basicranium. He noted that "During our investigations, my colleagues and I noticed that the shape of the bottom of the skull, or basicranium, is related to the position of the larynx. This is not surprising, since the basicranium serves as the roof of the upper respiratory tract." It probably was flat in the Australopithecines but, we find that it begins to curve in early *Homo spp.* Based on comparison of the basicranium of *H. erectus* and modern man, Laitman suggests that the location of the larynx in *H. erectus* was probably about midway down the throat compared to modern humans and in *H. habilis* it was likely even higher in the throat - more akin to the Australopithecines.

Laitman (in Leaky 1992) found that the Australopithecines had a typical ape like basicranium and would, therefore, be unable to produce the universal vowel sounds typical of human speech patterns. *Homo erectus* would likely be capable of producing speech and would not be able to produce some of the vowel sounds like those in "boot, father, and feet" and would have a level of speech equivalent to about a six year old modern human child. Leakey believes that *Homo habilis* "had a degree of spoken language ability considerably in excess of ape vocal communication. Fully developed spoken language like ours? No. I [Laitman] believe language abilities emerged gradually in the human career and were part of an evolutionary package built around the hunting-gathering way of life, something novel in the world of primates."

"Other anatomical evidence suggests that the anatomical requirements for fully articulate speech were probably not complete until late in the evolution of Homo. For example, the hypoglossal canal is much larger in humans than in great apes, suggesting that the hypoglossal nerve, which innervates the tongue, is also much larger in humans, perhaps reflecting the importance of tongued gestures in speech. The evidence suggests that the size of the hypoglossal canal in early australopithecines, and perhaps in Homo habilis, was within the range of that in modern great apes, while that of the Neandertal and early H. sapiens skulls was contained well within modern human range, although this has been disputed. A further clue comes from the finding that the thoracic region of the spinal cord is relatively larger in humans than in nonhuman primates, probably because breathing during speech involves extra muscles of the thorax and abdomen. Fossil evidence indicates that this enlargement was not present in the early hominins or even in Homo ergaster, dating from about 1.6 million years ago, but was present in several Neandertal fossils (Corballis 2011)."

"All humans have essentially the same version of a gene known as FOXP2 - a gene so important that it has swept through the population and become universal. By analyzing the variations in FOXP2 genes possessed by people around the world, Pääbo was able to fix a date, though rather roughly, for the time that all humans acquired the latest upgrade of the FOXP2 gene. It was fairly recently in human evolution, and certainly sometime within the last 200,000 years, he concluded. Language is such a complex faculty that it must be mediated by a large number of genes and have been the last in several stages (Wade 2006)." The FOXP2 gene may have been the last of the genes recruited to the language function – the first step in the development of modern speech. "Societies with two kinds of people, of greatly differing language abilities, may have existed during the evolution of language. As each new variant of gene arose, conferring some improvement in language ability, the carriers of the gene would leave more descendants. When the last of these genes – perhaps FOXP2 – swept through the ancestral human population, the modern faculty of language was attained (Wade 2006)."

We mentioned earlier that the only hyoid bone found to date came from Neandertal Man. Fagan (2010) poses the question "Was Neandertal Man capable of fluent speech? "Were they able to pass on information among their group and others?" Laitman (in Leaky 1992) notes that "Only one Neanderthal hyoid bone has been recovered, on a sixty-three-thousand-year-old skeleton from the Kebara cave, in Israel, but it is lower in the throat than that of *Homo sapiens*. The dimensions of the hypoglossal canal, which carries the nerves that run from the brain to the tongue, are equivalent in Neanderthals to those of the hypoglossal canal in modern humans, while the dimensions of the canal in the thoracic vertebrae, which house the nerves for controlling the diaphragm and breathing, are also similar. Thus we know that Neanderthals had the same motor control over their tongues and breathing as we do. They also appear to have had the same sound-perception structures as people today, as well as some of the hardware for speech. The bones of two of the Neanderthals at El Sidron have yielded the FOXP2 gene, which contributes to speech and language ability by acting on both the brain and the nerves that control facial muscles." With all that explanation, you must believe that Neandertal was capable of speech very much like we are today. Fagan (2010) goes off on a tangent at this point. Like many others before him, he has a hard time thinking of Neandertal Man as other than a large stupid oaf.

Fagan notes that "For all the anatomical features, there are reasons to believe that Neanderthals did not possess fluent speech. For one thing, the Neanderthals lived in small, intimate groups [and small intimate groups have no need to speak to each other?]. They shared knowledge and experience about their environment and their tiny social world, always with a common purpose." Of course they did! That's what language is all about! He proposes that their language consisted of humming - a theory proposed by Mithen (in Fagan 2010) called "Hmmmmm" or "holistic, manipulative, multi-modal, musical, and mimetic." Mithen notes that Neandertals could express all the information they needed to share through singing - not with words - but simply by humming. In Mithen's words "They didn't have much to say that had not either been said many times before or could not have been said by changing the intonation, rhythm, pitch, melody, and accompanying gestures of their widely understood, simple utterances." You have to ask two simple questions at this point: How did they say it many times before without language and by who was it said many times before if they could only hum? Mithen makes this entire discussion even more comical by proposing that deep within caves in southern France "Neanderthals danced and performed ... in total darkness ... and sung familiar rhythms by the light of a fire or flickering brands. Sounds would have echoed off the walls and resonated in the dark space, as their shadows ebbed and flowed against the walls." This is reminiscent of the hallucinogenic behavior of the "hippy movement" of the 1970's.

My answer to the questions posed by Fagan at the beginning of this discussion is emphatically yes! They did have fluent speech and were able to convey information among themselves.

***

We've seen through the fossil evidence that there was a significant increase in brain size from *Australopithecus africanus* to *Homo habilis* to *Homo erectus*. This tremendous increase is reflected in the tool kit, known as the Acheulean Industry, of *H. erectus*. Tool use would develop from two different roots: foraging and aggression. Kingdon (2003) combines these two in a new way of making a living he calls "tool-assisted, niche-stealing". "It was a change that let them cross a technological threshold, which took them into a new relationship with the environment." He provides a long laundry list of tools which could "trap, snare, or net their prey ... [or they could] ambush, excavate, expose, entice, corral, hook, spear, preserve, or contain a steadily enlarging range of food types." One of the key components of this development had to include an ability to teach others the fairly elaborate processes needed to produce this level of sophistication.

There are two areas of interest to us at this point: the "handedness" of the tool makers and the standardization and apparent increase in complexity of the tools themselves. A less direct indicator of possible language skills lies in the left hemisphere dominance of the brain which is associated with "right-handedness" in humans. This "right-handedness" could be taken as an artifact of language. If early *Homo spp.* can be shown to be predominately right handed, the implication is that they just might possess the initial vestiges of language. Leaky (1992) notes that "most early stone knappers - presumably *Homo habilis* - were right-handed. This surely implies that left hemisphere dominance had already emerged by this time, a significant fact in the evolution of language." It must be said that Leaky admits that he really wants *H. habilis* to have language skills and this may be a bias in his thinking on the subject. He notes "My belief, I have to admit, was based largely on intuition, not on hard data." Also, it is important to note that a Broca's area has never been found in the Australopithecines and that they were likely, just as the chimps, to not be either right- or left- handed.

"Linguistically, most left-handers are not mirror images of the righty majority. The left hemisphere controls language in virtually all right-handers (97%), but the right hemisphere controls language in a minority of left-handers, only about 19%. The rest have language in the left hemisphere (68%) or redundantly in both. In all of these lefties, language is more evenly distributed between the hemispheres than in the righties, and thus the lefties are more likely to withstand a stroke on one side of the brain without suffering from aphasia. There is some evidence that left-handers, though better at mathematical, spatial, and artistic activities, are more susceptible to language impairment, dyslexia, and stuttering. Even righties with left-handed relatives (presumably, those righties possessing only one copy of the dominant right-bias gene) appear to parse sentences in subtly different ways than pure righties (Pinker 2009)."

In this early stage of human development tools became more standard in form so that choppers became more like choppers and scrappers became more like scrappers. Though the tools appear to be more standardized and more refined, the implication that early humans were able to do a more varied array of jobs may not be true. Glynn (in Leaky 1992) noted that "It's not necessarily true that the increase in complexity reflects an increase in the number of tasks performed with stone tools, nor are the fancy tools necessarily more efficient in an engineering sense." When Glynn was questioned about why tool makers should go through the effort of making more refined tools he stated "My intuition is that we see in the stone tools the reflection of changes that were affecting culture as a whole. Probably more of all behavior, involved complex rule systems. In communications this presumably meant a more elaborate syntax and an extended vocabulary; in social relations, perhaps larger numbers of defined categories, obligations, and prescriptions, in subsistence, increasing bodies of communicable knowhow." This harkens back to our discussion on rules for kinship relationships to endure and tends to validate the emergence of kinship during this period.

***

"Every human species that preceded *Homo erectus* - that is, every species prior to about 2 million years ago - was distinctly like an ape in many respects: in anatomy, in important aspects of their social lives, in their economic lives (subsistence), and in their mental lives. It was as if, metaphorically, *Homo erectus* carried a banner that proclaimed, 'This way lies humanity' (Swisher et. al. 2000)! "Don't take this to mean that man was predestined to be man. Remember that evolution is a game of chance.

"The number of 'firsts' that *Homo erectus* can claim speaks volumes to the notion that a new kind of creature had arrived under the sun. *Homo erectus* was the first human species with a large brain, a large human-shaped (as opposed to ape-shaped) body, and long lower limbs and relatively short upper limbs. These physical differences contributed to the species' ability to be the first to expand its range beyond Africa [we'll see later that this statement need not be the case], where the human family had been born, and to establish homelands in Asia and Europe. It was the first to obtain a significant amount of meat in its diet through hunting, not just through scavenging. No previous human species had routinely established what might be called home bases, temporary campsites in the endless round of a nomadic lifestyle. No previous species could claim to manufacture complex stone tools into complex, preconceived shapes such as hand axes. *Homo erectus* was the first human species to use fire. It may have been the first time in human history that human babies had an extended childhood, similar to ours, a period of care and attention after weaning and before maturity when they could not survive without their parents. *Homo erectus* individuals may also have for the first time experience that was critical to learning complex cultural behaviors (Swisher et. al. 2000)." While these traits were certainly a part of *Homo erectus* and it pays to point them out, remember that they were building slowly over time - they didn't just happen to *Homo erectus*. We see many of these same traits beginning to show forth in *Homo habilis* and hints of many of them in Australopithecus spp.

A narrative provided by Glenn Isaac (in Leakey 1992) provides a very good summary of a day in the life of early man at the end of this period.

Three of the males leave camp at first light to check the snares they set up the previous day. Simply constructed from strips of soft bark and sharpened sticks, the snares are often highly effective, trapping the leg of a passing animal that happens to snag them. The three males carry long sharpened sticks, as much for self-defense as for spearing a fleeing prey animal. It is not easy to bring down animals at a distance with such implements. Guile and persistence are the real weapons of these hunters.

Meanwhile, some of the mature females of the band are preparing to spend the morning foraging. Soft animal skins skillfully slung and tied around the shoulder do double service as papoose and carrier bag. After a few hours of work the females are laden with fruits, nuts, and succulent tubers, enough to sustain the entire band for the day. Like the men, the women also carry long sharp sticks for protection. They have shorter ones, too, comfortable to grip, with which they dig for tubers. The foragers' skill is in knowing what fruits are ready for the picking and what wispy surface clues indicate the presence of nutritious tubers below.

Back at camp a couple of women and a man engage in idle chatter keeping an eye on the young child who did not accompany the hunting or foraging group. Yesterday, while he was stealthily tracking a young antelope, the man had slipped down a slope and gashed his leg on a jagged piece of lava. The hunt temporarily interrupted, his brothers tended first aid. One of them searched for a clump of sansevieria, a succulent plant that grows around the lake. From some broken fronds, he twisted the juices, letting them drip on the open wound. The brother knew that unless he applied this natural medicine, the wound would become very red and the young man might die. Another stripped some thorns from a nearby acacia and began arranging them across the wound, piercing both sides of the gash. Thin strips of bark looped on alternate sides of the thorns drew flesh together. First aid completed, the brothers continued on their way. Today, although the wound is sore, it looks clean, with little redness. The sansevieria worked, of course, he indicates to the women.

One of the women has been sitting at the edge of camp, deftly striking flakes from a lava cobble, the debris littered around her. She is now working strips of bark, softening them so that they will be supple and strong for making snares and tying skins. Another is using sharp flakes for whittling wood, making digging sticks. They wonder what the foragers - the hunters and gatherers - will bring back later that day.

The women, as always, can be relied on to bring back enough to keep hunger at bay; they represent the stable element in the economy. This day their haul is varied and plentiful, including some bird's eggs, probably flamingo. Soon, the hunting band can be heard returning, and from the noise they are making the camp knows that there will be meat to eat tonight. A large antelope was caught in one of the snares but had broken free, damaging its leg in the process. The hunting band spent much of the day tracking it and, was able to kill the animal as it lay exhausted near the shore. While some of the men were dismembering the animal, another hunter noticed vultures near the lake edge and went to investigate. He found a carcass. Tomorrow the band will go back to see whether anything is worth scavenging. Today the antelope fulfills all their needs.

As always when meat comes into camp, there is great excitement: anticipation of the feast to come and reconstructions of the chase, sometimes more than a little embellished to add to the drama. One of the men looks for lava cobbles suitable for making sharp flakes, and soon has enough to butcher the antelope leg. Meanwhile, one of the children has speared a catfish in the nearby stream. As they share the fruits of the day's efforts, they agree that it is a good place to stay for a few days. Darkness falls, and lightning can be seen in the distant hills, too far for the thunder to reach the little camp.

There's constant awareness of other, similar bands in the region, some of which contain relatives and potential mates. Young females in our group, when they reach maturity, will move to one of these other bands, and a network of relatedness and alliance will be built. Occasionally, the bands from distant parts that arrive in the region are sources of tension and some fear, physical aggression is possible when no alliances exist.

After some days at the camp, our band knows it is time to move on, not least because the storms are getting bigger. Thunder can now be frequently heard, and the rain in the hills is already swelling the stream. Soon it will flood the banks. The band can put off the move no longer, they leave their camp, now a scatter of bones, broken stones, catfish heads, abandoned animal skins and tendons, ends of tubers too bitter to eat, strips of bark, and half-whittled sticks. The band will seek higher ground.

Grandfather's Tale

This tale is taken from the Cherokee Indians of North Carolina.

The First Fire

In the beginning of the world, there was no fire. The animal people were often cold. Only the Thunders, who lived in the world beyond the sky arch, had fire. At last they sent Lightning down to an island. Lightning put fire into the bottom of a hollow sycamore tree.

The animal people knew that the fire was there, because they could see smoke rising from the top of the tree. But they could not get to it on account of the water. So they held a council to decide what to do.

Everyone that could fly or could swim was eager to go after the fire. Raven said, "Let me go. I am large and strong."

At that time Raven was white. He flew high and far across the water and reached the top of the sycamore tree. While he sat there wondering what to do, the heat scorched all his feathers black. The frightened Raven flew home without the fire, and his feathers have been black ever since.

Then the council sent Screech Owl. He flew to the island. But while he was looking down into the hollow tree, a blast of hot air came up and nearly burned out his eyes. He flew home and to this day, Screech Owl's eyes are red.

Then Hooting Owl and Horned Owl were sent to the island together. But smoke nearly blinded them, and the ashes carried up by the wind made white rings about their eyes. They had to come home, and were never able to get rid of the white rings.

The Little Snake swam across to the island, crawled through the grass to the tree, and entered it through a small hole at the bottom. But the smoke and the heat were too much for him, too. He escaped alive but his body had been scorched black. And it was so twisted that he doubled on his track as if always trying to escape from a small space.

Big snake, the climber, offered to go for fire, but he fell into the burning stump and became black as Little Snake. He has been the great blacksnake ever since.

At last Water Spider said that she would go. Water Spider has black downy hair and red stripes on her body. She could run on top of the water and she could dive to the bottom. She would have no trouble getting to the island.

"But you are so little, how will you carry enough fire?" The council asked.

"I'll manage all right," answered Water Spider. "I can spin a web." So she spun a thread from her body and wove it into a little bowl and fastened the little bowl on her back. Then she crossed over to the island and

through the grass. She put one little coal of fire into her bowl and brought it across to the people.

Ever since, we have had fire. And the Water Spider still has her little bowl on her back.

Chapter IV.
The Road to Modern Man
(1.0 million years ago to 100 thousand years ago)

In each great region of the world the living mammals are closely related to the extinct species of the same region. It is therefore probable that Africa was formerly inhabited by extinct apes closely allied to the gorilla and chimpanzee; and as these two species are man's nearest allies, it is somewhat more probable that our early progenitors lived on the African continent than elsewhere.

Charles Darwin, The Decent of Man

I fully subscribe to the judgment of those writers who maintain that of all the differences between man and the lower animals, the moral sense of conscience is by far the most important. It is the most noble of all the attributes of man, leading him without a moment's hesitation to risk his life for that of a fellow-creature; or after due deliberation impelled simply by the deep feeling of right or duty, to sacrifice it in some great cause.

Charles Darwin, The Decent of Man

There is hardly any need to argue that the powers of thought, imagination, invention, and control - the brain as mind, in other words - are the salient aspect of human evolution.

Peter J. Wilson, Man the Promising Primate: the Conditions of Human Evolution.

We left our story at the end of the last chapter with *H. ergaster* widely distributed throughout north and east Africa and with *H. erectus* about to migrate out of Africa to Asia. Recall from our previous discussion that the most likely line of evolution ran from *Australopithecus sediba* to *Homo habilis* to *Homo ergaster*. Some believe that *H. ergaster* and *H. erectus* are the same species. Others believe that *H. ergaster* evolved as a separate line from *H. erectus* and gave rise ultimately to *H. sapiens sapiens*. Our interest in this chapter is to answer the question, "Who were these descendants of *Homo ergaster/Homo erectus* leading to fully modern man?"

Anthropologists refer to the ancestral species of *Homo sapiens sapiens* as an "enigmatic guild of fossil humans sometimes referred to as 'Heidelbergs' or *Homo heidelbergensis;* these are treated by others as subsets of *Homo sapiens* and are confusingly known as 'Archaic *Homo sapiens*,' 'Archaic Moderns,' and more colloquially known as 'middlebrows.' It has been suggested that they derived from a second late excursion of ergaster-like humans out of Africa and that their main innovation was an increase in brain size that brought them within the range of modern humans. Alternatively, they could have risen within Eurasia (Kingdon 2003)."

In fact, there may be a continuum of evolutionary traits grading one to another making it difficult, if not impossible, to identify individual species. After all, it's this continuum that we would most expect from natural selection among a widespread group of people. Kingdon (2003) observes that the "scatter of Archaic humans from the Horn to the Cape, spanning more than a half a million years, suggests a single long-lived lineage." He adds that there is "... perhaps, the existence of a single, widely distributed gene pool. Fossils showing 'intermediate' features raise long-acknowledged problems in defining where *Homo ergaster* (or *Homo erectus*) ends, where *Homo sapiens* begin, and where intermediates such as *H. heidelbergensis* or *H. antecessor* fit in. "Are we vaguely more than a million years old, or is there some rubicon for our emergence as wholly modern humans? The fossil record unambiguously hints at some continuity for modern-looking humans over the last million years; yet there is genetic evidence for a sudden break in this apparent continuity, and its causes are the subject of much speculation (Kingdon 2003)."

The ancestral home of *Homo ergaster/H. erectus* was, in all likelihood, in Ethiopia in northeast Africa. We know that by 1.0 *Mya Homo erectus* was widely distributed throughout Eurasia and eastern and southern Africa. We also know that during this period "Ice Ages" were more evenly distributed over the million year period with periodicity of about 100,000 years. And, that stadial (cool) and inter-stadial (warm) periods were common. These wide swings in climate, from a geological perspective, would have a significant effect on the type and distribution of various habitats globally. "It is after this that many populations of *H. erectus*, particularly but not exclusively, living away from the tropics would have begun to experience severe climatic downturns that would have been interspersed with periods of milder climate. The break-up of some of these populations, and the extinction of a number of them, would have become a real possibility. Although the extinction of populations of proto-humans and humans had been a feature of the early part of our evolution, the attrition caused by regular climatic cycles with the Ice Ages was a new phenomenon. The geographical expansion of *H. erectus*, beyond any boundaries ever reached by humans before, was rudely brought to a halt. The mid-latitude belt, with its southern prong into East Africa, which had once been the subtropical woodland home of the Miocene apes and later the wooded savannah home of the proto-humans and *H. erectus*, was now dissected by inhospitable deserts, treeless steppes, and snow covered peaks (Finlayson 2009a)." These changes in habitat would mean that "many kinds of plants and animals suffered from recurring isolation. *H. erectus* and his descendants would not have been immune to this turmoil (Finlayson 2009a)."

"This change in climate would have had two important effects on humans. On the one hand, populations would have become repeatedly isolated with the result that genetic differences would have started to appear between them. On the other hand, the stresses on the marginal populations would have intensified the need for innovation. Climate change had produced new kinds of apes in the Miocene, new kinds of proto-humans in the Pliocene, and probably *H. erectus* at the start of the Pleistocene. Now it was about to start to work on populations of *H. erectus* and it is no surprise that we detect an increase in brain size among these people at this time (Finlayson 2009a)."

"Like *erectus, heidelbergensis* probably evolved from *ergaster*, and in its anatomy and its geographic distribution, it is a plausible common ancestor for the Neanderthals (*Homo* [*sapiens*] *neanderthalensis*) who appeared subsequently in Europe and for modern humans (*Homo sapiens*) who evolved later in Africa (Klein and Edgar 2002)."

"Rather than try to lump all the fossils from this critical but shadowy period of our evolution into one or two species and lineages, I [Finlayson] see them as regional variations on a theme. The theme is the adaptation of *H. erectus* to a rapidly changing world. Not all made it. Sometimes populations met and exchanged genes; at other times they were isolated from each other and went their separate ways; and at other times some populations could have expanded into the geographical area of others and may have overrun them. Some of the fossils may have belonged to the precursors of *Homo neanderthalensis* or *Homo sapiens* but others may have belonged to lineages that later went extinct or merged with others after brief periods of isolation (Finlayson 2009a)." It is in this turmoil that we find the emergence of *Homo heidelbergensis*.

"The ancestors of the Neanderthals lived and evolved from 127,000 to 780,000 years ago, in the Middle Pleistocene. The oldest fossil of the period is a jaw about half a million years old found at Mauer, near Heidelberg. The fossils from Sima de los Huesos, a cave site also in the Sierra de Atapuerca, stand out among the evidence of the Neanderthals' 'grandparents,' and they constitute the largest collection of human fossils not only in Europe, but in the whole world. The species name *Homo heidelbergensis* can be applied to all of them ... (Arsuaga 2001)." Another cave in Spain, Gran Dolina, where some eighty fossils were found to postdate those at Sima de los Huesos are attributed to a new species that Arsuaga (2001) calls *Homo antecessor.*

"By the time we reach *H. heidelbergensis* we find a powerful predator capable of dispatching large game by ambush hunting, well established on the scene. This powerful hunter then faced an evolutionary dilemma: become light and capable of ranging widely or keep a bulk that gave the edge in a world of close-quarter hunting of large grazers. Those that stuck to the tough build did well in habitat mosaics for a very long time but as climate deteriorated between 70 and 20 thousand years ago, many vegetated areas opened up and they all went extinct along with the large animals with which they had co-evolved. They included the Neanderthals in Europe and Asia, proto-ancestors elsewhere, and *H. erectus* in Africa.

"Much of the evolution from *H. erectus* to proto-Ancestors and beyond took place in a water-limited world, one in which water was the dominant limiting factor on the productivity of the environment. The ability to seek out ephemeral and widely scattered sources of water and to track the seasonal flushes of grass would have been high on the list of priorities. Once again evolution was active in the margins of the habitat mosaics that were the core territories of the tough *H. heidelbergensis* and some of their proto-Ancestor descendants. As increasingly large areas of this world came under the influence of rainfall-drought regimes and seasonal grasslands, humans that had survived in what had previously been marginal areas got lucky and thrived (Finlayson 2009a)."

"The Neanderthals, and their *H. heidelbergensis* ancestors, were not the cold-adapted humans of Ice Age Europe as they have often been portrayed; instead they grew up in the woodlands and savannahs, often around wetlands, of the mid-latitude belt of Eurasia alongside large herbivores. These were highly rich and productive environments that offered a wide range of foods and opportunities for collecting them. At times when mild climate conditions allowed, they tracked these habitats northwards and reached the British Isles and Germany. Like the large herbivores of temperate woodlands, further east their range was restricted to the southern fringe between the mountains of Eastern Europe and Asia to the south and the treeless Russian plains to the north.

"*Homo heidelbergensis* would have been at home in varied landscapes of trees and water but not in dense, impenetrable, jungle. In these environments these people would have been familiar with the animals of the temperate woodland mosaics, many of which would have been very large. Unlike the animals of the open plains, these ones probably did not live in large herds. They would have been, instead, widely dispersed across the landscape but they would have accumulated at higher density around favored places such as sources of fresh water where grassland and woodland animals would come together. It comes as no surprise that many of the *H. heidelbergensis* sites across north-western Europe, in places like Boxgrove, Hoxne, and Pakefield in the British Isles or Mauer, Schoningen, Miesenheim, and Bilzingsleben in Germany, are lakeside or riverside sites close to grassland and open woodland. These sites were places of active hunting, scavenging, and butchery of large animals. In Schoningen, beautifully preserved 400-thousand-year-old wooden spears testify to the hunting prowess and techniques of these people and show that wood was a raw material used by them (Finlayson 2009a)."

There are a number of options at this point for displaying our lineage from *Homo ergaster*. The first would be to draw a straight line from *Homo ergaster* to *Homo sapiens sapiens* with a side branch budding off about 270,000 to 440,000 years ago to *Homo sapiens neanderthalensis* ignoring the whole issue of the Heidelbergs altogether. The second option would be to break the Heidelbergs as a group into the separate species such as *H. antecessor* and *H. heidelbergensis*. Or, lastly, we could have Heidelbergs as the immediate descendent of *Homo ergaster* and then the line passing through a group called proto-ancestors to *H. sapiens sapiens* with a separate line to *H. sapiens neanderthalensis*. "The removal of East African fossils from *erectus* to *ergaster* would be trivial if we accepted the once common notion that *erectus* was directly ancestral to *Homo sapiens*, for *ergaster* would then be simply an early stage of *erectus*. Fossils that date from after 500,000 years ago, however, now indicate that *sapiens* evolved in Africa while *erectus* continued on largely unchanged in eastern Asia. In form and geologic age, *ergaster* is well positioned to be ancestral not only to *erectus* but also of *sapiens* … (Klein and Edgan 2002)." This is the simplest approach to the lineage and best suited to our discussion. My suspicion is that there was one continuously evolving species from *H. ergaster* to *H. sapiens sapiens* and that a bottleneck was formed sometime between 100,000 and 200,000 years ago that isolated the species and forced the evolution of *H. ergaster* to modern man. Figure 4-1 gives my best shot at sorting out who's who! Note I've left the line open for the Rudolfs for determinations by future anthropologists and archeologists based on new discoveries.

"… [T]he people who lived in Africa between 130,000 and 50,000 years ago may have been modern or near-modern in form, but they were behaviorally similar to the Neanderthals. Like the Neanderthals, they commonly struck stone flakes or flake-blades (elongated flakes) from cores they had carefully prepared in advance; they often collected naturally occurring pigments, perhaps because they were attracted to the colors; they apparently built fires at will; they buried their dead, at least on occasion; and they routinely acquired large mammals as food. In these respects and perhaps others, they may have been advanced over their predecessors. Yet, in common with both earlier people and their Neanderthal contemporaries, they manufactured a relatively small range of recognizable stone tool types; their artifact assemblages varied remarkably little through time and space (despite notable environmental variation); they obtained stone raw material mostly from local sources (suggesting relatively small home ranges or very simple social networks); they rarely if ever utilized bone, ivory, or shell to produce formal artifacts; they buried their dead without grave goods or any other compelling evidence for ritual or ceremony; they were relatively ineffective hunter-gathers who lacked, for example, the ability to fish; their populations were apparently very sparse, even by historic hunter-gatherer standards; and they left no compelling evidence for art or decoration (Klein and Edgar 2002)."

At this point, we'll briefly introduce Neandertal Man in the context of our lineage. The commonly accepted line from *H. heidelbergensis* leads to *H. sapiens neanderthalensis*. Once again, genetics provides us with a date range for this separation. Green et. al. (2010) found that "Assuming that human-chimpanzee average DNA sequence divergence was 5.6 to 8.3 million years ago, this suggests that Neandertals and present-day human populations separated between 270,000 and 440,000 years ago ..." I'll give species status to *H. sapiens neanderthalensis* over *H. neanderthalensis* based on the recent findings of Svante Pääbo that "Up to 4 percent of the DNA of people today who live outside of Africa came from Neandertals, the result of interbreeding between Neandertals and early modern humans (Wong 2010b)." Green et. al. (2010) note that "...when a colonizing population (such as anatomically modern humans) encounters a resident population (such as Neandertals), even a small number of breeding events along the wave front of expansion into new territory can result in substantial introduction of genes into the colonizing population as introduced alleles can 'surf' to high frequency as the population expands."

Based on mitochondrial DNA analysis, Svante Pääbo concluded that there was no contribution of mtDNA from Neandertals to modern humans. However, nuclear DNA analysis from a 38,000 year old Neandertal proved differently. They believe that the interbreeding took place in the Middle East between 50,000 and 80,000 years ago. Green et. al. (2010) confirms these findings. "Comparisons of the Neandertal genome to the genomes of five present-day humans from different parts of the world identify a number of genomic regions that may have been affected by positive selection in ancestral modern humans, including

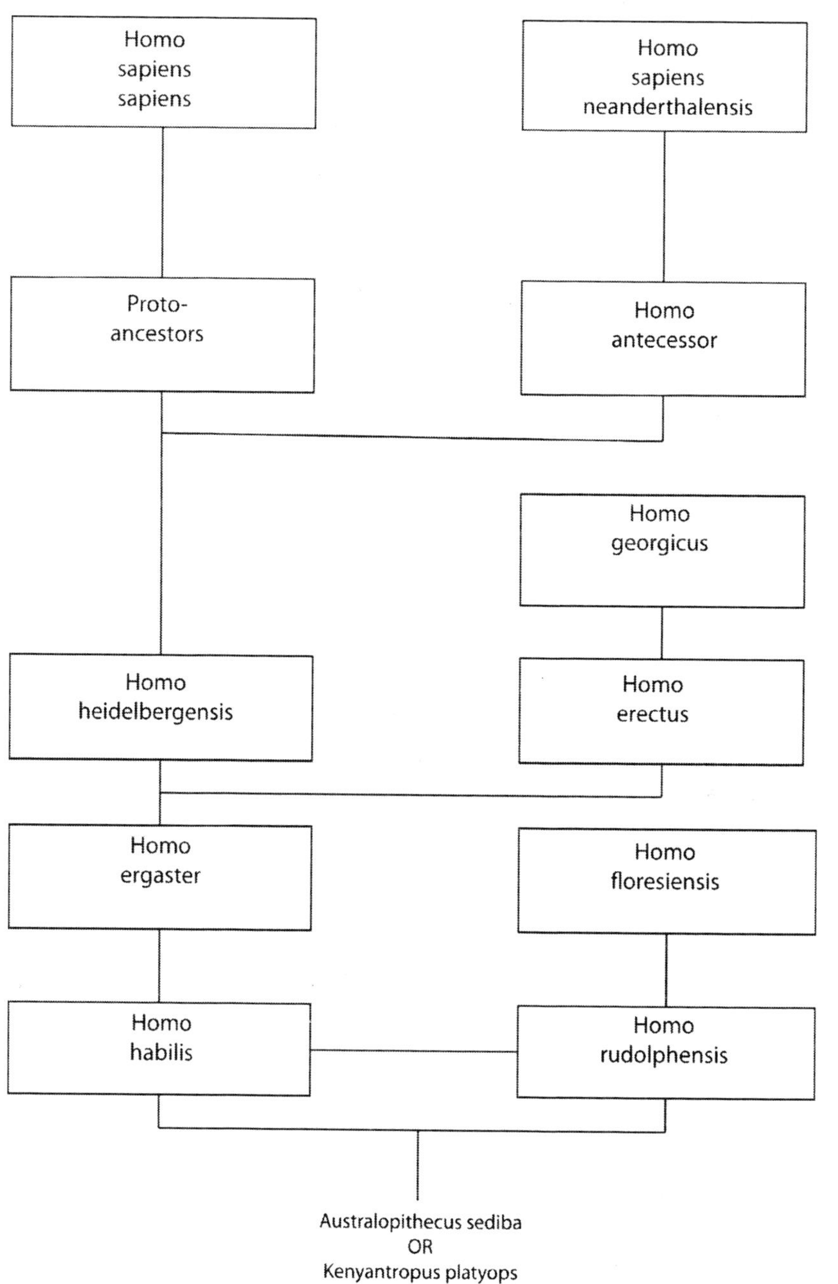

Figure 4-1. A proposed lineage of *Homo spp.* In part from Kingdon 2003 and Finlayson 2009a

genes involved in metabolism and cognitive and skeletal development. We show that Neandertals shared more genetic variants with present-day humans in Eurasia than with present-day humans in sub-Saharan Africa, suggesting that gene flow from Neandertals into the ancestors of non-Africans occurred before the divergence of Eurasian groups from each other." Because Neandertal mtDNA has not been found in modern populations, the interbreeding would likely have been between a male *H. sapiens neanderthalensis* and a female *H. sapiens sapiens.* It is possible that mating occurred between a female *H. sapiens neanderthalensis* and a male *H. sapiens sapiens* with only males being produced and surviving. If there were females produced in this union they did not survive to reproduce or their mtDNA would have been passed on to future generations. We'll see a little later in the discussion that mtDNA is not passed on to the next generation by the male. *H. sapiens sapiens* then spun off from *H. heidelbergensis* about 200,000 years ago. We'll hear more from both *H. sapiens sapiens* and *H. sapiens neanderthalensis* in the next chapter.

***

The major development throughout this evolutionary period was an increase in brain size leading to and caused by major leaps forward in culture and particularly in the development of self-consciousness and language. "At the level of the whole brain, the remark that there has been selection for bigger brains is, to be sure, common in writings about human evolution (especially among paleoanthropologists). Given that premise, one might naturally think that all kinds of computational abilities might come as a by-product. But if you think about it for a minute, you should quickly see that the premise has it backwards. Why would evolution ever have selected for sheer bigness of brain, that bulbous, metabolically greedy organ? A large-brained creature is sentenced to a life that combines all the disadvantages of balancing a watermelon on a broomstick, running in place in a down jacket, and, for women, passing a large kidney stone every few years. Any selection on brain size itself would surely have favored the pinhead. Selection for more powerful computational abilities (language, perception, reasoning, and so on) must have given us a big brain as a by-product, not the other way around (Pinker 2000)!"

"Modern humans are uniquely reliant on culture as a means of adaptation. Like all organisms, human populations evolve genetically, and, like most other animals, we also adapt using individual learning. Humans, however, acquire a greater proportion of their adaptive information by imitation, teaching, and other forms of cultural transmission than do other creatures. Culture clearly allows humans to adapt much more rapidly than non-cultural species. ...the shift from nomadic, egalitarian, food-gathering, societies to sedentary, stratified agricultural ones occurred over the period of a few thousand years. While these changes occurred slowly compared to the diffusion of technical innovations in the modern world, they were much faster than any analogous major adaptive change in non-human organisms (Boyd and Richardson 1996)." Boyd and Richardson (1996) also noted this rapid adaption can lead to:

1. "More within-species variation in characters that affect adaptation to local ecological conditions, and wider ecological range;

2. Greater within-species variation in social structure;

3. More group beneficial social behavior; and,

4. Stylistic differences of local groups."

"Information on how to handle the present so as to survive into the future is necessarily gleaned from the past. Non-random survival of DNA in ancestral bodies is the obvious way in which information from the past is recorded for future use, and this is the route by which the primary database of DNA is built up. But there are three further ways in which information about the past is archived in such a way that it can be used to improve future chances of survival. These are the immune system, the nervous system, and culture. Along with wings, lungs and all the other apparatus for survival, each of the three secondary information-gathering systems was ultimately prefigured by the primary one: natural selection of DNA. We could together call them the four 'memories'.

"The first memory is the DNA repository of ancestral survival techniques, written on the moving scroll that is the gene pool of the species. Just as the inherited database of DNA records recurrent details of ancestral environments and how to survive them, the immune system, the 'secondary memory', does the same thing for diseases and other insults to the body during an individual's own lifetime. This database of past diseases and how to survive them is unique to each individual and is written in the repertoire of proteins that we call antibodies – one population of antibodies for each pathogen (disease-causing organism), precisely tailored by past experience with the proteins that characterize the pathogens. Like many children of my generation, I [Dawkins] had measles and chickenpox. My body 'remembers' the 'experience', the memories being embodied in antibody proteins, along with the rest of my personal database of previously vanquished invaders. I have fortunately never had polio, but medical science has cleverly devised the technique of vaccination for planting false memories of diseases never suffered. I shall never contract polio, because my body 'thinks' it has done so in the past, and my immune system database is equipped with the appropriate antibodies, 'fooled' into making them by the injection of a harmless version of the virus. Fascinatingly, as the work of various Nobel Prize-winning medical scientists has shown, the immune system's database is itself built up by a quasi-Darwinian process of random variation and non-random selection. But in this case the non-random selection is selection not of bodies for their capacity to survive, but of proteins *within* the body for their capacity to envelop or otherwise neutralize invading proteins.

"The third memory is the one we ordinarily think of when we use the word: the memory that resides in the nervous system. By mechanisms that we don't yet fully understand, our brains retain a store of past experiences to parallel the antibody 'memory' of past diseases and the DNA 'memory' (for so we can regard it) of ancestral deaths and successes. At its simplest, the third memory works by a trial-and – error process that can be seen as yet another analogy to natural selection. When searching for food, an animal may 'try' various actions. Though not strictly random, this trial stage is a reasonable analogy to genetic mutation. The analogy to natural selection is 'reinforcement', the system of rewards (positive reinforcement) and punishments (negative reinforcement). An action such as turning over dead leaves (trial) turns out to yield beetle larvae and woodlice hiding under the leaves (reward). The nervous system has a rule that says. 'Any trial action that is followed by reward should be repeated. Any trial action that is followed by nothing, or, worse, followed by punishment, for example pain, should not be repeated'.

"But the brain's memory goes much further than this quasi-Darwinian process of non-random survival of rewarded actions, and elimination of punished actions, in the animal's repertoire. The brains' memory … is, at least in the case of human brains, both vast and vivid. It contains detailed scenes, represented in the internal simulacrum of all five senses. It contains lists of faces, places, tunes, social customs, rules, words. You know it well from the inside, so there is no need for me to spend my words evoking it, except to note the remarkable fact that the lexicon of words at my disposal for writing, and the identical, or at least heavily overlapping, dictionary at your disposal for reading, all reside in the same vast neuronal database, along with the syntactic apparatus for arranging them into sentences and deciphering them.

"Furthermore, the third memory, the one in the brain, has spawned a fourth. The database in my brain contains more than just a record of the happenings and sensations of my personal life – although that was the limit when brains originally evolved. Your brain includes collective memories inherited non-genetically from past generations, handed down by word of mouth, or in books or, nowadays, on the internet. The world in which you live is richer by far because of those who went before us and inscribed their impacts on the database of human culture: Newton and Marconi, Shakespeare and Steinbeck, Bach and the Beatles, Stephenson and the Wright brothers, Jenner and Salk, Curie and Einstein, von Neumann and Berners-Lee and, of course, Darwin.

"All four memories are part of, or manifestations of, the vast superstructure of apparatus for survival which was originally, and primarily, built up by the Darwinian process of non-random DNA survival (Dawkins 2009)."

"In non-cultural organisms most of the variation in social structure appears to occur among species. In contrast, there is a great deal of variation in social organization among groups within the human species. Moreover, much of this variation cannot be related to ecology in any simple way. Culturally related societies occupying different environments often have similar social systems, and culturally unrelated societies in similar environments often have very different social systems. Human's exhibit larger, more complex, and more co-operative societies than most other social organisms. Social complexity and degree of co-operation are suggestive that human social arrangements are more group-beneficial than those of other animals. Finally, the human species is subdivided into symbolically marked groups, such as ethnic groups, that are indicated by dialect, clothing, decoration of tools, and many other stylistic characters. Moreover, ethnic and other symbolically marked groups are also often distinguished by economic and ecological differences and variation in social structure (Boyd and Richardson 1996)."

This is a good point to introduce a relatively new science called memetics. Dawkins coined the term meme – a unit analogous to the gene for cultural evolution. "Memes such as tunes, ideas, and stories spread from brain to brain and sometimes mutate in the transmission. New features of a meme that make its recipients more likely to retain and disseminate it, such as being catchy, seductive, funny, or irrefutable, will lead to the meme's becoming more common in the meme pool. In subsequent rounds of retelling, the most spreadable memes will spread the most and will eventually take over the population. Ideas will therefore evolve to become better adapted to spreading themselves (Pinker 2009)." The jumping from brain to brain requires a "... process which, in the broad sense, can be called imitation. Of course, memes could only come into existence when the genes had provided brains that were capable of imitation – and the nature of those brains must have influenced which memes took hold and which did not. However, once memes had come into existence they would be expected to take on a life of their own (Blackmore 1999)."

"Genes are instructions for making proteins, stored in the cells of the body and passed on in reproduction. Their competition drives the evolution of the biological world. Memes are instructions for carrying out behavior, stored in brains (or other objects) and passed on by imitation. Their competition drives the evolution of the mind. Both genes and memes are replicators and must obey the general principles of evolutionary theory and in that sense are the same. Beyond that they may be, and indeed are, very different – they are related only by analogy. This means that memetic selection drives the evolution of ideas in the interests of replicating the memes, not the genes. This is the big difference that separates memetics from most previous theories of cultural evolution (Blackmore 1999)."

"The whole point of memetics is to treat the meme as a replicator in its own right, operating entirely for the benefit of its own selfish replication. If there is no second replicator, and you are a committed Darwinian, then somehow or other everything must come back to the genes – to biological advantage. If there are two replicators (or more) then there will inevitably be conflicts of interest – circumstances in which the interests of the genes pull in one direction and those of the memes in the opposite direction. These examples are very important for memetics because they would not be predicted by a purely genetic theory. If they occur, they prove that we need a theory of memes – or at least a theory involving some kind of second replicator (Blackmore 1999)."

"Memes fulfill the role of replicator because they exhibit all three of the necessary conditions, that is, heredity (the form and details of the behavior are copied), variation (they are copied with errors, embellishments or other variations), and selection (only some behaviors are successfully copied). This is a true evolutionary process. Effective memes will be those that cause high fidelity, long-lasting memory. Memes may be successful at spreading largely because they are memorable rather than because they are important or useful (Blackmore 1999)."

As noted earlier, selection for more computational abilities led to larger brain size. We see that the average brain size took a pretty good jump from 631 cc for *H. habilis* to 800 cc for *H. ergaster* (Table 4-1 and Figure 4-2). This represents a 25 percent increase in a short period of time. There is a steady climb in brains size from *H. ergaster* to *H. sapiens sapiens* to about 1350 cc - a nearly 70 percent increase. Neandertal Man had an even larger brain on average at 1450 cc and the diminutive Hobbit, *H. floresiensis* comes in at 417 cc - reminiscent of the Australopiths. The Hobbit is only known from Indonesia. Its immediate ancestors are thought to be either *H. habilis* or *H. rudolfensis.* While the Hobbit was likely born into the Pleistocene (2.0 *Mya*), she managed to eke out a living almost into the Holocene. She was still with us 17,000 years ago. Morwood and Oosterzee (2007) emphasize that the "Hobbit's brain ... underwent the most extraordinary evolution yet seen in any hominid. While it retained primitive traits, it had uniquely enlarged frontal and temporal lobes – precisely those areas concerned with higher cognitive activities, such as initiative and planning. Her brain may have been small, but it was normal, and the front part was restructured in a way that meant LB1 was probably smart, had language and could plan ahead." We'll have more to say about this little guy or gal in the next chapter.

Table 4-1. Estimated average cranial capacity (brain size) and EQ among hominids

Species in ascending order of age	Estimated average cranial capacity (cubic centimeters)	Encephalization Quotient (EQ)
Homo floresiensis	417	
Homo habilis	631	4.0
Homo georgicus	690	
Homo rudolfensis	575	
Homo ergaster	800	
Homo erectus - early	900	5.5
Homo erectus - late	1100	6.1
Homo antecessor	1000	
Homo heidelbergensis	1200	
Homo sapiens neanderthalensis	1450	
Homo sapiens sapiens	1350	7.6

"Through language, or, more precisely, through reflective thought and imagery, the human mind creates an internal model of the world that is uniquely capable of representing - and coping with - complex practical and social challenges. Inner thought, not outer communication, was the facility upon which natural selection worked ... Language was its medium - and, at the same time, it is an efficient tool for communication. In this context, it is impossible to separate the evolution of language from the evolution of introspective consciousness, a human facility that philosophers have wrestled with for millennia. Both language and consciousness, anthropologists are coming to believe, evolved in the increasingly intense nexus of our ancestors' social and economic lives. Individuals need to be able to understand, and predict, the behavior of others in their group, and the most efficient way to do that is to create a complex model of social interaction, partly by being intensely aware of one's own behaviors and motivations. Language and introspective consciousness can combine to do that. This has come to be known as the social brain hypothesis (Swisher et. al. 2000)."

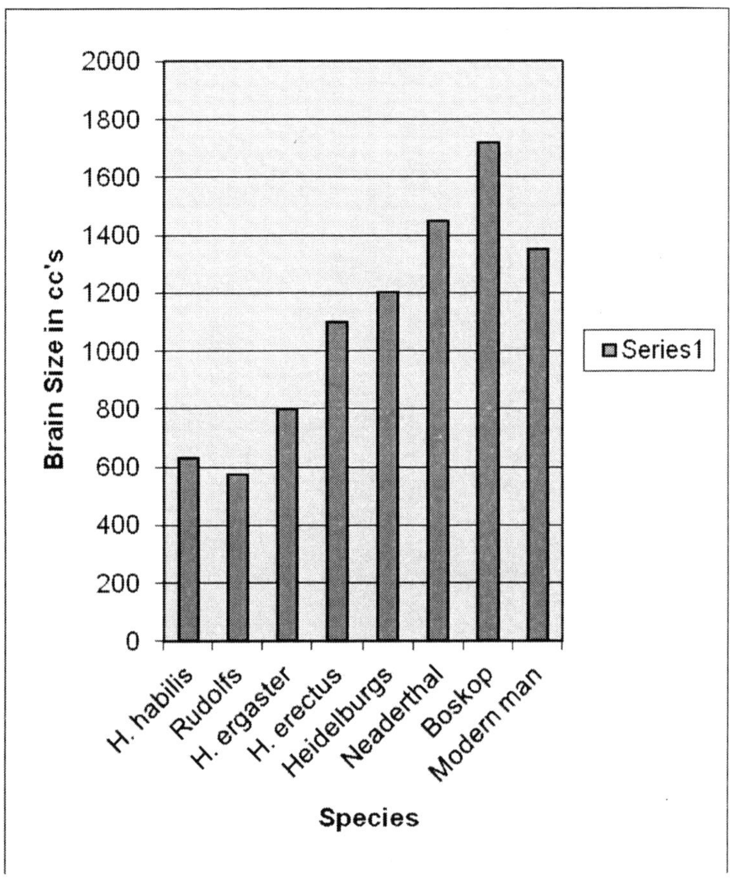

Figure 4-2. Average brain size by species

"For foraging societies, the functional usefulness of a more developed language would have centered on better planning and coordination of all jobs linked with subsistence. Integral to planning in large groups is a greater differentiation of tasks for specific larger objectives. This would have been less a matter of permanent or fixed roles (which decrease flexibility) than the development of a wider range of structured activities, such as building traps, weaving long nets and organizing battues or war parties. Indeed more fighting could have been one outcome of larger populations bred from improved efficiency and larger harvests. More frequent defense was, in turn, an incentive to develop better weapons. Acknowledged as being dependent on social organization for their success, such activities would have been intelligently organized in order to increase the entire group's effectiveness as a sort of combine harvester. Introducing flexibility of role also implies a big increase in the complexity of languages that had to describe, discuss and direct the diverse activities that the group shared in. This may not have been an innovation in principle but it was a big advance in practical efficiency (Kingdon 1993)."

"We may think of bigger brains as good things; big brains make smarter animals, so surely evolution wants to increase brain size. But brains are expensive. Every cell in your body, including brain cells, requires energy to operate. The reason we eat is to extract nutrients from other living things. We convert them into chemicals that fuel our cells like gasoline fuels a car. And it turns out that brain cells are the most expensive cells in your body, requiring approximately twice as much energy as other cells. Part of the cost is the expensive constant rebuilding of brain cells. Most cells in your body break down and are replaced over time, but the cells in your brain, with precious few exceptions, do not regenerate, and thus they have to engage in more laborious processes of in-place reconstruction. Just as it can be more expensive to extensively renovate an existing house than it would have been to build a new one from scratch, the upkeep on brain cells takes a great deal of energy (Lynch and Granger 2008)."

It is not enough just to know that brain size increased during this 1.5 million year period. More importantly, what caused brain size to increase so dramatically? What were the demands placed on man to drive brain size to increase? And, what were the feedback networks in human behavior that reinforced increased brain size driving it to expand exponentially (Figure 4-3) over such a short period of time? Why are we so much brainier than all the other mammals? Recall from our previous discussion that the combination of tool making and using and increased protein intake was a stimulus to increasing brain size. And, this stimulus was probably very important in the emergence of *Homo habilis* and possibly even *H. ergaster/H. erectus*. But, it's not enough to explain the continued expansion of the brain of *Homo sapiens* with the consequent increase in EQ.

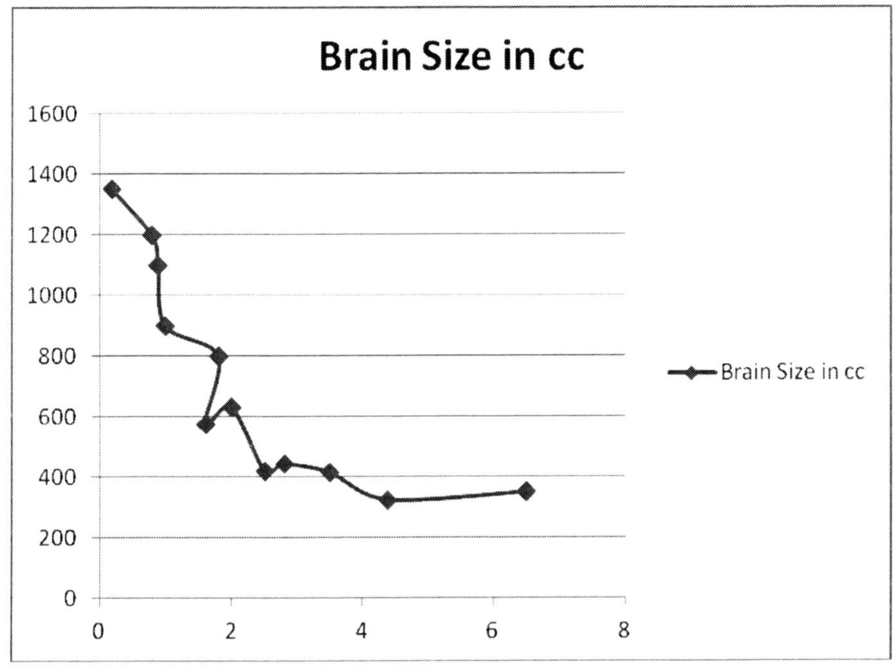

Figure 4-3. Relationship of brain size over time (millions of years)

People today have a brain that ranges between 1100 and 1500 cc with an average of about 1350 cc. There are exceptions on both ends of this scale. "At two extremes, satirist Jonathan Swift had an apparently giant brain of roughly 1900 cc while equally noted writer Anatole France reportedly had a brain size that barely topped 1000 cc. Geniuses are no exception. Einstein's brain reportedly measured an average and undistinguished 1230 cc (Lynch and Granger 2008)." However, from the archeological record we know that there were a vast array of ancestors - Neanderthals and other proto-ancestors - with much larger brains than ours. Table 4-2 (taken from Lynch and Granger 2008) provides a list of representative fossils with brain size in excess of 1350 cc.

Table 4-2. A representative list of fossils with brains size in excess of 1350 cc's (from Lynch and Granger 2008)

Name	Location	~size (cc)	`age (Kya)
Idalta	Ethiopia	1450	150
Omo II	Ethiopia	1435	130
Singa	Singa, Sudan	1550	130
Skhul V	Mt Carmel, Israel	1520	115
Skhul IV and IX	Mt Carmel, Israel	1590	115
Qafzeh	Nazareth, Israel	1565	95
Qafzeh	Nazareth, Israel	1508	90
Border Cave	Natal, S. Africa	1510	75
Shanidar	Zagros, Iraq	1600	60
Amud	Wadi Amud, Israel	1740	45
Cro-Magnon I	Dordogne, France	1600+	30
Brno I	Brno, Czech	1600	25
Fish Hook	Skildergat, S. Africa	1600	12
Zhoukondian	China	1500	10
Wadjak	Java, Indonesia	1550	10
Tuinplass	Pretoria, S. Africa	1590	?
Boskop	Transvaal, S. Africa	1717+	?

The range in brain size in this group is 1435 to 1740 cc with an average of just over 1470 cc. Only two of these fall within the range of today's folks – Idalta at 1450 cc and Omo II at 1435 cc. So it was not just Neanderthal Man that had a larger brain than us but a number of our early ancestors.

Because large brain size is expensive, then the hypothesis is "... that each brain size increase during primate evolution must have been strongly selected for, i.e., there must have been some strong behavioral improvement that made the brain increase advantageous in the fight for survival. Lynch and Granger (2008) think otherwise. They "... posit quite a different hypothesis: that brains increase for biological reasons - which may be largely accidental - and that behaviors follow this increase. As we have seen, it is relatively straightforward to posit how brain increase could arise from random genomic variation, but the question immediately arises how such a variation would be sustained in light of the added expense of a larger brain. This is where behavioral arguments often arise, and may indeed fit. It is not that a 'need' or a 'pressure' for a particular behavior, from sociology to linguistics, gave rise to a big brain. Rather, a big brain got randomly tossed up onto the table, and once there, it found utility. A randomly enlarged brain can find a previously-unexpected behavioral utility, and that utility may be sufficient to entail selection of the new brain size, despite its increased cost." This big brain looking for a purpose is rather counter intuitive and the reverse of current thinking as noted earlier. The evolution to modern man and a reduced brain size may indicate that the cost of the large brain became too high and that a smaller, more efficient, brain would perform better over the long term.

"As the brain grew increasingly larger during the course of human evolution, towards this peak, the need for it to manage and process data efficiently became greater until a point was reached that may have required the reorganization of its systems of data management ... The development of the cerebellum in the Ancestors may have been related to having to cope with an increasingly complex cultural and social environment. It would have provided the necessary computational efficiency to work smoothly in the new environment of increasing population density and social and cultural complexity (Finlayson 2009b)."

This cultural complexity may have led to a smaller brain through the process of the "taming" of our species. One of the side effects of the taming of a species, like the wolf or fox, is a smaller brain. Ridley (2003) suggests that the decline in brain size over the last 15000 years is associated with the "... taming of the human because of their movement towards living in 'civilized' human settlements. Ridley (1993a) proposed the "theory that the human intellect itself is a product of sexual rather than natural selection, for most evolutionary anthropologists now believe that big brains contributed to reproductive success either by enabling men to outwit and out scheme other men (and women to outwit and out scheme other women) or because big brains were originally used to court and seduce members of the other sex." One of the side effects of the taming of a species, like the wolf or fox, is a smaller brain.

We used to believe that our brain was a fixed organ incapable of change and that each area of the brain had a single function. This concept is known as localization. We now know that the brain is plastic. With this knowledge a whole new concept of neuroplasticity has developed and that this neuroplasticity can produce changes in brain structure. Doige (2007) notes that "Our brains are modified by the cultural activities that we do – be they reading, studying music, or learning new languages. We all have what might be called a culturally modified brain, and as cultures evolve, they continually lead to new changes in the brain. Our brains are vastly different, in fine detail, from the brains of our ancestors. In each stage of cultural development … the average human had to learn complex new skills and abilities that all involve massive brain change … Each one of us can actually learn an incredibly elaborate set of ancestrally developed skills and abilities in our lifetimes, in a sense generating a re-creation of this history of cultural evolution via brain plasticity.

"So a neuroplastically informed view of culture and the brain implies a two-way street: the brain and genetics produce culture, but culture also shapes the brain. Sometimes these changes can be dramatic." It may simply be that our brains became more efficient and thus need not have continued to increase in size to adapt to the new and changing cultural settings.

Both birds and mammals evolved from the reptiles. The avian brain was based on the visual system while that of the mammals was based on olfaction. In both, the brain increases in proportion to the body size by about a factor of 3 compared to their ancestors the reptiles. This represented the first time brain size deviated from body size. "The primitive olfactory system in fish and reptiles has neurons arrayed within it in layers or sheets, like a set of blankets laid over other brain structures. In these ancient animals, this olfactory system is referred to as 'pallium,' or cloak. When the first mammals developed, it is primarily this pallial structure that greatly expanded. The mammals grew and it extended until it covered much of the surface of their brains. The new structure is referred to as the cortex. The original reptilian olfactory cortex is transferred to mammals more or less intact, and all the rest of the mammal brain is the new cortex. We refer to the old, olfactory part of the cortex as 'paleocortex,' [you may have heard the term "lizard brain"] while the more recent parts of the cortex are called 'neocortex.' ...as the brain grows, most of the growth occurs in the neocortex (Lynch and Gardner 2008)."

Write (1994) notes that ... "the human brain [is] a 'triune' brain whose three basic parts recapitulate [repeat] our evolution: a reptilian core (the seat of our basic drives), surrounded by a paleo-mammalian brain (which endowed our ancestors with, among other things, affection for offspring), surrounded in turn by a 'neo-mammalian' brain. The voluminous neo mammalian brain brought abstract reasoning, language, and, perhaps, (selective) affection for people outside the family."

The mammalian cortex takes on a significantly different structure from its ancestors. Instead of being layered, like blankets stacked up together, it was uniform throughout. As it grew larger, this uniformity remained. As it expanded over the brain it took over the different tasks of vision, hearing and touch. The big difference in the cortex is that the neurons are scattered throughout the cortex in what is called random-access circuits. It's these circuits that the brain uses to integrate all the signals from the various senses.

"Perhaps the single most crucial feature of random-access circuits is this: they can be modified by experience. The more a particular set of connections are activated, the more they are strengthened, becoming increasingly reliable responders. The way they do this is worthy of a book in its own right - but suffice it to say that it is one of those instances in biology in which a broad swath of observations all fit together with amazingly tight coordination. When a mouse, for example, sniffs an odor, she sniffs rhythmically, about five times per second. It's not voluntary: she's biologically wired to sniff at this rate. The receptor cells in the nose, and then in the olfactory bulb and then the olfactory cortex, are all activated at the same frequency. When the activation reaches the cortex, something remarkable happens: the synaptic connections in the cortex contain a biological machine that can permanently amplify the signal, strengthening the synapses; and that machine is selectively activated by the precise rhythmic pattern of activation triggered by sniffing. In other words, when the animal is actively exploring her environment, the resulting brain activity causes synaptic connections to strengthen, enabling her brain cells to respond more strongly to this odor in the future. When she explores, she sniffs; and when she sniffs, she learns. All the components, from the submicroscopic world of proteins, voltages, and exotic chemistries, are all tightly linked to this rhythmic activation pattern. The mechanism for this had its origin more than half a billion years ago, in the rhythmic tail movements of primitive fish that enable swimming. When you're learning a new telephone number, you're engaging some of the same biological processes used by a hagfish slithering through the water. Biology uses and reuses its inventions, retaining and adapting them to new uses.

"The ability to strengthen synaptic connections is a simple form of learning. Strengthening a connection simply locks in a particular pattern in the brain, and that responding pattern becomes the brain's internal 'code' for whatever is being sensed. Your experience of a chocolate chip cookie is not a photograph or a recording; not a camera image, but an internal construct, a cortical creation.

"Once the cortex developed from the olfactory precursor circuits, it became independent; cortex no longer had to operate solely on olfaction, its original mode. In mammals it became the engine that analyzes olfactory odors, and visual images, and auditory sounds, and the sensation of touch, and a great deal more. And as we've said, how the nearly uniform structures of the cortex can end up doing the very different jobs of the different senses is another of the major mysteries of the brain.

"Images and sounds are converted into internal random-access codes much like olfactory codes. As these codes are created, they can be readily transmitted downstream to any other brain area, all of which now use the same internal coding system. And, using those connections, two different senses can be directly hooked together: smell of the chocolate chip cookie and its shape; its taste; the sound when it breaks. Using the shared cortical design, every sense acquires the same capability. The sound of a song can remind you of the band that plays it, the cover art on their CD, the concert where you saw them play, and whom you were with. All these senses participate in perceiving the event, creating memories of the event, and retrieving those memories (Lynch and Granger 2008)."

"The random-access circuits become a template for the explosive growth of cortex as mammal brains grew ever larger. Not only do these cortical circuits now operate not just olfaction but also vision, touch and hearing; they also generate the rest of the mental abilities in the mammalian brain. Our most sophisticated cognitive abilities are still based on that ancient design. Adding more of these same structures generates new animals with new mental capabilities (Lynch and Granger 2008)."

In the brain there are four major targets of the olfactory cortex: striatum, amygdala, hippocampus, and thalamus. "Each is very different from the others - different circuit structure, different connections, different functions (Lynch and Granger 2008)." The striatum "contains circuits that activate, and others that suppress the hindbrain muscle system. Through these two networks, a message sent from the cortex to striatum can initiate a 'go' signal or a 'stop'." Thus an odor recognized in the cortex can set the animal in motion (e.g., in response to food) or cause it to freeze (sensing a predator). The amygdala too can be understood in part by noting its outputs. It sends massive connections to a small region called the hypothalamus, a set of regulating structures that virtually runs the autonomic system of your body. The hypothalamus operates your endocrine glands (testosterone, estrogen, growth hormones, adrenaline, thyroid hormone, and many others), and generates simple primitive behaviors that are appropriate to these hormones. The amygdala largely rules the hypothalamus ...

A closely related function of the amygdala is its evocation of strong emotional responses. Where striatum and amygdala provide movement and emotion, the hippocampus is central to the encoding of memory (Lynch and Granger 2008)." The hippocampus is likely a "... temporary repository of memories which subsequently move onward to final, permanent cortical storage sites. When the hippocanthus detects a novel, unfamiliar contingency, it triggers a signal that alerts the cortex to store the new information. Without the hippocanthus, the novelties won't be detected, and won't be stored. The fourth and last target of the cortex is by no means the least: it is a connection to a large two-part system, consisting of another part of the cortex (frontal cortex) and part of the thalamus. This thalamo-cortical connection starts small, but as the brain grows large it becomes ever more important, until in big-brained mammals, thalamo-cortical circuits are the keystone of the brain (Lynch and Granger 2008)."

The cortex and these four structures are the major engines of the human brain. "All this anatomy produces a surprisingly straightforward, and surprisingly complete, picture of how odors [and, vision, hearing and touch] guide behavior. Cues are recognized by a primary cortex which then in parallel distributes signals to regions that initiate movement (striatum), intensify or weaken the movements (amygdala), detect anomalies during the search (hippocampus), associate the cue with objects (hippocampus again), and organize actions in appropriate behavioral sequences (frontal striatal-thalamo-cortical loops). The other sensory systems - vision, audition, and touch - follow the same basic pathways in connecting the outside world to useful responses (Lynch and Granger 2008)."

So how does all this structure work? There are a number of key ingredients. Remember that the original olfactory system is now an olfactory (both smell and taste), vision, hearing, touching system. Each of these individual senses sends "point-to-point" signals to the brain. The striatum connects to the thalamo-cortical loops just like the olfactory cortex does, thus creating even larger loops: striatal-thalamo-cortical loops, and amygdala and hippocampus in turn connect to the striatum, thus insinuating [inserting] themselves into that large striatal-thalamo-cortical loop (Lynch and Granger 2008)." The visual, hearing and touching systems have a similar behavior as the olfactory system. "... in vision, cells in the eye send their axons to a group of neurons in the thalamus, which in turn send their axons to the primary visual cortex, and throughout these initial stages, the image on the retina is projected, point-to-point, all the way to the cortex. The same process holds for hearing and touch. Those primary sensory areas then send connections into association areas that lose their point-to-point organization and acquire random-access organization. From there, connections go to the striatum, the amygdala, and the frontal thalamo-cortical system ... Critically, the association cortices also project to each other, and the larger the brain, the more of these connections there are (Lynch and Granger 2008)."

"These random-access circuits became a template for the explosive growth of cortex as mammal brains grew ever larger. Not only do these cortical circuits now operate not just olfaction but also vision, touch and hearing; they also generate the rest of the mental abilities in the mammalian brain. Our most sophisticated cognitive abilities are still based on that ancient design. Adding more of the same structures generates new animals with new mental capabilities (Lynch and Granger 2008)."

"The system is thus hierarchical - lower, simpler levels send messages to higher, more abstract levels, and receive feedback in return. Associational regions that are dominated by inputs from vision tend to connect to each other, but also connect to areas dominated by auditory input, as well as to areas whose input is not determined either by vision or sound alone. Go downstream far enough and there will be a region where information on face and voice can be combined into a single brain code (Lynch and Granger 2008)." You will instantly recognize the person this hierarchical system is detecting and analyzing.

"We have so far focused on association cortex in the back part of the cortex. This houses the pathways that collect input from sensory areas - vision, hearing, touch - and integrate them. There remains a vast territory, perhaps a quarter of the entire cortex, at the front of the brain, under your forehead, labeled the frontal association fields. That forehead tells the story: only humans have them. In most animals, the head sweeps aerodynamically back from the nose, because they have much smaller frontal fields. The larger the brain along the evolutionary ladder, the disproportionately bigger the frontal fields get. In primates, and most noticeably in us, those frontal areas become enormous.

Outputs from the frontal cortex provide more clues to its nature. There are three primary output paths. The first goes to the motor cortex, which contains a point-to-point map of the body's muscles. You want to move your left foot? An output signal moves from the frontal cortex to the premotor and motor cortex, triggering brainstem and muscle activity. The second pathway from frontal cortex connects to the striatum, which in turn projects right back to the thalamo-cortical circuits, thereby creating the closed loop: cortex to striatum to thalamus to cortex. The third and final pathways are two-way connections from frontal cortex to the sensory association cortex areas and back again. Taken together, these connection pathways explain three primary roles of the frontal cortex: planning movements, timing them, and coordinating internal thought patterns (Lynch and Granger 2008)."

"The third and final set of connections, from frontal cortex to association regions in sensory cortex, connect a planning region with areas that we described as responsible for internal feats such as combining information about a face and a voice. These provide the final ingredient needed to coordinate complex behavioral sequences. Unified information about our sensory perceptions allow us to walk smoothly, to learn to pronounce words to sound the way we want them to, and to parallel park our cars. The system does more than just organize actions, though; it also enables the organization of thoughts. Since the frontal cortex can assemble long temporal sequences, we can construct and reconstruct memories of past episodes. The system lends itself to creative use: frontal cortex essentially has its pick of all the vast amounts of sensory material stored throughout association cortical regions (Lynch and Granger 2008)."

"Although the specialized 'front end' circuits of the brain, with their point-to-point designs, specialize in their own particular visual or auditory inputs, the rest of the brain converts these to random-access encodings in association areas throughout [the] cortex. ...these areas take initial sensory information and construct grammars. These are not grammars of linguistic elements; they are grammatical organizations (nested hierarchical, sequences of categories) of percepts - visual, auditory, and other. Processing proceeds by incrementally assembling these constructs, and hierarchically passing them through long brain pathways, successively modifying them at each processing station in the path (Lynch and Granger 2008)."

If you aren't fully confused by now, stand by! There's one last point to be made and it's not terribly simple. "We now introduce the final crucial point about cortical circuits: they generate projections both forward, to downstream areas, and backward, back to their inputs. This occurs in the thalamo-cortical and in the cortico-cortical circuits: messages are sent in both directions, implying that the initial processing of an input - a sight or a sound - becomes altered by the downstream conception of what that sight or sound may be. Our 'higher level' processing actually modifies our initial processing. Perception is not pure and direct; it is affected by our learned 'expectations' (Lynch and Granger 2008)."

Let me break from all this scientific description of brain structure and function and tell a hypothetical story to try to put it all together. You've been out in the garage all morning working on your cherished 1948 Ford pickup. You've dropped the transmission and put it on the work bench. You're tired and dirty and decide this is enough for one day and go into the mud room to wash up. As you enter the house your nose picks up a pleasant smell - one you've experienced in the past. You're brain picks up this signal and interprets it as baking and you realize that while you were working on your cherished 1948 Ford pickup, your husband was in the kitchen baking! What!? It could happen! Use your imagination! (More on imagination later). Your cortex further examines the smell and sends a signal backward to the input that now distinguishes the baking as either cookies or cake and eliminates such things as pies and bread. Again your brain analyzes the smell in lieu of these new circumstances and verifies that, yes indeed, it is cookies. But what kind of cookies? Back to the cortex for more historical information looking only at the range of cookies. This time you know for sure! It's chocolate chip cookies! While all this is going on - a fraction of a second - your memory conjures up images of chocolate chip cookies. Salivating yet? You see a tray full of warm chocolate chip cookies with one missing. You've been known to snatch them up before they've cooled. You see yourself breaking the cookies in half and the warm melted chocolate streaming out between the two halves. You feel the soft texture and taste the delicious mixture of dough and chocolate. You begin to salivate. You skip the wash up routine and rush straight into the kitchen. The euphoria is broken when your husband tells you you're tracking dirt all over the house and that you need to go back to the mud room and wash up and take off those dirty cloths - which of course includes that brand new pair of jeans you just bought because all your other jeans have been ruined by grease stains!

"In the origin of the brain, the traditional view is that the 'point-to-point structures in our brain arose first, and the random access 'association' areas arose later in evolution (Lynch and Granger 2008)." Lynch and Granger (2008) propose an alternative hypothesis: "... the neocortex began with an overall olfactory design, which humans retain and expand in our huge association cortical areas; the more highly specialized sensory regions for visual, auditory, and touch senses were initially small and were added to over evolutionary time. Both modes of processing have critical uses, but ... it is the association areas that continued to grow explosively as the mammal brain expanded."

If we graph brain weight against body weight, similarly as we did to determine EQ, but this time only include primates we come up with a similar graph. This time we find that the human brain is 2.3 times as large as expected compared against all other primates. And, if we look at a graph of the three major components of the brain among all primates we get a graph similar to the one in Figure 4-4. The three brain areas plotted are the ancient midbrain, the diencephalon, and the neocortex. Looking at the graph we find that each of these areas get larger almost exactly as expected for all primates (humans are on the far right of the graph). If we look at the slope of each of the lines on the graph, we see that the neocortex increases at a higher rate than the diencephalon which increases at a higher rate than the ancient midbrain. So as we go from the primitive primates (lower left) to the humans (upper right) we find that the neocortex increases proportionately more than the rest of the brain. Recall it's the neocortex where all the action is. It is the neocortex that makes us smart. Lynch and Granger (2008) note that "a bigger cortex has within it much more association cortex than sensory regions. This gives different relative intelligence for different brain size."

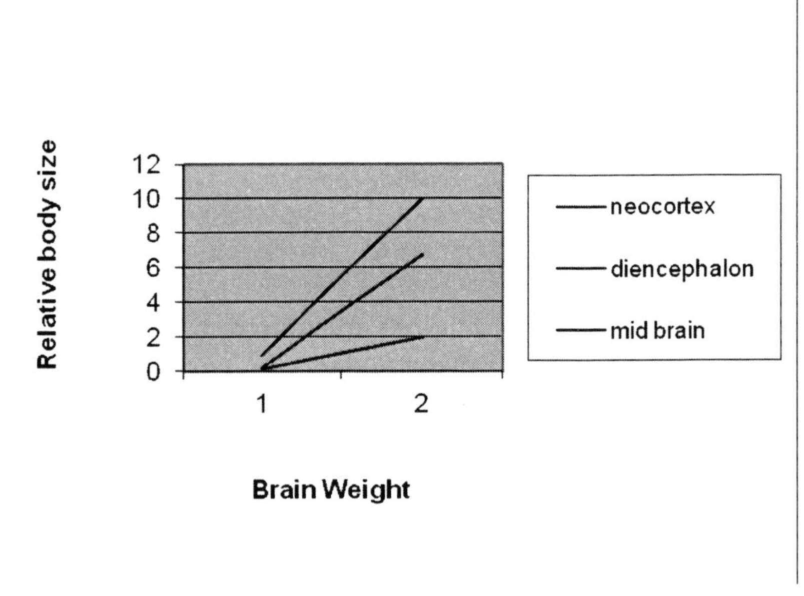

Figure 4-4 Relative brain weight vs. relative body size in primates

Let's return to Table 4-2 and look at a representative sample of fossils whose brains size exceeds 1350 cc's - the current human average. We find an individual at the bottom of the table, Boskop, from South Africa that has a brain size in excess of 1717 cc - more than 25 percent larger than modern humans. Also, there are two Neanderthals (Skhul V and Skhul IV, IX). The brains of both of these individuals are about 15 percent larger than modern humans. Were they smarter than us? There's absolutely no reason to believe if they were added to our graph in Figure 5 that they wouldn't just extend all three lines beyond those of the modern humans. Lynch and Granger (2008) did just that and estimated that Boskop's cortex would have been more than 50 percent larger than ours. Therefore, yes they would have been smarter than us by about the same amount that we were smarter than *Homo erectus*. So why are we here and they are not? I don't know. Maybe like Neanderthals they just painted themselves into a corner? Or, maybe like Neanderthals they live today in the genes of modern man?

***

I woke up one morning at 3:00 am. My mind was racing. Ideas were quickly popping in and out of my mind. Should I try to fall back to sleep or get up and write? I couldn't make up my mind. Finally I got up and wrote. And this is what my mind conjured up!

My mind must be a figment of my imagination!
But, it takes a mind to imagine!
Can I make up my mind?
Can I change my mind?
Can I help you make up your mind?
Can I change your mind?
I imagine so!

Is my mind in my heart?
Is my mind in my soul?

I imagine my mind is in my brain, among all the dendrons, axons, neurons and synapses and things.
But, I can't be sure.

Can I hold my mind in my hands?
Can I dissect it into its various components?
Can I determine the function of the various pieces and parts?
I imagine not!

If I can't hold my mind in my hands,
How can I stretch my imagination?
How can I let my imagination wander or run wild?
How can I be out of my mind?
How can I be in my right mind?
How can I put mind over matter?
I just can't Imagine!

I can alter my mind.
I can blow, twist, warp, and expand my mind.
I can be mindful or mindless.
I can be open-minded, closed-minded or narrow-minded.
I could go on, but it might just strain my imagination.

But wait, if it takes a mind to imagine.
And, I can imagine all this.
Then my mind must not be a figment of my imagination!
It must, in fact, be real!
Imagine that!

The purpose of this little mind exercise is two-fold. First, is to introduce the next topic and, second, to acknowledge that no animals, other than humans, are capable of this type of mind exercise let alone have the ability to write it down. The incremental shifts in brain size just didn't make it bigger it made it substantially more powerful. "Constructing models is something the human brain is very good at. When we are at sleep it is called dreaming; when we are awake we call it imagination or, if it is exceptionally vivid, hallucination (Dawkins 2008)."

"The human mind probably started its development about 800,000 years ago with the Heidelbergs and reached a peak around 45 -50,000 years ago in *Homo sapiens*. It is during this period "that we see for the first time multipart tools; animal bones punctured with holes to fashion musical instruments; burials with accoutrements, suggesting beliefs about aesthetics, and the afterlife; richly symbolic cave paintings that capture in exquisite detail events of the past and the perceived future; and control over fire, a technology that combines our folk physics and psychology and allowed our ancestors to prevail over novel environments by creating warmth and cooking foods to make them edible (Hauser 2009)."

"The mind is a system of organs of computation, designed by natural selection to solve the kinds of problems our ancestors faced in their foraging way of life, in particular, understanding and outmaneuvering objects, animals, plants, and other people. The summary can be unpacked into several claims. The mind is what the brain does; specifically, the brain processes information, and thinking is a kind of computation. The mind is organized into modules or mental organs, each with a specialized design that makes it an expert in one arena of interaction with the world. The modules' basic logic is specified by our genetic program. The operation was shaped by natural selection to solve problems of the hunting and gathering life led by our ancestors in most of our evolutionary history. The various problems for our ancestors were subtasks of one big problem for their genes, maximizing the number of copies that made it into the next generation (Pinker 2009)."

"The computational theory of mind ... says that beliefs and desires are *information*, incarnated as configurations of symbols. The symbols are the physical states of bits of matter, like chips in a computer or neurons in the brain. They symbolize things in the world because they are triggered by those things via our sense organs, and because of what they do once they are triggered. If the bits of matter that constitute a symbol are arranged to bump into the bits of matter constituting another symbol in just the right way, the symbols corresponding to one belief can give rise to new symbols corresponding to another belief logically related to it, which can give rise to symbols corresponding to other beliefs, and so on. Eventually the bits of matter constituting a symbol bump into bits of matter connected to muscles, and behavior happens. The computational theory of mind thus allows us to keep beliefs and desires in our explanations of behavior while planting them squarely in the physical universe. It allows meaning to cause and be caused (Pinker 2009)."

"... [T]he human brain uses at least four major formats of representation. One format is the visual image, which is like a template in a two-dimensional, picture-like mosaic. Another is phonological representation, a stretch of syllables that we play in our minds like a tape loop, planning out the mouth movements and imaging what the syllables sound like. This string-like representation is an important component of our short-term memory, as when we look up a phone number and silently repeat it to ourselves just long enough to dial the number. Phonological short-term memory lasts between one and five seconds and can hold from four to seven 'chunks.' (Short-term memory is measured in chunks rather than sounds because each item can be a label that points to a much bigger information structure in long-term memory, such as the content of a phrase or sentence). A third format is the grammatical representation: nouns and verbs, phrases and clauses, stems and roots, phonemes and syllables, all arranged into hierarchical trees.

"The fourth format is mentalese, the language of thought in which our computational knowledge is couched. When you put down a book, you forget almost everything about the wording and typeface of the sentences and where they sat on the page. What you take away is their content or gist. (In memory tests, people confidently 'recognize' sentences they never saw if they are paraphrases of the sentences they did see.) Mentalese is the medium in which content or gist is captured … Mentalese is also the mind's lingua franca, the traffic of information among mental modules that allows us to describe what we see, imagine what is described to us, carry out instructions, and so on. This traffic can actually be seen in the anatomy of the brain. The hippocampus and connected structures, which put our memories into long-term storage, and the frontal lobes, which house the circuitry for decision making, are not directly connected to the brain areas that process raw sensory input (the mosaic of edges and colors and the ribbon of changing pitches). Instead, most of their input fibers carry what neuroscientists call 'highly processed' input coming from regions one or more stops downstream form the first sensory areas. The input consists of codes for objects, words, and other complex concepts (Pinker 2009)."

"Natural selection cannot directly endow an organism with information about its environment, or with the computational networks, demons, modules, faculties, representations, or mental organs that process the information. It can only select among genes. But genes build brains, and different genes build different brains that process information in different ways. The evolution of information processing has to be accomplished at the nuts-and-bolts level by selection of genes that affect the brain-assembly process.

"Many kinds of genes could be the targets of selection for better information processing. Altered genes could lead to different numbers of proliferative units along the walls of the ventricles (the cavities in the center of the brain), which beget the cortical neurons making up the gray matter. Other genes could allow the proliferative units to divide for different numbers of cycles, creating different numbers and kinds of cortical areas. Axons connecting the neurons can be re-routed by shifting the chemical trails and molecular guideposts that coax the axons in particular directions. Genes can change the molecular locks and keys that encourage neurons to connect with other ones. As in the old joke about how to carve a statue of an elephant (remove all the bits that don't look like an elephant), neural circuits can be sculpted by programming certain cells and synapsis to commit suicide on cue.

Neurons can become active at different points in embryogenesis, and their firing patterns, both spontaneous and programmed, can be interpreted downstream as information about how to wire together. Many of these processes interact in cascades. For example, increasing the size of one area allows it to compete better for real estate downstream. Natural selection does not care how baroque the brain-assembly process is, or how ugly the resulting brain. Modifications are evaluated strictly on how well the brain's algorithms work in guiding the perception, thought, and action of the whole animal. By these processes, natural selection can build a better and better functioning brain (Pinker 2009)."

The incremental shifts in brain size over time from *Australopithecus africanus* to *Homo sapiens sapiens* may have led to the rubicon for our emergence: massive differences in computational power (Hauser 2009) of the human mind. Hauser (2009) notes that "This rearranging, deleting and copying of universal genetic elements created a brain with four special properties. Together these distinctive characteristics ... constitute ... our humaniqueness."

These traits are:

1. "Generative computation, the ability to create a virtually limitless variety of 'expressions,' be they arrangements of words, sequences of notes, combinations of actions, or strings of mathematical symbols. Generative computation encompasses two types of operation, recursive and combinatorial. Recursion is the repeated use of a rule to create new expressions. Think of the fact that a short phrase can be embedded within another phrase, repeatedly, to create longer richer descriptions of our thoughts - for example, the simple but poetic expression from Gertrude Stein: 'A rose is a rose is a rose'."

2. The minds capacity for the promiscuous combination of ideas. "We routinely connect thoughts from different domains of knowledge, allowing our understanding of art, sex, space, causality and friendship to combine. From this mingling, new laws, social relationships and

technologies can result, as when we decide that it is forbidden [moral domain] to push someone [motor action domain] intentionally [folk psychology domain] in front of a train [object domain] to save the lives [moral domain] of five [number domain] others."

3. The use of mental symbols. "We can spontaneously convert any sensory experience - real or imagined -into a symbol that we can keep to ourselves or express to others through language, art, music, or computer code."

4. Engaging in abstract thought. "Unlike animal thoughts, which are largely anchored in sensory and perceptual experiences, many of ours have no clear connection to such events. We alone ponder the likes of unicorns and aliens, nouns and verbs, infinity and God."

Hauser (2009) did an interesting experiment in nonlinguistic abilities for counting. There are two of these abilities known to occur in apes. The first is precise and limited to the numbers 1 through 4. The second is unlimited in scope but is centered around ratios. For instance, an animal that can distinguish between 1 and 2 can also distinguish between 2 and 4, 16 and 32, etc. Hauser (2009) found a third system in rhesus monkeys, "one that may help us understand the origins of the human ability to mark the difference between singular and plural. This system operates when individuals see sets of objects presented at the same time - opposed to individuals presented serially - and causes rhesus monkeys to discriminate one from many but not many from many food items." He proposes an exercise for us humans to try. He notes that something peculiar happens when our linguistic system meets up with our ancient computational system. "To see how, try this exercise: for the numbers 0, 0.2 and -5, add the most appropriate word: 'apple' or 'apples.' If you are like most native English speakers, including young children, you selected 'apples'. If you are surprised, good, you should be. This is not a rule we learned in grammar school - in fact, strictly speaking, it is not grammatically correct. But it is part of the universal grammar that we alone are born with. The rule is simple but abstract: anything that is not '1' is pluralized."

***

It is this very same generative computational power that leads to language. There are two aspects of language that I will discuss. The first is the "how" of the evolution of language, which is genetically, not culturally, determined. It is known as "universal grammar". The second is the "why" of the evolution of language and the cultural aspects of sociability. The third piece of the puzzle will be a discussion of language families in relation to the peopling of the world. "Language is not a cultural artifact that we learn the way we learn to tell time or how the federal government works [or doesn't work]. Instead, it is a distinct piece of the biological makeup of our brains. Language is a complex, specialized skill, which develops in the child spontaneously, without conscious effort of formal instruction, is deployed without awareness of its underlying logic, is qualitatively the same in every individual, and is distinct from more general abilities to process information or behave intelligently. For these reasons some cognitive scientists have described language as psychological faculty, a mental organ, a neural system, and a computational module. But I [Pinker] prefer the admittedly quaint term 'instinct' (Pinker 1994)."

"Children plainly have a language organ in their brains ready and waiting to apply the rules. They infer the basic rules of grammar without instruction, a task that has been shown to be beyond the power of a computer unless the computer has been endowed with some prior knowledge.

"From about the age of one and a half until soon after puberty children have a fascination with learning language and are capable of learning several languages far more easily than adults can. They learn to talk irrespective of how much encouragement they are given. Children do not have to be taught grammar, at least not of living languages that they hear spoken; they divine it. They are constantly generalizing the rules they have learned in defiance of the examples they hear (such as 'persons gived' rather than people gave'). They are learning to talk in the same way that they are learning to see, by adding the plasticity of vocabulary to the preparedness of a brain that insists on applying rules. The brain has to be taught that large animals with udders are called cows. But to see a cow standing in a field, the visual part of the brain employs a series of sophisticated mathematical filters to the image that it receives from the eye – all unconscious, innate, and un-teachable. In the same way, the language part of the brain knows without being taught that the word for a large animal with an udder is likely to behave grammatically like other nouns and not like verbs.

"The point is that nothing could be more 'instinctive' than the predisposition to learn a language. It is virtually un-teachable. It is hard-wired. It is not learned. It is – horrid thought – genetically determined. And yet nothing could be more plastic than the vocabulary and syntax to which that predisposition applies itself. The ability to learn a language, like almost all the other human brain functions, is an instinct for learning (Ridley 1993a)."

Pinker (1996) characterizes language as follows: "First, language is part of human biology, not human culture. Secondly, language is best thought of as a neural system, a computational module, and a mental organ. Thirdly, language is complex computationally, neurologically, and genetically. The implications are that language is as old as modern humans, that it did not necessarily appear at the same time as tools, art, or other cultural products, and that it did not fully evolve as a byproduct of something else." He adds that "Surely language would have been most likely to evolve among organisms that had something to say, and complex technologies were most likely to have developed among groups that could efficiently pass on knowledge from member to member and generation to generation." He defines language as a system that evolved by way of conventional Darwinian evolution. "Specifically, selection for neural circuitry enabling efficient communication as an unlimited set of messages of a certain kind (basically, hierarchical propositions involving human actions, beliefs, desires, and obligations; objects and their rough relative locations, motions, and forces; and the durations and relative times of events and states)." He notes that "mental grammar or psychogrammar is the neural system that allows us to speak and understand the possible word sequences of our native tongue. A grammar for a specific language is obviously acquired by a human being during childhood, but there must be neural circuitry that actually carries out the acquisition process in the child, and this circuitry may be called the *language faculty* or *language acquisition device*. An important part of the language faculty is *universal grammar*, an implementation of a set of principles or constraints that govern the possible form of any human grammar."

The memory structures being built inside our brains have a recognizable organization. Sequences of categories are constructed at each processing stage along the path "These are nested hierarchically, such that a single category at one stage may itself be part of another entire sequence of categories. These structures can be expressed in terms of grammars - just like those learned in school. This is a computational formalism, a code that captures the crucial characteristics of these hierarchical brain representations (Lynch and Granger 2008)." The code is quite simple. It is sentence (S) equals noun phrase (NP) plus verb phrase (VP): S = NP + VP. Now the noun phrase has the following components: an optional determiner (a, an, the), an adjective (also optional) and a noun. The verb phrase contains a verb and an optional adverb. So the possibilities for the noun phrase are Det Adj N | Det N | N and for the verb phrase are V | V Adv. If we diagram this out we have the structure in Figure 4-5.

Given all these possibilities we can construct an infinite number of sentences of infinite length. So the brain code for grammar is quite simple and can construct an incredibly complicated language system.

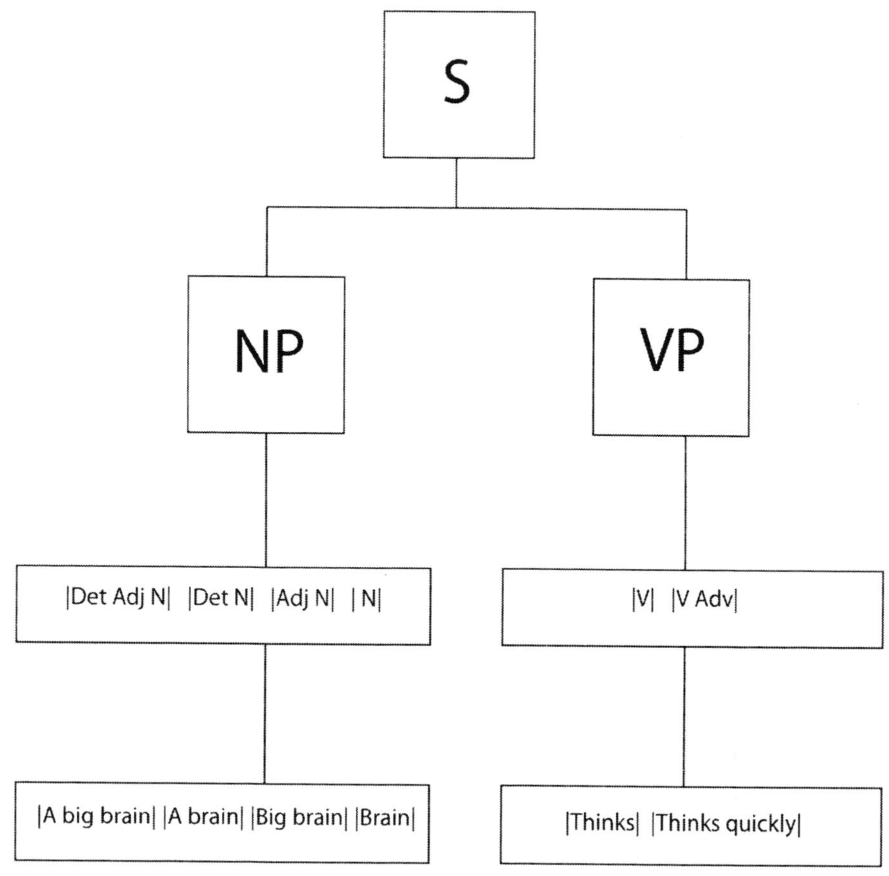

Figure 4-5 Grammar Structure

It's very similar to the DNA system. Four simple nucleotides can construct hundreds of thousands of proteins and string them together to make us.

"A grammar is a set of 'rewrite rules,' in which sequences of categories are rewritten to substitute category names with category members. Indeed, it is this ability that gives language its power and range: from a finite set of elements (words), grammars can create infinite numbers of strings (sequences) of those words - such as sentences. We can string new words together to make entirely new sentences under the sun, even meaningless ones that nonetheless are readily recognized as grammatically sound. Linguistic grammars are 'generative,' that is, they can be used to specify multiple instances of sentences without end. The generative properties of human language are what set it apart from other animal communication systems. The question has often been framed in these terms: how do generative grammars arise in human brains? It is not obvious how these linguistic structures could arise from the presumably 'lower' perceptual and motor processing in which other parts of the brain engage. If an innate bias related to sequences of categories of vocal utterances (speech) led, in larger-brained organisms, to a downstream bias for certain sequences of categories of assembled speech sounds (words), then this may at least in part account for this much-studied but still elusive characteristic of innate language capability (Lynch and Granger 2008)." The English language and the majority of other languages use the syntax of subject - verb - object or SVO described above. There are, of course, several other possibilities. The structure SOV "... is used by a few languages like Japanese, Korean and Tibetan. The syntaxes VSO and VOS are used by about 15 percent of all languages (Wells 2002)."

"The way language works, then, is that each person's brain contains a lexicon of words and the concepts they stand for (a mental dictionary) and a set of rules that combine the words to convey relationships among concepts (a mental grammar). The fact that grammar is discrete combinational system had two important consequences. The first is the sheer vastness of language. The second consequence of the design of grammar is that it is a code that is autonomous from cognition. A grammar specifies how words may combine to express meanings; that specification is independent of the particular meanings we typically convey or expect others to convey to us. Thus we all sense that some strings of words that can be given common-sense interpretations do not conform to the grammatical code of English (Pinker 2009)."

Language develops in humans from the moment of birth and possibly while still in the womb. "Babies continue to learn the sounds of their language throughout the first year. By six months, they are beginning to lump together the distinct sounds that their language collapses into a single phoneme, while continuing to discriminate equivalently distinct ones that their language keeps separate. By ten months they are no longer universal phoneticians but have turned into their parents; they do not distinguish Czech or Inslekampx phonemes unless they are Czech or Inslekampx babies. Babies make this transition before they produce or understand words, so their learning cannot depend upon correlating sound with meaning. That is, they cannot be listening for the difference in sound between a word they think means bit and a word they think means beet, because they have learned neither word. They must be sorting the sounds directly, somehow tuning their speech analysis module to deliver the phonemes used in their language. The module can then serve as the front end of the system that learns words and grammar.

"During the first year, babies also get their speech production systems geared up. First, ontogeny recapitulates phylogeny. A new born has a vocal tract like a nonhuman mammal. The larynx comes up like a periscope and engages the nasal passage, forcing the infant to breathe through the nose and making it anatomically possible to drink and breathe at the same time. By three months the larynx has descended deep into the throat, opening up the cavity behind the tongue (the pharynx) that allows the tongue to move forwards and backwards and produce the variety of vowel sounds used by adults. Not much of linguistic interest happens during the first two months, when babies produce cries, grunts, sighs, clicks, stops, and pops associated with breathing, feeding, and fussing, or even during the next three, when coos and laughs are added. Between five and seven months babies begin to play with sounds, rather than using them to express their physical and emotional states, and their sequences of clicks, hums, glides, trills, hisses, and smacks begin to sound like consonants and vowels. Between seven and eight months they suddenly begin to babble in real syllables like ba-ba-ba, neh-neh-neh, and dee-dee-dee. The sounds are the same in all languages, and consist of the phonemes and syllable patterns that are most common across languages. By the end of the first year, babies vary their syllables, like neh-nee, da-dee, and meh-neh, and produce that really cute sentence-like gibberish.

"Why is babbling so important? The infant is like a person who has been given a complicated piece of audio equipment bristling with unlabeled knobs and switches but missing the instruction manual. In such situations people resort to what hackers call frobbing – fiddling aimlessly with the controls to see what happens. The infant has been given a set of neural commands that can move the articulators every which way, with wildly varying effects on the sound. By listening to their own babbling, babies in effect write their own instruction manual; they learn how much to move which muscle in which way to make which change in the sound. This is a prerequisite to duplicating the speech of their parents.

"Shortly before their first birthday, babies begin to understand words, and around that birthday, they start to produce them. Words are usually produced in isolation; this one-word stage can last from two months to a year. For over a century, and all over the globe, scientists have kept diaries of their infants' first words, and the lists are almost identical. About half the words are for objects: food (juice, cookie), body parts (eye, nose), clothing (diaper, sock), vehicles (car, boat), toys (doll, block), household items (bottle, light), animals (dog, kitty), and people (dada, baby). There are words for actions, motions, and routines, like up, off, open, peekaboo, eat, and go, and modifiers, like hot, all gone, more, dirty, and cold. Finally there are routines used in social interaction, like yes, no, want, bye-bye, and hi – a few of which, look at that and what is that, are words in the sense of listemes (memorized chunks), but not, at least for the adult, words in the sense of morphological products and syntactic atoms. Children differ in how much they name objects or engage in social interaction using memorized routines.

"Around eighteen months, language takes off. Vocabulary growth jumps to the new-word-every-two-hours minimum rate that the child will maintain throughout adolescence. And syntax begins, with strings of the minimum length that allows it: two. Here are some examples: All dry. I sit. No pee. More cereal. Other pocket. Mail come. Our car. All messy. I shut. See baby. More hot. Boot off. Airplane all-gone. Papa away. All wet. No bad. See pretty. Hi Calico. Siren by. Bye-bye car. Dry pants.

If we divide language development into somewhat arbitrary stages, like Syllable Babbling, Gibberish Babbling, One-word Utterances, and Two-word Strings, the next stage would have to be called All Hell Breaks Loose. Between the late twos and the mid-threes, children's language blooms into fluent grammatical conversation so rapidly that it overwhelms the researchers who study it, and no one has worked out the exact sequence. Sentence length increases steadily, and because grammar is a discrete combinational system, the number of syntactic types increases exponentially, doubling every month, reaching the thousands before the third birthday (Pinker 1994)."

Pinker (1994) provides a series of facts "that refute many popular assumptions, have strong implications about the relevant evolutionary forces, and offer some suggestions about where to look further" to develop a complete theory of language evolution. I'll briefly summarize these facts.

1. Language is found in all human societies present and past.

2. There is no correlation between grammatical complexity and cultural complexity.

3. The existence of universal properties of grammar is uncontroversial but also uninformative about the biological basis of language.

4. There is no difference among racial or ethnic groups in their ability to learn language.

5. There is fairly poor correlation between typology and historical relatedness of language, except over very short periods of time.

6. Language is found universally across individuals within a society regardless of IQ or social status.

7. No formal instructions, lessons, or tutelage, is necessary for language acquisition to take place successfully.

8. Language acquisition in children is not propelled by attempts to improve communicative power.

9. By the age of 3 years, children obey many or all language universals and avoid many tempting errors.

10. Children can create a language themselves in a single generation.

Recall our discussion of the Brodman area in the last chapter. The Broca's area is divided into two parts - Brodman area 44 and 45. "It is now evident that area 44 is involved not only in speech, but also in motor functions unrelated to speech, including complex hand movements, and sensorimotor learning and integration. In the course of human evolution, then, it seems that vocalization must have been incorporated into the system, which explains why language can be either vocal, as in speech, or manual, as in signed languages (Corballis 2011)." If we look more closely at the Broman area we find what are known as mirror neurons. The mirror neurons were discovered in monkeys in an area of the brain designated as F5 which is homologous to the Broca's area in humans. This overlap of the two systems "... has led to the notion that language grew out of the mirror system itself... (Corballis 2011)." The mirror neurons "... form a subset of a class of neurons that are active when the monkey makes an intentional movement of the hand, such as reaching for a peanut." In addition "some of these neurons also responded when the monkey observed another individual (such as a researcher) making the same movement (Corballis 2011)." The mirror neurons are often referred to as the "monkey see, monkey do" neurons. "... [T]he mirror system in the monkey is in essence a system for understanding action. That is, the monkey understands the actions of others in terms of how it would itself perform these actions. This is the basic idea underlying what has been called the motor theory of speech perception, which holds that we perceive speech, not in terms of the acoustic patterns it creates, but in terms of how we ourselves would articulate it (Corballis 2011)."

"Indeed the mirror system is now well documented in humans, and involves characteristics that are more language-like than those in the monkey. For example, in the monkey, mirror neurons respond to transitive acts, as in reaching for an actual object, but do not respond to intransitive acts, where a movement is mimed and involves no object. In humans, in contrast, the mirror system responds to both transitive and intransitive acts, and the incorporation of intransitive acts would have paved the way to the understanding of acts that are symbolic rather than object-related (Corballis 2011)."

"The first step from manual gesture to speech may have been the incorporation of facial gestures. Even in the monkey, manual and facial gestures are closely linked neurophysiologically, as well as behaviorally. Some neurons in area F5 fire when the animal makes movements to grasp an object with either the hand or the mouth. An area in the monkey brain that is considered homologous to Broca's area is involved in control of the orofacial musculature – though not of speech itself. These neural links between hand and mouth may be related to eating rather than communication, perhaps involved in preparing the mouth to hold an object after the hand has grasped it, but later adapted for gestural and finally vocal language. The link between hand and mouth suggests that early communicative gestures may have involved facial gestures as well as gestures of the hands. Indeed, speech itself retains a strong visual component (Corballis 2011)."

"The final act, then, was the incorporation of vocalization. Part of the reason for this may have been that facial gestures increasingly involved movements of the tongue that are invisible, and activation of the vocal cords simply allowed these invisible gestures to be accessible. Speech might be described as facial gesture half swallowed, with sound added. Along with the incorporation of vocal sound, the vocal tract itself changed, and control of the tongue was enhanced, enabling a wider diversity of sounds (Corballis 2011)."

Pinker (1994, 2009) discusses the "how" of language development and provides an insight to language evolution. Now we need to look at the "why" of language evolution. Corballis (2011) notes that memory and foresight may be keys to the evolution of language. Humans have the ability to remember specific events in the past, foresee possible future ones, and even invent fictional ones. We possess two basic types of memory – episodic memory – memory of actual events that occurred in the past – and, semantic memory – the storehouse of all our knowledge. And, episodic memory may combine with semantic memory which is called autobiographical memory. "... [T]he storage of episodic memories depends on semantic memories that are already in place – one can scarcely record a visit to a restaurant without already knowing what a restaurant is, and what happens there – but are then related to the self in subjectively sensed time. This allows the actual experience of the event to be stored separately from the semantic system. In this way, episodic memories could not be stored in the absence of semantic memory, which is perhaps why our childhood episodic memories do not begin until the semantic system is well established, by around age four or five (Corballis 2011)." Recall that Pinker (1994) tells us that "All Hell Breaks Loose" in language use in children from about the late 2's to the late 3's. Now at age 4 or 5 memory begins to play a definitive role.

Corballis notes that "Humans pass through four developmental stages – infancy, childhood, juvenility, and adolescence. Each of these stages contributes differently to acquisition of language. "During infancy, lasting from birth to age two and a half, infants proceed from babbling to the point that they know that words or gestures have meaning, and can string them together in two-word sequences" as noted previously. And, he notes that in "childhood that seems to be especially critical to the emergence of grammatical language and theory of mind. Childhood seems to be the language link that is missing in great apes and the early hominins, which may account for the fact that, so far at least, great apes have not acquired recursive grammar. But it is also during childhood that theory of mind, episodic memory, and understanding of the future emerge. Childhood may be the crucible of the recursive mind. During the juvenile phase, from age 7 to around 10, children begin to appreciate the more pragmatic use of language, and how to use language to achieve social ends. The final stage is adolescence, which Locke and Bogin suggest is also unique to our own species, and sees the full flowering of pragmatic and social function, in such activities as storytelling, gossip, and sexual maneuvering."

"Recursion holds the key to that difference in mind, underlying human characteristics as language, theory of mind, and mental time travel. It was not so much a new faculty, though, as an extension of existing faculties. Animals communicate, but in the course of human evolution recursive principles were added, allowing us to construct and understand an unlimited number of possible messages. Animals may have some awareness of the mental states of others, but the recursive principle extends this so that one may know that others know what one knows, providing for extended empathy and cooperation. It also introduces enhanced opportunities for deception and exploitation, which drove theory of mind to deeper levels of recursion and Machiavellian intrigue. Animals have memory, but our forebears added a recursive principle that allowed them to insert past episodes into present consciousness, and generate potential future episodes. The generation of episodes led also to storytelling and fiction (Corvallis (2011)." Corvallis (2011) suggests that "... the recursive principle evolved in these facets of social life, allowing our forebears to bond, and to share information within social groups, but also to compete, and to indulge in ever-escalating forms of warfare between groups."

"Although recursion was critical to the evolution of the human mind, ... it is not a 'module,' the name given to specific, innate functional units, many of which are said to have evolved during the Pleistocene. Nor did it depend on some specific mutation, or special kind of neuron, or the sudden appearance of a new brain structure. Rather, recursion probably evolved through progressive increases in short-term memory and capacity for hierarchical organization. These in turn were probably dependent on brain size, which increased incrementally, albeit rapidly, during the Pleistocene. Recursion "... allows us to travel mentally in time, inserting consciousness of the past or future into the present consciousness. Recursion is also the main ingredient distinguishing human language from all other forms of animal communication (Corballis 2011)."

One of the design features of language is called displacement. Displacement is "... the capacity to refer to events removed from the here and now [to travel mentally in time]. Language may therefore have evolved primarily to enable humans to share their memories, plans, and stories, enhancing social cohesion and creating common culture (Corballis 2011)." More basic to the sharing of human memories however may be the simple need to increase social interaction. About two-thirds of all conversations appear to be matters of social import. "Who is doing what with whom, and whether it's a good or bad thing; who is in and who is out, and why; how to deal with a difficult social situation involving a lover, child or colleague (Dunbar 2010)." Dunbar 2010 notes that "... we seem to be obsessed with gossiping about one another. Even the design of our minds seems to reinforce this.

Psychologists have developed a concept called Theory of Mind. "Having a Theory of Mind means being able to understand what another individual is thinking, to ascribe beliefs, desires fears and hopes to someone else, and to believe that they really do experience these feelings as mental states. We can conceive of a kind of natural hierarchy: you can have a mental state (a belief about something) and I can have a mental state about your mental state (a belief about a belief). If your mental state is a belief about my mental state, then we can say that 'I believe something to be the case'. These are now usually referred to as orders of 'intentionality'. Thinking about mental states in this way yields the following rough hierarchy.

"Machines such as computers have zero-order intentionality: they are not aware of their own mental states. Presumably, we also have zero-order intentionality when we are in a coma, and most insects and other invertebrates are also probably zero-order intentional beings. Ever since Descartes produced his immortal aphorism *Cogito ergo sum* ('I think, therefore I am'), we have known about first-order intentional states (I believe something to be the case). After that, we run into the beginning of an infinite regression: I believe that you believe that I believe something (third order intentionality); I believe that you believe that I believe that you believe something (fourth order intentionality), and so on. For obvious reasons, the higher orders of intentionality are often referred to as 'mind-reading'.

"There is good reason to believe that humans are capable of keeping track of, at most, six orders of intentionality, and after that they probably have to see it written down. In the words of the philosopher Dan Dennett" 'I suspect [1] that you wonder [2] whether I realize [3] how hard it is for you to be sure that you understand [4] whether I mean [5] to be saying that can recognize [6] that I can believe [7] you to want [8] me to explain that most of us can keep track of only about five or six orders [of intentionality] (Dunbar 2010)."

***

Dunbar (2010) found a direct correlation between the size of the neocortex in primates and group size – a measure of their social complexity. She used group size as a measure of social complexity for two reasons. "Firstly, it is one of the few things the field-workers invariably count and give solid numerical values for. It was therefore easy to obtain data on group size for a large number of primate species. Qualitative observations (like saying that one species is more complex than another) aren't very helpful when it comes to testing hypotheses rigorously, because it is easy for us to deceive ourselves into seeing what we want to see. The other reason was that, in one very important respect, social complexity does increase with group size."

Dunbar (2010) notes that "...what makes primate social groups quite different from those of other species is the fact that monkeys and apes are able to use very sophisticated forms of social knowledge about each other. They use this knowledge about how others behave to predict how they might behave in the future, and then use these predictions to structure their relationships." In addition she noted that "All such societies are characterized by tiers of social organization that are increasingly inclusive. At the bottom of the hierarchy are the temporary overnight camps of 30-35 people, some five or six families in all. These are essentially ecological groupings, the temporary coming together of a variable number of families who find it convenient to pool resources for a while and collaborate in hunting and food-finding. At the top of the scale is the largest grouping, the tribe, typically numbering some 1500 – 2000 people. The tribe is the linguistic group, the set of people who speak the same language (or in the case of the more widely spoken languages, the same dialect). Between these two levels can sometimes be found the groupings of about 500 people commonly referred to as 'mega-bands'. Below them, smaller groupings can sometimes be discerned, often referred to as clans. These turn out to average almost exactly 150 and to be very much less variable in size than any of the other kinds of groupings mentioned above."

"In fact, 150 turns out to have even more interesting properties. It is, for example, roughly the number of living descendants (including all the wives, husbands and children) that you would expect an ancestral people to have produced after four generations at the kind of birth-rates conventionally observed in hunter-gatherer and peasant societies. What's interesting about this is that a pedigree, or family tree, five generations deep takes you back to grandma's grandma, or as far back as any living member of the group can remember as a matter of personal experience. In other words, as far back as anyone can vouch for particular relationships: only within the circle of individuals defined by those relationships can you specify who is whose cousin, and who is merely an acquaintance." Recall that the smallest groups had a group size of about 30 to 35 people – similar in size to our primate cousins. This value was also used as the baseline in Dunbar's scheme for neocortex size – which she assigned a value of 1. Human brains are about 4 times larger (a ratio of 4:1) than our primate cousins and therefore have a predicted group size of about 120 to 140 – a pretty good fit with reality.

Dunbar (2010) notes that there is a "... well-established principle in sociology suggesting that social groupings larger than about 150 – 200 become increasingly hierarchical in structure. Small social groups tend to lack structure of any kind, relying instead on personal contacts to oil the wheels of social intercourse. But with more people to coordinate, hierarchical structures are required. There must be chiefs to direct, and a police force to ensure that social rules are adhered to."

She goes on to make the argument that "We do seem to use language in establishing and servicing our relationships. Could it be that language evolved as a kind of vocal grooming to allow us to bond larger groups than was possible using the conventional primate mechanism of physical grooming? Language does have two key features that would allow it to function in this way. One is that we can talk to several people at the same time, thereby increasing the rate at which we interact with them. If conversation serves the same function as grooming, then modern humans can at least 'groom' with several others simultaneously. A second is that language allows us to exchange information over a wider network of individuals than is possible for monkeys and apes. If the main function of grooming for monkeys and apes is to build up trust and personal knowledge of allies, then language has an added advantage. It allows you to say a great deal about yourself, your likes and dislikes, the kind of person you are; it also allows you to convey in numerous subtle ways something about your reliability as an ally or friend."

"Language has an additional benefit invaluable in these circumstances. It allows us to exchange information about other people, so short-circuiting the laborious process of finding out how they behave. For monkeys and apes, all this has to be done by direct observation. I may never know that you are unreliable until I see you in action with an ally, and that opportunity is likely to occur only rarely. But a mutual acquaintance may be able to report on his or her experiences of you, and so warn me against you –especially if they share a common interest with me. Friends and relations will not want to see their allies being exploited by other individuals, since a cost borne by an ally is ultimately a cost borne by them. If I die helping out a scoundrel, my friends and relations lose an ally, as well as everything they have invested in me over the years. Language thus seems ideally suited in various ways to being a cheap and ultra-efficient form of grooming (Dunbar 2010)."

Dunbar (2010) sums up her argument this way: "In a nutshell, I am suggesting that language evolved to allow us to gossip." Corballis (2011) agrees. He notes that "… it is the social function of language, its role in storytelling, gossip and group bonding that was probably critical to the evolution of language in the first place."

It is no wonder, then, that religion and language are closely linked. "Stories about the supernatural morph naturally into religion. Bryan Boyd points out that religious conviction derives less from doctrine than from stories. Like other religious works, the Bible tells us stories of such supernatural incidents as virgin birth, walking on water, or rising from the dead. As Boyd notes, evolution will favor belief in falsehood; it motivates more adaptive behavior than does belief in a truth. One falsehood that is perhaps encouraged by tales of the supernatural is the notion of life after death and subsequent oblivion, and the belief is no doubt fortified if expressed in stories that are widely shared, more generally, the adaptiveness of stories about the supernatural may derive from their sheer power to spread throughout a culture, engendering social cohesion (Corballis 2011)."

"Although religious rituals usually involve dance and music, they are also very verbal, since the sacred truths have to be stated. If so, religion, at least in its modern form, cannot pre-date the emergence of language. If religion had to await the evolution of modern, articulate language, then it ... emerged shortly before 50,000 years ago (Wade 2006)." There's no real good reason religion had to wait for modern language to form. We see a need for a set of behavioral rules with the first inclination for the need for parental investment. Modern "... religion, language and reciprocity are the comparatively recent elements of the glue that holds human societies together. Religion began as a mechanism for a community to exclude those who could not be trusted. Later, it grew into a means of encouraging communal action, a necessary role in hunter-gatherer societies that have no chiefs or central authority. It was co-opted by the rulers of settled societies as a way of solidifying their authority and justifying their privileged position (Wade 2006)."

***

We find that genetics, linguistics, and archeology working together provide a better and more accurate view of our evolution from this point in our story on to the end. The science of genetics has made a huge contribution to our understanding of the evolution of mankind. As noted in Chapter II, geneticists were able to establish a reasonable date for the rendezvous with our and the chimpanzees common ancestor. Their next task focused on when we diverged from *Homo ergaster.* And now their task is to determine a common ancestor for modern man and, where the most likely place of origin might be?

There are two key features of mitochondrial DNA that make it particularly useful in genetics research. First mutations in mitochondrial DNA (mtDNA) occur about 20 times more often and on a more regular basis than in nuclear DNA. Therefore it is more reliable as a "genetic clock". The second feature is that it is passed down from generation to generation by the mother. The father has mtDNA but it does not combine with the egg as part of fertilization – the mitochondria are located in the tail of the sperm which is lost on contact with the egg. So geneticists using mtDNA are able to track mutations back in time from mother to grandmother to great-grandmother ... until they find a single great-grandmother that we all share - as it turns out our 10,000 great-grandmother.

The very first findings using mtDNA found that all humans whether Africans, Eskimos, Hawaiians, or Vietnamese, or any other person have very few mutational differences between them -*tous parents, tous differents* - all related, all different. Second, when geneticists compared the lineages of all of the DNA they came up with two branches - one consisted solely of Africans and a second branch consisted of all the remaining people of African origin and everyone else in the world. This single limb that connected these two branches showed that we all have a common root in Africa. And, lastly, there are slightly more mitochondrial DNA mutations among the Africans compared to non-Africans implying their roots are a little older. When they did the arithmetic they found a common ancestor between 142,500 and 285,000 years ago. Other studies have yielded dates of 220,000 years ago, 200,000 years ago and 143,000 years ago. This great-grandmother is often called "African Eve" or "mitochondrial Eve".

Fagan (1990) asks the question "Did Eve exist?" He, of course, is referring to the "genetic" and not the "biblical" Eve. If one accepts the genetic theory of African origins for modern humans, was there actually an Eve, a woman of flesh and blood rather than just a genetic ancestor? In the sense that there was once a single person one could point to as the identifiable, single ancestor of *Homo sapiens*, the answer is no. Scientific reality is very different, for genes are the products, not of a single individual, but of interacting populations of males and females. There may, indeed, once have been a single Eve, but we can never hope to identify her in person. Her identity is submerged in those of thousands of other archaic humans, who were living south of the Sahara around 200,000 years ago." Stephen Jay Gould (in Fagan 1990) notes that "They [geneticists] have demonstrated the underlying unity of all humankind, a contingent fact of history that we often forget. Despite our external differences of skin color, hair form, and size, all modern humans have a relatively recent, common ancestry in Africa."

We can use the same basic principles for nuclear DNA that we did for mtDNA using the Y chromosome of the male to trace back in time the emergence of a proverbial "genetic Adam". We find that all the Y chromosomes of the males living today coalesce in the Y of a single man. Wells (2002) notes that "...the oldest splits in the ancestry of the Y-chromosome occurred in Africa. In other words, the root of the male family tree was placed in Africa - exactly the same answer the mtDNA had given us for women. The shocker came when a date was estimated for the age of the oldest common ancestor. This man, from whom all men alive today ultimately derived their Y-chromosomes, lived 59,000 years ago." This means that all modern humans were living in Africa before about 60,000 years ago. We can ask the same question of Adam that Fagan (1990) asked of Eve. "Did Adam exist"? And, the answer is no for the same reason that there was not a single Eve. The mitochondrial Eve and the genetic Adam surely did not know each other. In fact, in all likelihood, they didn't live in the very same time period or at the same location in Africa.

"The reason we don't find male lineages coalescing at the same time as Eve is because of early human sexual behavior. In most traditional societies, a few men do most of the mating – think of chiefdoms and warlords, for instance. Some men never get to have children, while others have more than their fair share. This is known as the *variance in reproductive success*, and it is higher for men than women, which means that women have more equal opportunities to have children. Since women are passing on their mtDNA, the result is that – in general – mtDNA lineages have a more equal chance of being passed on to the next generation than Y-chromosome lineages. This behavioral quirk tends to reduce the *effective* population size of the Y-chromosome, since not every man will pass on his Y, while most women will pass on their mtDNA. Since genetic drift acts more quickly in small populations, changes in the lineage composition of the Y-chromosome pool occurred more rapidly. Over time, this meant that Y lineages were more likely to be lost than mtDNA lineages. The result is that the deeper Y lineages were lost over the past 170,000 years, and the only ones left date to around 60,000 years ago (Wells 2007)."

***

I mentioned earlier that there were two human natures (two genders) – male and female. The differences are due to the hormone testosterone. "There are two periods when testosterone levels rise in male children: in the womb, from about six weeks after conception, and again at puberty. The effect of this double-barreled burst of testosterone on the male brain is dramatic. The first dose produces a baby that is mentally different from a girl baby from its first day on the planet. Baby girls are more interested in smiling, communicating, and people, boys in action and things. Shown cluttered pictures, boys select objects, girls people. Boys are instantly obsessed with dismantling, assembling, destroying, possessing, and coveting things. Girls are fascinated by people and treat their toys as surrogate people. Hence, to suit their mentalities, we have invented toys that suit each sex. We give boys tractors and girls dolls. We are reinforcing the stereotypical obsessions that they already have, but we are not creating them (Ridley1993a)."

"And the brain itself begins to show strange differences. Brain functions become more diffuse in girls, whereas they take up specific locations in the heads of boys. The two hemispheres of the brain become more different and more specialized in boys. The corpus callosum, which connects the two, grows larger in girls. It is as if testosterone begins to isolate the boy's right hemisphere from colonization by verbal skills from the left.

"These facts are far too few and unsystematic to be regarded as anything more than hints of what actually happens, but the role of language acquisition must be critical. Language is the most human and therefore most recent of our mental skills – the one we share with no other ape. Language seems to come into the brain like an invading Goth, taking the place of other skills, and testosterone appears to resist this. Whatever actually happens, it is an indisputable fact that at the age of five, when they first arrive at school, the average boy has a very different brain from the average girl (Ridley 1993a),"

"Puberty strikes a young man like a hormonal thunderbolt. His testicles descend, his voice breaks, his body becomes hairier and leaner, and he begins to grow like a weed. The cause of all this is a veritable flood of testosterone from his testicles. He now has twenty times as much of it in his blood as a girl of the same age. The effect is to develop the mental photograph laid down in his head by the womb's dose and to make his mind into that of an adult man.

"Asked about their ambitions, men from six different cultures replied with much the same answer. They wanted to be practical, shrewd, assertive, dominating, competitive, and self-controlled. They sought power and independence above all. Women from the same cultures wanted to be loving, affectionate, impulsive, sympathetic, and generous. They sought to serve society above all (Ridley 1993a)."

So why, in the first place, is there a need for sex? Ridley (1993a) provides an extensive review of the possible theories. The one that seems to hold the most promise is that sex is used to combat the threat of parasites. "Organisms need sex to keep their genes one step ahead of the parasites. Men are not redundant after all; they are women's insurance policy against her children being wiped out by influenza and smallpox (if that is a consolation). Women add sperm to their eggs because if they did not, the resulting babies would be identically vulnerable to the first parasite that picked their genetic locks." We have to back up to the reproduction process to get at the answer of how this is so. When sperm are produced, they carry a minimum payload – "... a nucleus cargo, a mitochondrial engine, and a flagellum propeller. The sperm-making cells go to great lengths to strip off the rest of the cytoplasm before the sperm is complete and re-digest it at some expense.

Even the propeller and engine are jettisoned when the sperm meets the egg; only the nucleus travels farther (Ridley 1993a)."

So why is this important? In the cytoplasm itself there may be infectious bacteria or viruses which you don't want to pass on. Ridley (1993a) notes that "[g]ender then, was invented as a means of resolving the conflict between cytoplasmic genes of the two parents. Rather than let such conflict destroy the offspring, a sensible agreement was reached: All the cytoplasmic genes would come from the mother, none from the father. Since this made the fathers gametes smaller, they could specialize in being more numerous and mobile, the better to find eggs. Gender is a bureaucratic solution to an antisocial habit."

Reproduction is a fairly simple process. In both the male and female the 23 pairs of chromosomes unwind. This forms, in the female, a cell (having only a single chromosome from each of the 23 pairs of chromosomes) that becomes the egg. In the male a similar cell is formed called the sperm. For an egg or sperm to form the reproductive cell must divide twice, a process called meiosis, to provide the requisite number of chromosomes. "Every gene normally consists of a strand of DNA and its complimentary copy closely entwined in the famous double helix. Special enzymes move up and down the strands, and where they find a break, repair it by reference to the complementary strand. DNA is continually being damaged by sunlight and chemicals. If it were not for the repair enzymes, it would quite quickly become meaningless gobbledygook.

"But what happens when both strands are damaged at the same place? This can be quite common – for example, when the two strands get fused together like a spot of glue on a closed zipper. The repair enzymes have no way of knowing what to repair the DNA to. They need a template of what the gene used to look like. Sex provides it. It introduces a copy of the same gene from another creature (outcrossing) or from another chromosome (recombination) in the same creature. Repair can now refer to a fresh template (Ridley 1993a)."

Once these two cells – egg and sperm – meet up they form a cell which becomes the initial cell that will begin to build a unique individual. If the male donates an X chromosome from the 23rd pair the female will give birth to another female. If, on the other hand, he donates a Y the female will give birth to a male. A second very complex process of cell division, mitosis, begins the process of cell differentiation that will build all the correct body parts in all the correct areas - hopefully. It's a very efficient process, since most of us turn out OK! Do mutations occur in these processes? You bet. Even identical twins may be slightly different from each other.

Are the mutations always bad? Not necessarily. Olson (2002) gives a pretty good rundown on the possible outcomes. "The fate of a mutation depends in part on how it affects the body. In that respect, mutations have much in common with lottery tickets. First of all, the vast majority of mutations are worthless or neutral, because they have absolutely no effect on our bodies. (For that matter, the vast majority of our DNA has absolutely no effect on our bodies. Just a small fraction serves any known function: some of the rest may have regulatory or structural functions, but its purpose remains obscure."

"Most mutations are neutral for another reason. DNA has a fair amount of redundancy built into it. Even if a mutation occurs in a functional region of DNA, the mutation may not change the underlying function. In that case a mutation can persist from generation to generation as a harmless variant (Olson 2002)." Going back to our example of the gene segment GCAATGGCC we might find that a mutation has occurred in the replacement of the first G with a C - now you have CCAATGGCC. If it happened in a nonfunctional part of the gene segment, it would have no effect. However, if a mutation occurred in the functional segment AATGGC (a C replaces a G to form AATCGC) it could be benign, because of redundancy, and have no effect on the function of the gene.

"Not all mutations are benign. The second most common category consists of those that harm an organism. In humans, most of these mutations are so damaging that they result in miscarriages. At least a fifth (and probably much more) of all conceptions end in miscarriages, and many of these result from genetic mutations that are fatal to a fetus. Other severe mutations do not exert their effects until after birth. Approximately 5 percent of children are born with serious physical problems, many of which are caused by genetic mutations. These abnormalities take an enormous psychological and economic toll on individuals and their families. One reason governments invest so much money in biomedical research, some of which can seem exceedingly abstract, is to understand and overcome these genetic disorders.

"The third category of mutations, by far the rarest and most mysterious, are those that benefit an organism by making it more likely to survive and reproduce in a particular environment. If one organism leaves more offspring than another, its mutations, including its beneficial mutations, will become more common in the next generation. As Charles Darwin realized, this is the process that has filled the world with such an amazing array of living things (Olson 2002)." This is the essence of natural selection.

Ridley (1993a) notes that "The old … model of sex, in which sex is an aid to faster evolution, implies that organisms would prefer to keep their mutation rate fairly high – since mutation is the source of all variety – and then do a good job of sieving out the bad ones. But, as Williams put it, there is no evidence yet found that any creature ever does anything other than try to keep its mutation rate as low as possible. It strives for a mutation rate of zero. Evolution depends on the fact it fails."

Cavalli-Sforza (2000) notes that "From the beginning of modern genetics, four evolutionary forces have been recognized: mutation, which produces new genetic types; natural selection, the mechanism which automatically selects the mutated types best adapted to a particular environment; genetic drift, the random fluctuation of gene frequencies in populations' and migration, sometimes called gene flow."

We've already covered mutation and natural selection. Genetic drift and migration are inter-related.

Genetic drift, a special case of migration, is the chance fluctuation of gene frequencies over several generations. A good example of genetic drift versus natural selection (taken from Cavalli-Sforza) involves the blood type of American Indians. They are about 100 percent O blood type. However, they came from a population in Asia where type O is represented in about 60 percent of the population and A and B the remaining 40 percent. Let's assume that a group broke from the original population and contained 75 percent O type. A generation or two later then, another group broke from this group and were 95 percent O type. And finally, another small group from this population broke off from the main population, and they were all O blood type. The new population would not have any representatives of A or B blood type. The evolutionary force directing these changes would be attributed to genetic drift. This is also known as the founders' effect. It usually occurs where the group that broke away is quite small. If on the other hand, this first group broke away from the main population and migrated to the America's with a mix of O, A and B blood types but, later only had O blood type in the new population then natural selection might be the evolutionary force at work. In the case of American Indians we know that they had an unusual disease that no other humans had - syphilis. The O blood type provides resistance to syphilis where A and B do not. Therefore, natural selection would select for O blood type and against A and B. Since we can't go back and do blood typing on the original American Indian founding population(s), we don't know if their predominant O blood type is due to genetic drift or natural selection. This type of migration, as in the case of the American Indians, involves colonization of new territory - the Americas. And, as we see from the blood type example, can create a very homogeneous population.

A second, and very important form of migration, involves marriage. There can be two effects on the genetics of a population through marriage. The first situation involves just one of the spouses moving away from his or her mother population. In this case the female usually moves to a neighboring population to marry. This interchange of "brides" between neighboring populations usually results in genetic homogeneity. That is both neighboring populations become more like each other since the genes are shared back and forth over time. Migration that involves spouses, in which one or both parents and the child move to a neighboring population or beyond without the cross-sharing of genes back and forth, is the most important form from a genetic point of view. This resettlement of the spouses and child can cause the new population center to become more heterogeneous by introducing a completely new set of genes into the neighboring population.

"In summary, natural selection is determined by the difference between the mortality and/or fecundity of different genetic types (also called 'genotypes'). Those genes that reduce mortality or increase fertility will increase in frequency in subsequent generations. The genotypes that increase mortality, especially among the young, or that reduce reproductive output tend to disappear from the population. The biological adaptation of an individual to the environment in which he [or she] lives is measured solely by his [or her] capacity to survive and reproduce. The process is completely automatic, and the 'survival of the fittest,' or more accurately the greater representation in future generations of those who have better chances of surviving and reproducing (i.e., are genetically fitter), is the cornerstone of natural selection (Cavalli-Sforza 2000)."

Mutation, natural selection, genetic drift, and migration would all be at work in determining the point where *H. ergaster* and Modern man would part ways. "Two preconditions might have been necessary for a tiny band of technologically oriented Moderns to emerge from one corner of Africa to take over the continent (and eventually the world). One is a relatively prolonged period of isolation. This is a problematic option for advanced, mobile humans in a land mass that is as singularly continental as Africa; but some degree of regional isolation could have been caused by an unusually severe environmental change or upheaval, reinforced by an epidemic or starvation. The second precondition could have centered on a decisive role for technology. If the ancestral Moderns were equipped with peculiarly appropriate tools and techniques to cope with the vicissitudes of their environment, they might have been able to maintain physical separation from other humans long enough for cultural barriers to arise. If it came to the point where such barriers became impregnable, human evolution could be said to have become 'self-made,' because both obstacles *against* and facilitators *for* gene flow did not so much derive from the environment as they were self-generated."

***

Studies indicate that within any ancient group of people mortality is greater in the very young and the very old. Once adulthood is attained, the chances for survival increase. They also note that up to one-third of the people in the group will be pre-adults - or more specifically, teenagers. Kingdon (2003) notes that this class of teenagers would have had a "proportional role in determining the day-to-day functioning of the group. It would be wrong to see non-adults as mere hangers-on, tagging along on the heels of the decision-taking adults; they must have been active participants in the overall foraging success of the group. The brighter and better the children were at communicating among themselves and with adults, at collecting food and keeping out of harm's way, the better their chances of overall survival." He also favors the selection process that favors "smart kids and loving mothers" over the conventional view of "sexual selection and mate choice". I've heard it said that at some point in time, God noted that He was actually unable to be in all places at all times. He, therefore, created mothers.

Kingdon's (2003) rationale behind his thinking follows the following line of reasoning. "In the real life of the prehistoric human, there would have been an unbroken continuum between an animal appetite for food in all its diversity; the context in which food was found (i.e., ecology); the ways in which food could be obtained (i.e., tool and techniques); and the social context in which the food was gathered, shared, and consumed (i.e., the group or 'society'). All the peculiarities of human thought or mentality, language, and social awareness must have arisen within that continuum, and it is an intellectual conceit to dismiss subsistence and tools and lift out social relations as the 'prime mover' in the development of human minds. Other animals have complex societies, but the facts that tools and techniques are the interface between modern humans and nature and that their effective use to harvest resources requires complex communication skills are differences of overriding significance.

"These differences are significant because tools are not separable from the technical procedures and mental planning that is used in their employment, from the ecological or food-gathering contexts in which they have been devised, or from the social cooperation that enhances their scope and range. The use, practice, learning, and rewards of technical expertise and skills in obtaining food exist as much in their ecological setting as they do in a social one, but the public display of competence in an ever widening repertoire of techniques must have been a ubiquitous and continuous preoccupation among prehistoric peoples. For any single individual, significant social status would have attended articulate displays of competence in the gathering of food or the protection of resources.

"Remember, too, that all foraging would have occurred in response to a rapidly changing calendar of food availability. These seasonal changes would have demanded versatility and good judgment as to when to switch from one food or one set of skills to another. Furthermore, a prime difference between the ecological niche of other animals and the 'niche-stealing' specialization of humans was the continuous expansion of food sources. Novelty was a continuous, not an intermediate, property of human foraging not only because food sources were changing by season (and being augmented as more niches were 'stolen') but also because there was always a class of youngsters that were learning how to survive on these resources for the first time. This learning was not innate for the food or the technique, and it would have been a continuous challenge - for both teacher and the taught - to articulate and occasionally improve skills. Such skills were 'autorewarding' in that both participants, sometimes the entire group, ended up with the prize of extra items of food."

Recall that there were two preconditions required by Kingdon (2003) of a prolonged period of isolation and a decisive role for technology to allow a small band of Moderns to emerge from a small corner of Africa. Kingdon (2003) speculates about where in Africa this bottleneck might occur. "The gap between 'emergence' and exit out of Africa carries a strong implication that the northern reaches of the continent were not prime habitat and also suggests that Moderns originated south of the Sahara. When Moderns did leave Africa, they would seem not to have colonized cooler habitats; that they were excluded from cool, dry habitats because these were already preoccupied by Neandertals or Erects could be part but not all of the explanation. If this aversion derived from their immediate place of origin, then North Africa and Ethiopia are even less likely sources for the earliest Moderns, as are the South African Uplands. If choice is further narrowed by eliminating both rain forests and deserts from consideration, the largest blocks of remnant territory became the savannas of eastern, western, and southern Africa." Indeed a bottleneck did occur isolating a population in South Africa somewhere between 195 and 123 thousand years ago. The climate changed significantly. During this period the planet entered a long glacial stage known as the Mann Isotope 6 (MIS6). During this cool dry period, the deserts would have expanded significantly making much of the African continent uninhabitable except for the southern coast. The effect was nearly the extinction of mankind. Genetic studies indicate that the human population dropped from an estimated 10,000 people to as low as a few hundreds of people. Probably not below 500 since this is the point where in-breeding would produce deleterious effects. The exact date and residual population numbers are unknown (Marean 2010). "What we do know is that this transition turned people with rather broad, long, and low braincases, with quite strong browridges (like the Florisbad and Jebel Irhoud remains), into individuals with higher, shorter, and narrower crania with smoother foreheads (like the Kibish or Border Cave fossils). Chins became a prominent facial feature even in children, though the rest of the face remained short, broad, and flat, with a wide nose and low, well-spaced eye sockets - which, in life, might have enclosed brown irises with epicanthic folds around them. The skeleton retained its lanky tropical shape, but bone thickness and muscle power was decreasing (Stringer and McKie 1996)."

Marean (2010) excavated a cave known only as PP13B "at a place called Pinnacle Point, a promontory near the town of Mossel Bay that juts out into the Indian Ocean." His team has "recorded a remarkable record of activities undertaken by the people who inhabited this area between approximately 164,000 and 35,000 years ago, hence during the bottleneck and after the population began to recover." In addition, they were able to lay out a plausible account of how these people lived during a climatic crisis. He also noted that "The remains also debunk the abiding notion that cognitive modernity: evidence of behavioral sophistication abounds in even the oldest archaeological levels at PP13B. This advanced intellect no doubt contributed significantly to the survival of the species, enabling our forebears to take advantage of the resources available on the coast." This, of course, verifies the thinking of Kingdon (2003) noted earlier.

The area surrounding this cave is known as the Cape Floral Region. It is a 90,000 square kilometer strip that contains an astonishing 9,000 plant species, some 64 percent of which live only there. Marean (2010) notes that "Of the vegetation groups that occur in this realm, the two most extensive are the fynbos and the renosterveld, which consist largely of shrubs. To a human forager equipped with a digging stick, they offer a valuable commodity: the plants in these groups produce the world's greatest diversity of geophytes - underground energy-storage organs such as tubers, bulbs and corms." Geophytes have a number of qualities useful to man: they are high in carbohydrates, they are located underground and less susceptible to other predators, they are low in fiber and therefore easily digestible, and they are adapted to dry conditions and would have been plentiful during this arid glacial period. A high source of protein would also have been available to these people in the form of shellfish. Marean (2010) notes that "The southern coast also has an excellent source of protein to offer, despite not being a prime hunting ground for large mammals. Just offshore, the collision of nutrient-rich cold waters from the Benguela upwelling and the warm Agulhas current creates a mix of cold and warm eddies along the southern coast. This varied environment nurtures diverse and dense beds of shellfish in the rocky intertidal zones and sandy beaches. Shellfish are a very high quality source of protein and omega-3 fatty acids. And as with the geophytes, glacial cooling does not depress their numbers. Rather, lower ocean temperatures result in a greater abundance of shellfish." A number of technological advances would have given these people a distinct advantage in this environment. Merean (2010) surmises "... that the people who lived at PP13B - which 164,000 years ago was located much farther inland, two to five kilometers (approximately 1.25 - 3.0 miles) from the water, because of lower sea levels - scheduled their trips to the shore using a lunar calendar of sorts, just as modern coastal people have done for ages." They also used fire to heat treat silcrete to make it easier to work. Silcrete is a fine grained rock that is superior to the quartzite readily available to the people at Pinnacle Point for making the fine bladelets that they used. However, silcrete is very difficult to work in its natural state. Marean (2010) describe the process as follows: "The process of treating by heat testifies to two uniquely modern human cognitive abilities. First, people recognized that they could substantially alter a raw material to make it useful - in this case, engineering the properties of stone by heating it, thereby turning poor-quality rock into high-quality raw material. Second, they could invent and execute a long chain of processes. The making of silcrete blades requires a complex series of carefully designed steps: building a sand pit to insulate the silcrete, bringing the heat up to 250 degrees Celsius, holding the temperature steady and then dropping it down slowly. Creating and carrying out the sequence and passing technologies down from generation to generation probably required language." The people at Pinnacle Point also had an artistic bent. They found and used red ochre, probably mixed with animal fat to make paint, for decoration of likely themselves and seashell ornaments. Marean (2010) notes that "Such decorations typically encode information about social identity or other important aspects of culture - that is, they are symbolic." If this is the case, it predates the art found in early populations in Europe by tens of thousands of years.

Marean (2010) sums up his findings in the following statement. "I suspect that a driving force in the evolution of this complex cognition was a strong long-term selection acting to enhance our ancestors' ability to mentally map the location and seasonal variation of many species of plants in arid environments and to convey this accumulated knowledge to offspring and other group members. This capacity laid the foundation for many other advances, such as the ability to grasp the link between the phases of the moon and the tides and to learn to schedule their shellfish-hunting trips to the shore accordingly. Together the readily available shellfish and geophytes provided a high-quality diet that allowed people to become less nomadic, increased their birth rates and reduced child mortality. The larger group size that resulted from these changes would have promoted symbolic behavior to express their social identity and build on one another's technologies, explaining why we see such sophisticated practices at PP13B."

Being adapted to these coastal environments, Modern humans would have moved up the east coast of Africa as local populations expanded. Their path out of Africa sometime after about 100,000 years ago would lead to the north and east across the Red Sea on a coastal path to Asia.

Grandfather's Tale

This tale is taken from the Natchez on the Gulf Coast.

Adoption of the Human Race

In the beginning, Moon, Sun, Wind, Rainbow, Thunder, Fire, and Water once met a very old man. This wise old man turned out to be Chief of the Sky Spirits. Thunder asked him, "Can you make the people of the world my children?"

"No, no, no!" Wise Old Man replied. "They cannot be your children, but they can be your grandchildren."

Sun asked Old Man, "Can you make the people of the world my children?"

"No, they cannot be your children" answered Old Man. "But they can be your friends and grandchildren. Your main purpose is to give plenty of light."

Moon asked, "Can you make the people of the world my children?"

"No, no, I cannot do that," Old Man replied. "The people of the world can be your nephews and friends."

Fire asked that the people of the world be made his children.

Wise Old Man replied, "No, I cannot give them to you to be your children, but the people of the world can be your grandchildren. You can be their warmth and give them fire to cook their food."

Wind asked the same question as the others. Wise Old Man told Wind, "No, no, the people of the world cannot be your children, but they can be your grandchildren. You can remove the bad air and all kinds of diseases from the people, and keep them healthy."

Rainbow wanted the people to be his children. "No, they cannot be your children," Wise Old Man explained. "You will always be busy preventing too much rain and floods upon the earth."

Water asked that human beings be made his children, but Wise Old Man answered, "No, the people of the world can never be your children. You shall give them long life."

Wise Old Man continued, "I have now told all of you the best ways to guide yourselves and what you can do to help the people of the world. You must always remember that these children of the human race are my children."

PART 2
ON CONQUERING THE PLANET

Chapter V.

Who Wants to Stay and Who Wants to Go?

(2.0 million years ago to 20 thousand years ago)

Wanderlust is innate and ubiquitous; it has carried hominids to verdant oasis and frigid tundra; it has occurred during times of warfare and famine as well as peace and opulence.

Jon Turk, In the Wake of the Jomon

Many psychologists and linguists now agree that spoken language is the loom on which some of the finer fabrics of consciousness are woven. The two qualities of the human mind are inextricably meshed with each other.

Richard Leakey, The Origin of Humanity

... after the power of language had been acquired, and the wishes of the community could be expressed, the common opinion how each member ought to act for the public good, would naturally become in a paramount degree the guide to action.

Charles Darwin, The Decent of Man

If we possessed a perfect pedigree of mankind, a genealogical arrangement of the races of man would afford the best classification of the various languages now spoken throughout the world; and if all extinct languages, and all intermediate and slowly changing dialects, were to be included, such an arrangement would be the only possible one.

Charles Darwin, The Decent of Man

Kun mena mana? Kun mena aqwa?" In pidgin it means "Who thinks we should stay? Who thinks we should go across the water?" This begins the story of man's settling of the Eurasian continent. Not a mass migration but a slow crawl over many millennia. "Our African Exodus was the greatest of all human journeys, a global endeavor that took our ancestors over every conceivable obstacle thrown up by nature: estuaries, deserts, mountain ranges, steppes and tundra, dense forests, fields of ice and snow, and sheer distance - the 9,000 mile length of the Americas being a prime example (Stringer and McKie 1996)." Finlayson's innovators will always be in the lead - the conservatives will perish. We will see that there are four separate stories to be told in the settling of Eurasia. The first involves *Homo erectus*, and possibly earlier ancestors, and their moving out of Africa about 1.8 *Mya* into Eurasia. The second covers the establishment of Neanderthals in Europe about 500 *Kya* and their cohabitation with proto-ancestors in the Middle East from about 130 - 80 *Kya*. The third is the movement of *Homo sapiens sapiens* out of East Africa and their spread to Asia, Australia and finally to Europe from somewhere between 100 - 40 *Kya*. And the last covers the story of Modern Man, Cro-Magnon Man, re-entering the Middle East and their replacement of Neanderthal Man over the period of about 40-25 *Kya*. The migrations to the Americas will be discussed in Chapter VI.

Before we can discuss the actual "migration" scenarios, we need to have a basic understanding of how the hunter-gather societies function and their evolution within the context of the environments they colonized and settled. Over the last 100,000 years humans have progressed from the basic family organization through to the establishments of large city states. How they defined, used and maintained territory in their individual ecological context is a large part of that evolution.

"A universal aspect of hunters and gathers is that these societies must maintain territory, or range, from which the essential resources to sustain life are derived. Within each territory, resources are not distributed uniformly. Certain resource concentrations play a more important role in subsistence activities than others. Because the harvest of these resources is essential for human survival, their distribution and the timing of harvesting them result in concentrations of human populations. Furthermore, collective efforts enable hunter-gathers to maximize their harvests. As a result of these cooperative activities, an economic surplus … [is] … generated that facilitate[s] human settlement for extended periods of time. These types of subsistence activities led to predictable settlement locales that were restricted to specific geographic locations for a specific duration of time.

"Although each group, or band, maintains a geographical range, or territory, there is 'communication between groups, including reciprocal visiting and marriage alliances, so that the basic hunting society consisted of a series of local 'bands' that were part of a larger breeding and language community'. The major limiting factor on band size is how many people can be supported, or carried, by the harvest of the resources within their territory. The maximum number of people that can be supported by the environment is called the carrying capacity, which is a general ecological term referring to the number of living things an environment can sustain. Human carrying capacity is determined by the resources that can be harvested within the territorial range given a group's level of technology.

"There are many controls on carrying capacity. For example, a technological innovation, such as a shift from simple spear fishing to using fish weirs to concentrate the fish and force them to swim past one spot or into a trap, could greatly increase a harvest and the number of people that might be supported. It is not only the total amount of resources within a given environment that is important, but also the ability of the people to perceive and harvest them. This is called the effective environment and is limited to the resources conceptualized and utilized by the band. The band's potential to harvest resources is dependent to a large degree on their specific level of technology and their knowledge of their territory's resources. [It is also reflective of their level of innovation within the band.] Environmental factors also affect carrying capacity, such as unusually cool summers that might alter the migration routes of caribou or high water that might reduce salmon harvests. Gradual and more efficient adaptation to the effective environment could conceivably take hundreds of years or more following colonization before the carrying capacity of a specific territory is reached.

"Considerable ethnographic evidence shows that social conflict, frequently over resources, causes groups of people to break away from the larger group and relocate. Anthropologists call this phenomenon fissioning. Given our knowledge of hunter-gathers, a hypothetical colonization model for the Americas [or any other lands] can be constructed. Very simply stated, as a group approached or exceeded the carrying capacity of their territory, fission occurred. Although adjacent bands might be able to absorb some of the splinter population, this would also create increased competition for resources. To successfully colonize and settle adjacent unoccupied territory, the splinter group separating from the parent band would have to be a viable breeding population containing at a minimum adult males, young adult females, and possibly children.

"This model is based primarily on environmental criteria and views colonization as a process characterized by a series of stages. The first stage is exploration by humans of unoccupied regions adjacent to a settled area. Exploration provides knowledge of the adjacent region's geography and resources, but the area remains unsettled by humans. Fission is the second stage: it requires that a viable breeding group break away from an existing parent population. The third stage, migration, is the actual process of the splinter group moving from the parent settlement to the known, but unoccupied, region to be colonized. The fourth stage is colonization. Colonization is the establishment of residential locales, or frontier settlements, in a territorial range that provides the resources essential to sustain life. The final stage is settlement, the process by which the colonizing population expands to fully exploit and occupy the effective environment of the new territory. Exploration may have been undertaken primarily by young adult males able to travel long distances unencumbered by children, pregnant females, and the elderly.

"As adaptation to the new territory becomes increasingly effective, a greater number of people are supported by available resources. Over time the population would eventually approach or exceed carrying capacity and fission would occur again. A subsequent migration would take place and an adjacent explored but unoccupied territory would be colonized. The absence of competition would tend to promote colonization of unoccupied territory (Dixon 1999)." So in this leap-frog manner humans slowly colonized new territory and gradually filled up every possible environmental niche around the world. Not all of these niches were equal as regards to quality and quantity of resources, so different strategies were needed to fully utilize them.

The early hunter-gathers out of Africa existed in a subsistence economy. "In the subsistence economy, the goal is not to maximize production but to minimize the effort expended in meeting household needs. A specific mix of strategies all exploited at the same cost level minimizes procurement costs for the households of a region. Such a mix should remain stable except where upset by changes in population, technology, or the environment. As one example of such change, the diets of the peoples of the prehistoric world gradually broadened to include an increasing range of foods as the landscape was gradually filled by hunter-gathers (Johnson and Earle 2000)."

There are four major factors associated with the subsistence economy that tend to keep populations low. "First, a chronic caloric deficiency lowers fertility; because of season cycles in food availability and limited storage capabilities, periods of food shortage were common. Second, a long nursing period delays renewed ovulation; since most wild foods are apparently not well suited for weaning young infants, nursing among foragers typically remains a child's main food source for the first two or three years. Third, intense physical exercise required for mobile foraging may lower female fertility. Fourth, because closely spaced children are an economic hardship in a mobile society, infanticide may have been used to space births. Although these factors no doubt operate differentially under different environmental conditions, the fertility of mobile groups is invariably low (Johnson and Earle 2000)." Low populations would tend to persist over time until resource utilization improved through changes in technology and/or environmental conditions increased carrying capacity of the territory or the people simply moved on to new territories. Once all environments are filled with people, new territories are no longer available for the taking. With no place to go, a local group, whose population is increasing, will be forced to compete with their neighbors.

"As a result of population increase, the subsistence economy must be intensified to provision greater numbers of people on the same resource base. Intensification does not come cheaply, but typically creates four kinds of problems, their relative importance varying according to environmental conditions. These are production risk, raiding and warfare, technology needs, and resource depletion. The solutions to these problems generally require increasing the economic integration of communities and the power of leaders.

"The first problem is production risk. As a landscape fills with people, the most desirable foods are soon depleted, and the less desirable foods, those that once served as buffers against starvation in bad years, come to be part of the regular diet. With fewer buffers and less food generally, the risk of starvation increases, and each household faces the need for a margin of security in food production against the possibility of lean seasons or years. Households can do some of this individually, through over production or private food storage. But at some point it becomes more efficient for the community to share costs of security. A classic form of risk management is community food storage; another is reciprocal arrangements between communities for visiting (and feasting) in lean times. A region with community storage or reciprocal feasting can sustain a larger population; but such arrangements call for leadership and create opportunities for control.

"The second problem is resource competition. At all ethnographically known economic levels, competition between families for prized resources occurs. At the family level, characterized by low population densities and dispersed resources, households tend to avoid competition by scattering and staying out of each other's way. With intensification, however, locally rich resources, such as fertile bottomlands, become even more precious, and improvements to the land, such as long-yielding tree crops, become more common. Such developments increase the benefits of the violent seizure of territory relative to the costs of violence. The general level of violence in a region accordingly rises, and small groups form alliances with other small groups for more effective defense of their resources. Effective defense also requires greater social integration and generates opportunities for control.

"The third problem, inadequate use of resources, refers to resources that may be used only if costly technologies are developed. As population rises in a resource area, and with it production risks, it becomes advantageous to invest in technologies that utilize resources that were ignored at lower population levels. For example, the efficient use of marine resources may require the construction of huge canoes or whaling boats; the use of lands for farming may require an irrigation system; in the far north, the abundance of food in summer may stave off hunger in winter only by means of large-scale technologies for harvesting and storage. These technologies are frequently beyond the capacity of a single family; they require the collaboration of households in a community and in due course come under the control of a manager.

"Fourth, the depletion of local resources brought on by population growth may increase the need for goods that cannot be produced locally but can be obtained in exchange for local goods. Trade can even out seasonal or annual shortfalls in production, and it can increase food production by making tools (e.g., axes) available in places poorly supplied with the raw materials for producing them. In both these ways trade in specialized goods increases the overall efficiency with which a population can be provisioned from limited resources, and thus the capacity to sustain a larger population on the same resource base, which is what intensification essentially is. But trade, especially long-distance trade, requires a knowledgeable head trader who can make decisions binding the trading group. And this, too, is an opportunity for control (Johnson and Earle 2000)."

With this in mind, Johnson and Earle (2000) examined the cultural evolution of society. We will meet many of these variants of societal evolution along the journey out of Africa and the settling of the diverse habitats around the world. Different people will react differently to the four problems of the subsistence economy defined by Johnson and Earle (2000). Societal evolution has not followed a straight line linear path from foragers to complex city-states. Instead, the concept of multilinear evolution states "... that social evolution can follow different courses depending on local history and ecology readily embraces the possibility that particular communities, having achieved a working solution to the problems posed by population and environment, need not evolve at all if conditions do not significantly change. No intrinsic perfecting tendency drives technology to ever-increasing levels of energy efficiency. Hunter-gathers can remain hunter-gathers indefinitely; horticulturalists and pastoralists, though having harnessed energy can remain small scale and egalitarian (Johnson and Earle 2000)."

Basic human instincts are centered around the quest for health and safety. Various societies through evolutionary time have developed different strategies for ensuring their health and safety. People have been willing to sacrifice some of their liberty in order to improve their overall life-style in order to share the resources within their and neighboring territories. "They participate in social groups that make this possible, and agree to certain restrictions on their own behavior in order to reap the benefits of society, including protection from hazards. But these restrictions can, under certain circumstances, include requirements to participate in ritual occasions, share wealth in communal distributions, and take orders from authorities. Thus the materialist focus of an ecological perspective does not remain on nutrients, shelter, and defense, but expands out into the entire social and cultural world, where many adaptive solutions are found to the problems faced by individuals in their efforts to achieve health and safety (Johnson and Earle 2000),"

There are three principle modes to exchange goods and services: reciprocity, redistribution, and exchange. "Reciprocity, in which individuals (or groups) of roughly equal standing engage in a customary give-and-take of equivalent values over time, characterizes the pattern of exchange typical of households, lineages, villages, and many other small social groups. Redistribution, an intrinsically hierarchical flow of goods into a center where they are controlled and then redistributed by some central authority, typifies the feasts and gift-exchanges of some Big Man systems and the centralization of larger-scale communities like chiefdoms, as well as most modern governments. Exchange, the market driven flow of goods and services under the regime of supply and demand, typifies the modern market economy (Johnson and Earle 2000)."

Johnson and Earle (2000) have identified three critical levels of socioeconomic integration: "(a) The Family-Level Group, including family/camp and the family/hamlet; (b) the Local Group, including acephalous [having no leader or chief] local group and Big Man collectivity; and (c) the Regional Polity, including the chiefdom and the state." They are clear to state that explicit dividing lines between these groups do not exist and they provide examples of current societies in each of these areas. Instead they note "These labels do not signify perfectly discrete levels or plateaus, to one or another of which all known cultures must be assigned; rather, they designate stations along a continuum at which it is convenient to stop and make comparisons with previous stations."

The earliest people out of Africa again lived in a subsistence economy and were certainly of the Family-Level Group. Johnson and Earle (2000) characterize this group as follows: "The family or hearth group is the primary subsistence group. It is capable of great self-sufficiency but moves in and out of extended family camps or hamlets opportunistically as problems or opportunities arise.

"The family/camp is characteristic of foraging societies of low density (less than one person per ten square miles). Camp groups of twenty-five to fifty persons typically form when resources are highly localized or when a group larger than the individual family is required for risk management or for a particular subsistence activity. The group can then dissolve into small segments consisting of single families (five to eight persons) that independently exploit low-density, dispersed resources. These societies are characterized by a simple division of labor by sex. Suprafamily leadership is ephemeral and context-specific, relating to immediate organizational requirements such as a hunting expedition requiring the participation of numerous families. Although homicide is fairly common, organized aggression (warfare) is not. Ceremonialism is ad hoc and little developed. A camp characteristically has a home range but does not claim exclusive access to this territory or strictly defend it against outsiders.

"The family/hamlet is characteristic of somewhat higher density societies (from one person per ten square miles to two per square mile) in which families cluster into a settlement group or hamlet (twenty-five to thirty persons) on a more permanent basis. The subsistence economy continues to rely heavily on wild foods, sometimes in conjunction with the beginnings of horticulture or herding. Storage is more prevalent. During the year individuals or families move out to exploit specific resources; from year to year, the hamlet re-forms and fragments as households change locations to minimize resource procurement costs. The hamlet does not form a clearly demarcated political group, and leadership continues to be context-specific and minimal. Ceremonialism is little developed. As with the family/camp, the hamlet's territory consists of undefended home ranges, and warfare is uncommon."

Later migrations may have included Local Groups in addition to the Family Level Groups. Johnson and Earle (2000) note that "Local groups of families, running to five or ten times the size of family-level groups, form around some common interest such as defense or food storage. They are usually subdivided along kinship lines into corporate lineages or clans. Depending on the extent of their common interests, these groups are either acephalous, village-sized units or larger groups integrated by regional networks of exchange headed by Big Men.

"The acephalous local group is typically found in societies with density greater than one person per square mile. The subsistence economy in most cases focuses on domesticated species, although in some cases wild resources, especially maritime resources, dominate. A frequent settlement pattern is a village of perhaps one hundred to two hundred people subdivided into clan or lineage segments of hamlet size (i.e., twenty-five to thirty persons). The local group forms a ritually integrated political group and may have a headman; but it typically fragments into its constituent kin groupings either seasonally or periodically as a result of internal disputes. Because of endemic warfare intercommunity relationships of various sorts are critically important for community security, but such relationships are contracted essentially on an individual, family-by-family basis. Ceremonialism is important for publicly defining groups and their relationships. Resources are held exclusively by kin groups, and territorial defense is common.

"The Big Man and his managed intergroup collectivity are found at higher but variable population densities in areas in which warfare between territorial groups has traditionally been intense. Subsistence is focused heavily on agriculture, pastoralism, or extremely productive natural resources. The local community of perhaps three hundred to five hundred people is a territorial division, typically containing multiple clan or lineage segments that either live together in a village or are dispersed throughout the well-defined territory of the group. The local group is represented by a Big Man, a strong, charismatic leader who is essential for maintaining internal group cohesion and for negotiating intergroup alliances. The Big Man is also important in risk management, trade, and internal dispute settlement, and represents his group in the major ceremonies that coordinate and formalize intergroup relationships. His power, however, is dependent on his personal initiative; if his followers desert him for a competitor, little may be left of the reputation he has tried to build for himself and his local group, or of the alliances he has contracted."

If we assume modern humankind has been evolving over the last 100,000 years, the majority of that time humans have been in a subsistence mode in a Family-Level Group setting. With the development of agriculture about 10,000 years ago, the Local Group setting with a need for warfare to protect land resources and capture women as mates became prominent. Pinker (2009) notes that "In primitive warfare, mobilization was more complete, battles were more frequent, casualties higher, prisoners fewer, and weapons more damaging. War is, to put it mildly, a major selection pressure, and since it appears to have been a recurring event in our evolutionary history, it must have shaped parts of the human psyche." He asks the inevitable question "Why would anybody go to war?" He also notes that the major reason men go to war is to obtain or keep women. "Access to women is the limiting factor on males' reproductive success. Having two wives can double a man's children, having three wives can triple it, and so on. For a man who is not at death's door, no other resource has as much impact on evolutionary fitness."

"Is warfare between local groups an outcome of human nature? Humans are phylogenetically prepared to be aggressive under some circumstances: they competitively seek to obtain food, shelter, and mates to support themselves and their offspring, and will defend these prizes against all comers. But it is not so clear that this explains warfare: family-level communities, in which these same tendencies apply, manage to live rather peacefully by comparison.

"Peace becomes impossible; however, as the subsistence economy intensifies. In social evolution, we have seen that the first response to resource competition – at very low population levels – is to scatter and avoid conflict. Men may fight one another over this resource or that, but the overall pattern is to maintain peace through disengagement. As landscapes fill in and opportunities diminish, competition inevitably rises, but the family's first thought is not to create a regional political structure to resolve conflict. Rather, applying conservative logic, families do what they have always done: try to locate the best resources before others, and, if they are scarce, occupy them exclusively. If competitors try to dislodge them, those in possession must dig in, especially if other resources are now also occupied by other families. The result is that as population rises, competition increases, and so does aggressiveness (Johnson and Earle 2000)."

It wasn't until about 5,000 years ago that the Regional Polity came into play. The Regional Polity then developed once major human dispersal had settled almost every available environmental niche in the world. More will be said about the Regional Polity in Chapter VII.

***

The earliest emergence out of Africa has been attributed to *Homo erectus* beginning about 1.8 *Mya*. The consensus thought on this migration was that "*Homo erectus* was the first human species that had long legs and a big brain, could make tools, and actively hunted in grassy savannahs for meat. This suite of features allowed it to migrate out of Africa and colonize Asia (Finlayson (2009a)." Let's look at each of the various pieces of this scenario. First, was it a migration? Probably not! Finlayson (2009a) notes: "Instead, it was a gradual expansion into new areas by the children and grandchildren ... The expansion was at the level of generation and not of individuals. So it would have been for the first proto-humans that would have gradually expanded into favorable habitats wherever these were." Did these first folks out of Africa need long legs, big brains and the ability to make tools and hunt? No! Finlayson (2009a) emphatically notes: "... I fail to see the relevance of, for example, long legs. What would have favored expansion would have been reproductive output and suitability of habitats. The first proto-humans to expand from the geographical core area in tropical Africa did not have to wait until they were Olympic marathon champions to move out of their ancestral home. They did not have to wait to be super-brainy either. How many different species of animals have traversed huge parts of the planet and occupied remote places? How many species of tree re-colonize former distant, haunts after an Ice Age? All that was necessary was that their requirements matched the environments that they were moving into. The vast majority did so without particularly stunning brains, tools, or a body that made them superb at long-distance migration." *Homo erectus* was well adapted to open, savannah-like, grassy woodland. This habitat "stretched right across a mid-latitude belt from West Africa to China, with a southern prong down to South Africa (Finlayson 2009a)." The habitat would have been available to any proto-human wanting to make use of these environments. The pace need only be about 3 Km (1.86 miles) per generation (20 years per generation – that's less than one tenth of a mile per year). It would not have been directed towards any geographic goal but simply a matter of moving to find food and short migrations when groups got too large for their habitat to support and split into smaller kinship groups.

There are three diminutive creatures that could throw a "monkey wrench" into the long-established scenario - *Australopithecus bahrelghazali, Homo georgicus,* and *Homo floresiensis* -that *H. erectus* was the first out of Africa. By all accounts, the presence of *Australopithecus bahrelghazali* in Chad 3.0 to 3.5 *Mya* doesn't seem to fit any scenario developed so far. He's out of place and out of time. He's supposed to be in East Africa with all the other Australopithecines. Then there's *Homo georgicus*. We find him in Georgia - the country of not the state of - 1.77 *Mya*. He's contemporaneous with *Homo erectus* and already "out of Africa". He appears way too small to be a descendent of *Homo* erectus. Wong (2009) notes that "The discovery of the Georgian remains dispelled that notion of a brawny trailblazer with a tricked-out tool kit, because they were on the small side for *H. erectus* and they made Oldowan tools, rather than the advanced Acheulean implements experts expected the first pioneers to make." How did they get there? Who were their descendants?

"Found in the 1990s, and then dated to 1.8 million years old, the Dmanisi hominids are currently the most ancient undisputed hominid fossils outside of Africa. At a stroke they almost doubled the accepted date for the first dispersal of humans out of Africa, which was thought to have occurred a little over a million years ago, based on *Homo erectus* finds in Java. The Dmanisi hominids predate the appearance of large-bodied, large-brained, humans. Some are small bodied (1.4 meters stature), small brained (ca. 600 cubic centimeters) and have primitive characteristics previously only seen in the earlier members of the genus *Homo* and australopithecines. They do have modern body proportions. They have been variously described as early Asian *Homo erectus*; earlier *Homo ergastor*; even earlier *Homo habilis*; a new species *Homo georgicus*; and a new subspecies *Homo erectus georgicus*. This assemblage can definitely be described as taxonomically challenging.

"Another problem raised by the Dmanisi hominids is that they are about the same age as the earliest identified species of *Homo* in Africa – and there is certainly no guarantee that the site represents the oldest hominid presence in Asia! Any older findings, and there will be even more reason to question the present disciplinary preoccupation with Africa as the one and only source of all stages of hominid evolution (Morwood and Van Oosterzee 2007)." I've saved the best for last - The Hobbit!

"And then there is LB1 [*Homo floresiensis* the first human fossil from Liang Bua cave]. On Flores, an oceanic island east of mainland Asia and always a difficult colonization prospect, we have found a hominid, LB1, who is smaller in stature, with a smaller brain and more primitive characteristics than any of the Dmanisi specimens. Her more apelike body proportions, for instance, are similar to *Homo habilis, Australopithecus afarensis* and particularly *Australopithecus garhi*, but differ from other hominid species for which proportions can be calculated. Endemic dwarfing goes only part of the way to explain her diminutive brain and body size. The ancestral population for *Homo floresiensis* almost certainly underwent size reduction on the island, but general biological principles indicate that proto-hobbits cannot have been a large-bodied hominid species such as *Homo ergastor* or *Homo erectus* – a conclusion supported by the range of primitive traits retained by LB1 and her kin (Morwood and Van Oosterzee 2007)."

Homo floresiensis, the Hobbit, is a strange mix of *Homo* and Australopithecine characteristics. She may be as old as 2.0 million years and managed to hold on to her spot on earth until 17,000 years ago. Researchers originally believed that LB1 and the other Hobbits, formally known as *Homo floresiensis*, were descendants of a human ancestor with essentially modern body proportions known as *Homo erectus* that shrank dramatically in response to the limited resources available on their island home. *Homo floresiensis* is a good example of how a new idea is met first with disbelief followed shortly thereafter by ridicule. Many scientists refused to accept her as a new species and fell to the old tactic of assuming she was simply a modern human with a disease resulting in a small body and brain. Some of these diseases include: Laron syndrome (a genetic disease that causes insensitivity to growth hormone), Microcephalic osteodysplastic primordial dwarfism type II (a genetic disorder whose victims have small bodies and small brains but nearly normal intelligence) and, Myxoedematous endemic evetinism (a condition that arises from prenatal nutritional deficiencies that hinder the thyroid). There was a similar situation with Neanderthal Man. In 1887 a fossil was discovered in the Neander Valley in Germany and named Neanderthal Man. Rudlof Virchow, who detested the concept of evolution, said that this fossil simply represented a modern human with a bad case of rickets. Of course he was proved wrong! This tactic didn't work when used to discount Neanderthal Man and it won't likely work for the Hobbit. But a new analysis suggests *H. floresiensis* is significantly more primitive than *H. erectus* and evolved either right after one of the earliest known members of our genus, *H. habilis* or right before it with *H. rudolfensis*. Either way the study implies that *H. floresiensis* evolved in Africa, along with the other early *Homo* species, and was already fairly small when the species reached Flores, although it may have undergone some dwarfing when it got there. So she was either the immediate descendent of *H. habilis* or *H. rudolfensis*. "*Homo floresiensis* and *H. georgicus* hint to us that they had not been confined to Africa at all but had, instead, spread far and wide. It is tempting to think that it was part of the same 3.5-million-year-old geographic expansion out from north-east Africa that not only took them south into South Africa and west to lake Chad but also north into western Asia and east as far as Indonesia. For now, this remains conjecture although it looks increasingly likely that it was small-brained proto-humans that, sometime between 3.5 and 1.77 million years ago (when we find them in Dmanisi), first ventured across the new grassy savannah habitats of Asia. They were well ahead of *H. erectus* (Finlayson 2009a)."

These savannah habitats were a ..."vast belt of seasonal, semi-arid to sub-humid warm habitats, with some woodland and open plains and a preponderance of grassy savannahs and lakes. Looked at from the perspective of this vast belt of habitat, the presence of small-brained proto-humans from Ethiopia to South Africa, were at least to Lake Chad, north to Dmanisi, and probably east to Flores, makes perfect sense. It means that at the start of the Pleistocene 1.8 million years ago, and before the first appearance of the larger brained and taller *H. erectus*, proto-humans were already living across a wide area of the Afro-Eurasian supercontinent. The evidence available strongly favors a region in the core of this habitat belt, present-day Ethiopia, for the origin of the proto-humans, but was it also the heartland of *H. erectus* (Finlayson 2009a)." We have evidence of *H. erectus* in East Africa 1.78 *Mya* and in Java 1.81 *Mya*. There doesn't seem to be anything in between. "All other sites that make claims to *H. erectus* contain fossils that are best classified as small-brained proto-humans or are classified based on tools only (Finlayson 2009a)." These tools could have been made by either *H. erectus* or the small-brained proto-humans. We know that "*Homo erectus* and small-brained proto-humans survived side by side in continental areas, as in Lake Turkana in Kenya, for up to half a million years: this shows that the two managed to avoid competing with each other and that there was no immediate obvious superiority of *H. erectus* over the others. Like so much of the human story, there were often several kinds around at the same time and evolution was not a neat progression from one kind to another (Finlayson 2009a)."

The taxonomy of the East Asian hominids was "tackled by Harvard evolutionary biologist Ernst Mayr, who lumped a number of specimens together, including the fossils from Chou Kou Tien and Java, into one species – *Homo erectus*. He also distinguished a number of geographical and chronological subspecies, including *Homo erectus pekinensis* from China; and Lower Pleistocene *Homo erectus robustus*, Middle Pleistocene *Homo erectus erectus* and Late Pleistocene *Homo erectus soloensis* from Java. The latter, from the sites of Ngandong, Ngawi and Smbungmacan along the Solo River, are a population with more modern traits. Mayr's taxonomic system had been widely accepted.

"A recent study by the Japanese paleoanthropologist Kaifu and his colleges of all hominid teeth and lower jaws recovered from Sangiran in Central Java, confirms that substantial changes took place in the local *Homo erectus* population over time. The study shows that two strikingly different hominid populations occupied the site separated by a gap of 400,000 to 600,000 years. The Lower Pleistocene *Homo erectus* population about 1.2 million years ago had some very primitive traits harking back to the earliest members of *Homo* in Africa. In contrast, those present at Sangiran during the Middle Pleistocene, from *around* 800,000 years ago, had traits similar to those of Peking Man, thus banishing the old idea that *Homo erectus* was some static stuck-in-the-mud hominid. Did one species evolve into the other, or was Java invaded by different *Homo erectus* populations in a world of mobile hominids? Like other mammals in Southeast Asia, *Homo erectus* seems to have had an evolutionary history as dynamic and complex as the landscape (Morwood and Van Oosterzee 2007)."

"So in the context of information on past environments, let's follow the imaginary dispersal of *Homo ergastor* out of Africa at, say, 1.75 million years ago. Moving unimpeded by great desert barriers, initially toward Israel, 'Ergaster' would immediately be confronted by local hominids, since there are stone artifacts in Israel, at Erq-el-Ahmar, dated by paleomagnetism to between 2.0 and 1.8 million years. Moving farther north, 'Ergaster' would have met the small Dmanisi hominids in their warm Pleistocene cul-de-sac.

"'Ergaster' also could have walked woodland and savanna environments to reach the fertile Indus Valley and the richly wooded foothills of the Himalayas, where stone tools dated to at least two million years indicate that hominids had already moved in. Continuing farther east and south, and by now 8,000 kilometers from Africa on the east coast of Asia 'Ergaster' may have met another smaller hominid species, one that resembled the small Dmanisi people and later gave rise to an endemic little people on Flores. Finally, at the very southeast corner of the Asian continent, 'Ergaster' would have reached Java to find an established population of large-bodied hominids with some very different, and more primitive, traits.

"*Homo ergaster* clearly was not the only player on the Asian stage in early Pleistocene. On their march east, would they have met other hominids traveling west? Was Asia always a passive recipient of hominids? Two-way traffic is well documented in other animals moving into and out of Africa. There is really no reason at all to assume that hominids always moved out of Africa into Asia. Given the lack of an obvious East African ancestry for *Homo ergaster,* could the species even have originated in Asia, evolved from one of the other hominid populations that had earlier migrated out of Africa? This would explain the sudden appearance of *Homo ergaster* in Africa better than invoking a short, undocumented process of evolution. In fact, the Dmanisi hominids may be 'close to the stem' from which the *Homo ergaster* lineage evolved, and at a time when the core of grassland expansion was Asia – and Africa was peripheral. The expansion of savanna, with a corresponding need for people to cover greater distances and place more reliance on hunting, may have provided the impetus for their increases in stature and brain size that occurred 1.8 million years ago. If so, was Asia the epicenter for some major developments in the human story? Maybe the little hominids from Dmanisi and Liang Bua, with their pre-*ergaster* traits, are the first real challenges to the 'Only Out of Africa' orthodoxy (Morwood and Van Oosterzee 2007."

***

Neanderthal Man emerged from *Homo heidelbergensis* who was well established in Europe. The ancestors of the Neanderthals lived and evolved from 127,000 to 780,000 years ago, in the Middle Pleistocene. The oldest fossil of the period is a jaw about a half million years old found at Mauer, near Heidelberg. The fossils from Sima de los Huesos, a cave site also in the Sierra de Atapuerca [Spain], stand out among the evidence of the Neanderthals' 'grandparents', and they constitute the largest collection of human fossils not only in Europe, but in the whole world (Arsuaga 2001)." Neanderthal Man was firmly ensconced in Europe from about 500,000 years ago. They were the lone occupants of Europe. Most of that time the weather was cold but they adapted to it and were quite successful in this environment. "Neanderthals ventured out of this small continent to populate the Middle East, and Southeast Asia. The entire epoch between 40,000 and 127,000 years ago is 'the Neanderthal period' (Arsuaga 2001)." Neanderthal Man lived from Wales in the northwest to Gibraltar in the southwest; from near Moscow in the north to Uzbekistan in the east. There is no indication that they ever inhabited the Old World of Africa, India or any other part of Asia.

"Although it is really just a peninsula of Eurasia, Europe is blessed with one of the most varied and abundant physical settings on earth. To the south it borders the placid waters of the Mediterranean, providing ready access to the Middle East and Africa. To the east it merges with the vast steppes of Asia, whose people have had a major influence on the continent. To the west it is bounded by the Atlantic, which serves not only as a route to the rest of the world but also as a critical buffer on climate. Because of the oceanic currents that flow across the Atlantic, Europe is remarkably warm despite its far northern location. This warmth influences even the most northern fringes of the continent, where the forests give way to subarctic plains.

"Internally, Europe is even more diverse. It has high mountains, fertile prairies, extensive forests, and vast swamps, and it is threaded by rivers that provide an easy way to move from one ecological niche to another. Because of its many bays and gulfs, Europe has much more shoreline per unit of area than does any other continent. Also, unlike Asia and Africa, it does not have large areas of homogeneous terrain, so people in nearby regions could develop quite different ways of life. Interactions among these groups, whether trade, cultural diffusion, or warfare, could then drive social change (Olson 2002)."

"The Neanderthals were not simply primitive versions of ourselves. They were not lesser humans with very limited mental faculties. They had many characteristics in common with us, by virtue of the long evolutionary history that we had shared up to the point where our two lines diverged. But the European branch was not stagnate after the time of that divergence. It continued to evolve and produce the Neanderthals, who developed their own distinctive characteristics, just as we did elsewhere. The Neanderthals were not living fossils. They did not belong to the past, and they were not anachronistic. In their particular epoch, they were just as 'modern' as our ancestors, the Cro-Magnons, were (Arsuaga 2001)."

Neanderthals preferred method of hunting was "ambush hunting". "There were two requirements to make ambush hunting successful. One was cover, which would allow the Neanderthals to get close to their prey. This was amply provided by the open woodland and savannah habitats in which they lived. The second was power, and they had plenty of that in their robust and muscular bodies, the legacy of *H. heidelbergensis. Homo heidelbergensis* was the product of several hundred thousand years of investment in the ambush hunting tradition within the temperate woodlands of Eurasia. The appearance of projectile technology among the Ancestors is often hailed as an advancement of technology, but the reality is that such technology would have been useless against powerful animals of Middle Pleistocene Eurasia. To down such animals needed power, guile, cooperation, and getting up close. That they did not shy away from contact is demonstrated by the injuries which they regularly sustained, comparable to those of modern Rodeo athletes. Neanderthals would have often stared at their prey right in the eye (Finlayson 2009a)."

Sometime between 100 and 130 *Kya* Neanderthal Man was at Tabun and proto-ancestors were at Skhul and Qafzeh in the Middle East. This period began during the last interglacial when the climate was warm and wet, and savannahs encroached into the Sahara Desert. The Middle East was connected to East Africa by the expansion of savannah and humans were able to move freely within the area. The climate oscillated between warm-wet and warm-dry periods during this 30 thousand year period with a peak of warm-wet about 125 *Kya*, a peak of warm-dry about 122 *Kya*, and a return to warm-wet with a peak at about 105 *Kya*. These changes in climate would cause habitats to expand and contract fairly regularly during this time frame. The Middle East was unique in that it was at a "junction between the mid-latitude belt ... and the southward prong that leads to the southern tip of Africa. Here three main kinds of environments - Mediterranean woodland, dry steppe, and subtropical desert - are in close juxtaposition and the area covered by each fluctuation changed as temperatures and rainfall changed. There are clear signals that the fortunes of the animals typical of each of these environments waxed and waned as they reacted to changing conditions (Finlayson 2009a)." This junction of habitats was at the far northern range of most African animals. We find that the majority of the animals hunted by both Neanderthal Man and our proto-ancestors were made up of animals adapted to life in these mixed environments. The main species available to hunters were adapted to Mediterranean woodlands but not dense forests, smaller numbers were adapted to dry steppe with a few adapted to subtropical desert.

There were times when Neandertal Man and proto-ancestors co-existed in the Middle East. Interestingly, Neandertal and proto-ancestors shared a common tool kit. So, why did Modern Man survive and Neandertal Man become extinct? John Shea (in Stringer and McKie 1996) notes that "If Neanderthals abilities to use stone technology are taken as a measure of their intelligence and adaptive ability, then the archeological record suggests few major differences between Neandertals and early modern humans [our proto-ancestors]. In Neanderthals our ancestors would have confronted hominids perhaps very different in appearance, but every bit as intelligent as themselves. Why we are here today and they are not, is one of the most intriguing questions in palaeoanthropology." The similar tool kits can be explained "... as a reflection of their common ancestry, about 150,000 years previously. Both had evolved separately since that period, but in those days of sluggish cultural innovation, their tool technology had progressed slowly and in parallel. And that, very simply, is why they share the same level of implements (Stringer and McKie 1996)." It was not the tool kits per se but the use these two individuals applied to their tools and how they used their environments that distinguished them from each other. Lieberman (in Stringer and McKie 1996) showed that "Clearly modern humans [our proto-humans] migrated between various winter and summer locations, while Neanderthals lived in sites the year round. And that is crucial - for its hard work to stay in the same place. The area becomes depleted of nuts, berries, tubers and vegetables, and the local game - even the gazelle - learn to avoid you. That is why hunter-gatherer tribes today move around all the time. Neanderthals, on the other hand, would have had to have travelled further and would have had to have worked harder to stay in the same place. And if you have to search for many hours for food, there would be no point in bringing back only a few berries or a couple of potatoes. You want a deer or a gazelle, a high protein shot in return for all your effort. It is the schlepp effect. You are not going to go schlepping about for long distances unless there is good reward. And in the Neanderthals' case that meant big game. They hunted more, but moved their homes less than modern humans. We moved more, but hunted less." Lieberman concludes that "The two species were living in the same environment but were exploiting it in different ways. Neanderthals were using it in a more resource intensive way compared with modern humans. They were working harder, and at earlier ages, than modern humans because they used the environment differently." The work of Shea and Lieberman indicate that Neandertal Man was "a species striving harder and harder to stay still. Lacking a less labor-intensive way to exploit their environment, they slogged towards hardship, big bones, athleticism, and extinction. But in *Homo sapiens* [our proto-ancestors], we can see patterns of a lighter, more effective touch upon the environment (Stringer and McKie 1996)."

By 80,000 years ago the populations of proto-ancestors that had been living in Skhul and Qafzeh either had gone extinct or had retreated back into Africa, leaving the Middle East in the hands of the Neandertals. We know that throughout this last glacial period there were times when the temperatures increased and times where it decreased above and below some average level. One of these extra cold spells occurred about 74,000 years ago with the eruption of Mount Toba on the island of Sumatra. This eruption represents the largest eruption in the past 450 million years. It's blast is estimated to be "4,000 times more powerful than that of Mt. Saint Helens and would have sent more than 1,000 cubic kilometers of dust and ash into the atmosphere, plunging the earth into years-long volcanic winters. Summer temperatures could have dropped by as much as twelve degrees centigrade [22 degrees F], while forests shrank, deserts spread, and in eastern Asia, a prolonged winter monsoon would have swept clouds of dust from inland deserts round the globe (Stringer and McKie 1996)."

***

We begin the last stage of our story, that of modern man, toward the end of the Pleistocene with the last of the great glaciations. This glacial period lasted from about 110,000 years ago to about 12,000 years ago. There was a period of cooling between 75 - 59 *Kya* and a second between 27 - 16 *Kya*. In between from about 50 - 44 *Kya* the climate was relatively mild but, certainly not as warm or mild as the previous interglacial. During the last maximum glaciations, rivers were blocked by ice and ice dams built up forming huge lakes behind them. When climate warmed and the ice dams melted, super-floods would have wiped out everything in their path. "These massive inland lakes would have been important barriers across a huge corridor of flat land that stretched from the Pacific coast of eastern Siberia, west all the way to the British Isles. This corridor would have been bound to the north by the ice sheets and to the south by the mountains of the Mediterranean, south-western and central Asia. The east-west corridor changed in character in response to the whims of climate. The overarching trend of climate deterioration between 50 and 30 thousand years ago favored the expansion of treeless habitats far and wide across this belt of plains. With increasing cold, tundra moved south ahead of the advancing ice sheets; with increasing aridity, steppe spread from its strongholds in the center of Eurasia. Where they met a new habitat, the steppe-tundra was born. It was a double expansion wave that swept east and west across this vast area. When it got warmer and wetter, the two habitats decoupled and gave way to trees. Moments of reprieve for wooded habitats got rarer and rarer, especially deep within the continent where warmth and moisture became scarce environmental commodities (Finlayson 2009a)."

"The period from 50 - 30 thousand years ago marked the high water for dry-steppe and prairie adapted animals such as the horse, steppe bison, and saiga antelope. The other groups of animals that thrived were the woolly mammoth, woolly rhino, reindeer, and musk ox. Some like the moose, giant deer, roe deer, red deer, aurochs, chamois, wild sheep and ibex were able to adapt to their new habitat and hang on through this period of cold. The large carnivores that survived were those able to eat plants and fruit and hibernate during the cold, dark, winters: brown and cave bears. The other carnivores were smaller species that did not depend on the large herbivores. The arctic fox reached right down to France and the red fox retreated. Lynxes and wild cats completed the panorama. But it was the predators that could follow the large herbivores across the steppe-tundra that thrived. Lions and spotted hyenas managed to make a living for a while but they were no match for the ultimate long-distance pursuit runner (Finlayson 2009a)."

"The wolf became the master of this open landscape. Only in the far north did it have a rival; the polar bear had evolved from a population of brown bears and became a specialized meat eater. Like the wolf, it ranged across great distances in pursuit of its prey but, unlike the wolf, it went to sleep at the height of the northern winter. The wolf and the bears together give us a picture of what was needed to be a successful hunter in these landscapes. Most important of all you had to be a marathon runner; energy-sapping sprinting was not a viable alternative. The ability to mix the diet, to ride bad periods by storing reserves as fat and reducing energy expenditure or by caching food, and hunting cooperatively in groups were optional extras. A hunter that could do a number of these things had the potential to become the super-predator of the treeless plains of Eurasia (Finlayson 2009a)."

From about 16 - 12.9 *Kya* the climate warmed with a brief period of *cooling*, known as the Younger Dryas between 12.9 and 11.5 kya. The end of the Pleistocene and the beginning of the modern Holocene is marked by a date of 12,000 years ago. It is these stadial and inter-stadial periods within the last glaciations that will mitigate the climate and ultimately manipulate the habitat that will drive the exodus of modern man from Africa across Asia and Australia and finally to Europe. We'll pick up the settling of the Americas in the next chapter.

***

At about this same time, as we saw in the previous chapter, modern humans were moving out of Africa. There were two possible routes taken - one route up, probably along the coast of the Red Sea, through the Middle East to northern India and finally to Asia; and, a second route either back down the eastern coast of the Red Sea or across the Red Sea along the Arabian and Indian coasts to Asia. Both are probable and even likely. There are proponents of each. I am a proponent of the coastal migration. We'll find a little later that genetics provides the best evidence of this route. From Asia they were able to colonize Australia and later Europe and the Americas.

Wells (2008) provides a good story depicting likely events that led up to this expansion out of Africa.

The sun is going down on the east Africa savannah and it is starting to get noticeably colder. You shiver, relieved that you and the other mem-

bers of the hunting band managed to kill a lame gazelle. The clan will eat well tonight. When you return to camp, everyone takes a simple stone cutting tool - sharp on one edge and blunt on the other - and butchers the animal. The tool ... is simple but effective. You make quick work of the sinew and bone and soon you are relaxing around the fire, watching the meat cook over the flames. A hyena howls in the distance and you begin to think about other things for the first time in hours.

As you mull over the day's hunt you are thankful that your luck has held out again - the animal herds do seem to be getting sparser. Of course you don't know it, but the African climate had been getting drier, and the resources to support the herds are simply not as common as they used to be. After dinner your mate brings your son to you. Although he is a strong, healthy child, he worries you because he seems so unlike the other children. For one thing, he has already learned to speak - at age two - while the other children do not do this until they are at least three. He also seems to be much better at making things than the other children in the clan, and enjoys playing games with the small pieces of stone that lie scattered around the camp. He seems much more emotional than the others, often erupting into violent temper tantrums that scare the other clan members. The strangest thing, though, is that he has begun to trace images in the dust that are similar to the animals that you bring back to camp. You find this especially frightening, and quickly rub them out when you see them. Others in the clan have noticed, though, and there have been some mutterings about his unusual behavior.

Time passes. As your son grows up you teach him to hunt and make simple tools, but his knowledge soon surpasses yours. He seems to have a magical ability to anticipate what the animals will do, which makes him a popular member of the clan, in spite of his odd behavior. At an early age -around fifteen -he becomes the accepted leader of your small group. Under his guidance your clan eats well and prospers. He fathers many children, and they too seem to be much cleverer than others in the group. Within a few generations all of the members of the clan can trace their ancestry to him. He becomes the "totemic ancestor" of the group - the founding father - and everyone descended from him is by definition a member of the group. Other clans denied the mysterious knowledge of animal behavior and superior tool-making ability that give his clan such an advantage on the hunt, either move away or are disbanded in raids organized by the clever ones. The women are taken by the raiders and are incorporated into the clan structure, but the men are usually killed or chased away. Soon there are too many clan members to live in one small territory, and in the ensuing arguments over access to food some of the young men take their mates and set off to find new territory. The process is repeated many, many times over the next few thousand years, until essentially every man in the region traces his ancestry to that first, clever child.

"When the ancestral human population dispersed across the world 50,000 years ago [and possibly earlier], evolution set in motion a grand experiment: each population in its fiercely guarded territory would develop in its own way. This development would be both cultural, leading to a vast family of different languages, religions and lifestyles, and also genetic, as members of each society respond to different climates, ecologies and social arrangements of their own making. Isolated on their separate continents, the far flung branches of the human family were to follow different trajectories as each adapted to the strange world that lay beyond the boundaries of their ancestral homeland (Wade 2006)."

"The earliest confirmed date of modern humans in Asia is in China, 67,000 years ago; and the first settlement of New Guinea and Oceana is believed to have been 50,000 to 60,000 years ago at the earliest, or at most somewhat later, 40,000 years ago. Did modern humans reach East Asia by land or perhaps by boat, along the southern coast? Probably both (Cavalli-Sforza 2000)." A waterborne migration seems likely ... Once modern humans reached the Arabian Peninsula they kept on going - around the Persian Gulf, along the shorelines of Iran and Pakistan, south along the Indian Coast, north again to the mouths of the Ganges, and finally into southeastern Asia (Olson 2002)." "There exist archeological estimates of the rate at which the re-colonization of northern Europe took place at the end of the last glaciation (around 13,000 years ago): it varied from .5 to 2 kilometers per year, not too different from that of farmers. But the major limiting factor may have been largely that of ice withdrawal, rather than that of human movement. It is possible that the coastal route allowed a faster movement. How long did it take to travel by the hypothetical coastal route from East Africa, a likely point of departure, to Southeast Asia (a necessary point of arrival, from which some may have continued north to East Asia along the coast of the Pacific, and others south to New Guinea and Australia)? We may venture a minimum guess of 10,000 years. Perhaps modern adventurers might try to repeat a small part of the journey in conditions similar to those of our ancestors. This could be very informative, even if today's coast and the availability of seafood must be different from what it was then. Assuming that it took that much time to go from East Africa to Southeast Asia, the average displacement that may have taken place would have been on the order of 50 to 60 kilometers per generation (2.5 -3.0 km/yr) [This amounted to about .5 to .75 miles per year]. This is about twice the rate of advance of farmers many thousands of years later. We are speaking of a mode of life of which there do not seem to exist any living or historical examples (other than perhaps in Borneo). I [Cavalli-Sforza] would describe it as nomadic fishing. But the model of demographic expansion would not be very different from that of Neolithic farmers, in the sense that there must be both migration and reasonably active reproduction for an expansion to take place. Or course, over the generations, some extended families or small groups may have settled or chosen to leave the coast and go into the interior, as others continued their almost random wandering along the coast. The settlement of Southeast Asia, New Guinea, and Australia led to the development of the Indo-Pacific and Australian linguistic families. Andamanese and other Negritos of Southeast Asia are the closest living descendants of Africans who were among the first settlers of Southeast Asia and Oceania (Cavalli-Sforza 2000)."

"Even if modern humans left no archeological record of their passage, they left a genetic record. Sometime more than 60,000 years ago a mutation occurred in the mitochondrial DNA of a woman living in eastern Africa (Specifically, a cytosine mutated to a thymine at position 10,400 of her mitochondrial DNA). Through the workings of genetic chance, the haplogroup defined by this mutation, known as haplogroup M, became common among the populations living in the area. A substantial portion of the people who made their way out of Africa and onto the Arabian Peninsula carried mitochondrial DNA from haplogroup M (most of the others had mitochondrial DNA in a related haplogroup labeled N). Today haplogroup M is found in the southern part of Arabia and in India and is widespread in Asia. But it is rare in the Middle East, which is another argument for an early migration route from the Horn of Africa rather than through the Sinai (Olson 2002)."

"China and Japan may have been settled before Australia, and may have been the first areas of development of the Dene-Sino-Caucasian family [of languages], which must have spread west through central Asia to Europe. The Na-Dene branch went to Siberia, from which it later (about 10,000 years ago) migrated to North America, after the Amerindians first colonized it (some 15,000 to 30,000 years ago) (Cavalli-Sforza 2000)."

***

"Forty thousand years ago, our ancestors occupied much of the Old World, including the cold tundra of Pleistocene Europe and Asia. The human brain was of modern size. These people used primitive tools, but these tools had persisted essentially without change for tens of thousands of years. People showed concern for their fellow humans by burying the dead and taking care of the handicapped and injured. Then long patterns of development that had begun 2 million years earlier began to flower, bringing language and communication, ritual and ideology, social organization, art and design, settlement and technology. Tools and methods began to vary from region to region and became rapidly more sophisticated. More implements were made from a greater variety of raw material than ever before. Bows, boats, buttons, fishhooks, lamps, needles, nets, spear-throwers, and many other items appear in the archaeological record for the first time. The dog was domesticated as a hunting companion and occasional source of food. With the dawning awareness of art and image-making, caves and artifacts were decorated with paintings, carvings and engravings. Sites from this period are larger and more common than those from previous periods, suggesting larger social groups, more economic activity, and perhaps rapid population growth. From almost any perspective, this period represents a dramatic change in human behavior.

"The conquest of new territory accelerated sharply at the same time. Having spread to North Africa and the Near East by at least 100,000 B.P., anatomically modern people advanced slowly across Europe and Asia until they reached China, Japan, and Australia around 40,000 B.P. In some areas the newcomers replaced, and in others interbred with, other humans such as the robustly built Neanderthals. At 40,000 B.P. the movement into uninhabited landscapes accelerated. These people explored new continents and brought with them the basic cultural foundations of early American culture.

"The causes of this remarkable migration and cultural exploration are a mystery. Some anthropologists suggest that a change in the brains of modern humans improved their linguistic ability, which led to more sophisticated cultures. Others emphasize the rapid change in people's social and economic organization, perhaps caused by some fundamental behavioral change. Either way, demographic changes and possibly food shortages compelled some humans around the world to abandon one way of life for another, and so invent new technologies and lifestyles.

"Still other researchers argue that the crucial element may have been human species' extreme sociability, which created an environment conducive to sophisticated tools, language, and artwork. Evidence of the extensive social relations among early *Homo sapiens* in Europe, for instance, is everywhere. Starting around 40,000 B.P., our ancestors left behind the remains of large campsites, suggesting that they occasionally gathered in large groups, possibly for ceremonial purposes. Richard Klein and others believe that the huge leap in the sophistication of hunting techniques and fishing technologies reveals the beginning of intense cooperation among group members. Discoveries of shell, bone, and flint hundreds of miles from their original sources indicate that modern humans had vast networks for exchanging goods.

"Cognitive skills, belief systems, and art probably played key roles in the development of a complex network of groups after people initially moved into new environments. Their learning capacity was obviously crucial to the way they adapted to new environments and developed new cultural patterns. But the belief systems of groups were probably equally important. These systems probably preserved some degree of community, an acceptance of the social system, and hence a general social solidarity, particularly in times of duress. Communal beliefs may also have functioned to store and transmit ancient knowledge as groups split up and entered new worlds (Dillehay 2000)."

"Central Asia shines through as a region of huge importance as a reservoir of genetic diversity; they [our ancestors] had more time to accumulate genetic novelties here than anywhere else in Eurasia and it is from here that people expanded across the plains, westwards toward Europe, and eastwards towards the Pacific coast of Asia and, eventually, into North America. From these humble beginnings on the edge of the steppe around 45 thousand years ago we find a thriving population of undisputed Ancestors out on the plains 15 thousand years later. Neanderthals are to be found nowhere in this scene. The remnants were confined to refuges further south. Difficult as the threads are to disentangle, the outcome of this long period of constant attrition of human populations caused by rapid climate change is the exclusive presence of an ever-expanding population of Ancestors on the Eurasian Plain. This plain was only marginally exploited by earlier peoples, including Neanderthals, so its invasion was not at the expense of others. People had broken through a new frontier where no one had been before. It truly was a giant leap for mankind (Finlayson 2009a)."

Finlayson (2009a) provides a narrative story depicting events of the time.

How do you deal with a flat and endless landscape where there are no trees from which you can stalk your prey or escape to if chased by a carnivore, no caves which you can shelter in, and no features that can prevent you from getting lost? It all looks the same and going into it can appear a scary and daunting proposition. Every so often, from the comfort of a cave up in the hills along the edge of the plains you see a herd of juicy reindeer, waiting to be taken, and the temptation becomes too great and the opportunity too good to be missed. You probably start off with forays, maybe at particular times of the year when the animals are most abundant or most vulnerable, but you never lose sight of the broken relief that is home. You invent new weapons and technologies that allow you to hunt in the open but you do not abandon the old ways completely. You still ambush the odd ibex or red deer back in the hills. They are your life insurance and you have become a sophisticated risk manager.

You have developed a mixed strategy, one of many similar developments that neighbors and distant cousins have been practicing for generations. Sometimes you meet these neighbors, or you spy on them, and you pick up new ideas; they do the same to you when you are not watching. Sometimes your neighbors are friendly and keen to cooperate, share technology, or exchange commodities.

Other times strange people who speak in an odd tongue enter the area but you are unsure of their motives. They are powerful and muscular so you keep a safe distance. They soon move on, apparently keener on the solitary red deer stags dispersed all over the hills than in the big herds of reindeer down below. You have seen them trying to catch the reindeer before and you have been impressed by their weapons, some of which you have copied, but they are awkward out in the open and they lose more animals than they get despite their inventiveness. As they pass, your attention turns once more to the hundreds of tiny specks on the flat lands down below.

Threads of smoke rise over a collection of strange domed structures crowded together close to a river meandering in an otherwise featureless landscape. We have travelled in an imaginary time machine back to 28 thousand years ago and a remote place somewhere north of the Black Sea. As dawn breaks wolves howl and bark, surprisingly from the area where the structures are situated. A man comes out from one of the domes that appear to have been made from animal hides. He carries a large piece of meat which he throws to the odd-looking wolves now quite visibly tethered. Noisy children emerge followed by women, probably their mothers. Soon the camp has come to life and people of all ages and both sexes are performing different activities. Some talk to each other while others just get on with the tasks. The children watch and play. The whole camp appears to behave like a giant superorganism, like a nest of ants. But these are not automated giant ants, they are people, and their daily routines are interspersed by large doses of improvisation and activities that have clearly been shaped by their experience of what has gone on the day before. We have stumbled across a community of Ancestors who will become known by their cultural achievements and, even though they started their career in the east, their culture will become known as the Gravettian from a French site where it was first recognized.

"The source of the Central Asian population carried a unique genetic marker that originated sometime between 40 and 35 thousand years ago. The people who spread westwards into Europe carried another identifier genetic mutation, one derived from this parent stock and which indicates a westward expansion around 30 thousand years ago (Finlayson 2009a)."

The "... behavioral novelties of the Gravettian culture can be understood without recourse to explanations that involve sudden changes in brain wiring which produced particularly brainy people. These people were already smart and so were many of their contemporaries, and predecessors, many of whom never made it. The innovations had instead to do with the pressures imposed by the new terrain in which they were living (Finlayson 2009a)."

"Many of the inventions that have been attributed to being modern can be simply understood as ways of dealing with this environment. Take the use of bone, ivory, and antler to make tools and weapons. These materials are quite typical of Gravettian sites. Major sources of antler were reindeer while mammoths contributed ivory and bone. These products would have been readily available to people living on the plains but not to those living in other environments without such animals. Their use reflects availability and also ingenuity on the part of people who could turn these into functional, and in some places also decorative, objects. For much of human history, people would have used wood as the prime raw material for tools and weapons. Rarely do we find evidence of such wooden objects as they perish more readily than stone but the available examples show that wood was already being used by *Homo heidelbergensis*. People finding themselves in an environment in which trees were a scarce commodity would soon have turned their attention to other organic materials, like bone, antler, and ivory, as substitutes.

"The places that the Gravettians and their kin frequented would have been cold for much of the year with the ground frozen for long periods. This frozen ground, the permafrost soil, would have acted as a kind of natural freezer that slowed down the decay of dead animals. As these rotted slowly, or were exposed by the thawing permafrost in summer, they would have become available to people. There must have been lots of carcasses and skeletons lying about in places and people soon found uses for these, particularly the remains of mammoths (Finlayson 2009a)."

The Gravettians would have learned to dig pits into the ground for storage of food in these natural freezers. "They had invented an economy in which surplus catered for risk. Storage then became a central element of this way of life. With it came the need to protect and curate the stored supplies. For parts of the year at least, those times when foods were kept within the Gravettian village, some people in the group would have been confined to its immediate vicinity. We can conclude two things from this observation: on the one hand some people's nomadic hunter-gatherer ways were curtailed and, on the other, there must have been a division of labor within the group. Quite possibly specialists at different crafts arose from this early separation of tasks among the group members (Finlayson 2009a)."

"The Gravettian people were the hallmark of human culture and technological achievement. Their origins were in the plains north of the Black Sea or even further east, and their genetic roots go back to those of Central Asian people who first learnt to tame the open steppe. With their skills and ingenuity the Gravettians managed to succeed where others had failed and they achieved this at a time when the climate was building up to the height of the last Ice Age. From 30 until 22 thousand years ago, almost as long as the later world of farmers, the Gravettians ruled the roost across the western Eurasian Plain and they managed to penetrate south into the Mediterranean Peninsulas, with pioneers even reaching the south-western extreme of Iberia. They probably had little contact with the Neanderthals - by the time they reached the last outposts of these people, the Neanderthals were long gone (Finlayson 2009a)."

"The Gravettians had a semi-nomadic lifestyle based on village life that would serve as the nerve center for the group. Hunting groups would spread out from the village on scouting trips to locate game, mostly reindeer and horse. They would return to the village at a prescribed time and share their information with each other. If they found game in enough numbers, they would organize a hunting party. Periodically, the village would move closer to the herds as game became scarce in their area. [They] ... must have had specialized terms and words related to their hunting first and foremost, and that the need for such terms must have driven the development of complex language. With it came other things intimately human: the wrong transmission or reception of information - error and deception. The permanence and size of some of the Gravettian camps does suggest that these people had taken the unprecedented step of amalgamating small bands of hunter-gatherers into large collectives. Perhaps this was happening only seasonally or in particular sites but it marked the beginning of a new world, one in which humans were able to adopt new ways of life simply by cooperating with each other in numbers. At the heart of this new way of living was the base camp, an early kind of village; what was new about this when humans had been gathering around camp fires for hundreds of thousands of years? The difference had to do with the environment they were in. This environment was a vast expanse of emptiness. If you found a herd of grazers you had hit the jackpot, but what did you do in the meantime (Finlayson 2009a)."

"In the absence of natural rock shelters, the people of the plains had to construct their own and the large bones and tusks of the mammoth made excellent superstructures that could be covered with animal hides to make tents. Some of these structures were partly dug into the ground and similar ones would have been made earlier, using wood, by people first venturing onto the plains, and probably by many others, including the Neanderthals, who had the capability of working wood. Tent building is a good example of a technology that probably developed gradually but which suddenly received a big boost as people came under pressure to find shelter from the icy winds of the plains. With a shortage of wood, but an abundance of large bones, tusks, and hides, architecture leapt forward (Finlayson 2009a)."

One of the consequences of this new lifestyle, "... and one that might have inadvertently held the key to the ultimate success of these people would have been a reduction of the interval of time between child births in women. In the fully nomadic lifestyle of the Neanderthals and many other groups of Ancestors, families would have needed to raise children to some level of independence before the next pregnancy. Such a strategy may have put a premium on precocity among children, something which appears to have had, on average, an accelerated rate of development relative to Ancestors' children: the dental development of a Neanderthal child aged eight at death was, for example, comparable to that of modern human children who are several years older.

"With a part of the community staying behind in the village, for at least part of the year, a woman would have been able to repeat pregnancy before her earlier child was fully independent. There would be less need for her to move long distances and there would have been other women, men, and grandparents about to help with the rearing. This cultural change would be expected to have subsequently led to positive selection of physiological attributes that optimized the new behavior. The pressure on child precocity was also released. This simple shift in strategy of landscape use would have catapulted a rapid growth in the population (Finlayson 2009a)."

***

"Between 40,000 and 30,000 years ago, when modern humans were replacing Neanderthals, the continent was significantly colder than it is today. The summers were warm, but the winters were brutal, with many weeks of subfreezing temperatures and heavy snowfall. The climate must have posed severe challenges to modern humans, whose long limbs are more adapted to the warmth of the tropics. Yet it also provided tremendous opportunities. During this period much of northern Europe consisted of vast subarctic grasslands. Huge herds of game wandered the plains, including reindeer, wild horses, bison, mammoths, and wooly rhinoceroses. Modern humans clearly organized their lives around these animals. They made their camps along the rivers, where herds of game descended from the high regions toward the coastal plains as winter approached. Many of their tools were designed to process the products of the hunt. The openings of caves where they lived often faced south, so that on cold days the people could be in the sun while working at their tasks (Olson 2002)."

"The Neanderthals 50 thousand years ago had already experienced the effects of the encroaching treeless habitats across the Russian Plain and Eastern Europe and their range had started to contract. By 40 thousand years ago their homeland had been pinned back to the Mediterranean, south-west France, and pockets around the Black Sea (Finlayson 2009a)." "There was likely a culture change occurring at this same time. The Mousterian culture developed in response to the changing climate and habitat conditions and was likely the sole purview of the Neanderthals (Finlayson 2009a)."

Cro-Magnon Man established his dominance in the Middle East about 30,000 years ago. At this time, we enter a new period coined by Jared Diamond as the "Great Leap Forward". Diamond (2006) defines this short moment in time as follows: "For most of the many millions of years since our lineage diverged from the apes, we remained little more than glorified chimpanzees in how we made our living. As recently as forty thousand years ago, Western Europe was still occupied by Neanderthals, primitive beings for whom art and progress scarcely existed. Then there came an abrupt change, as anatomically modern people appeared in Europe, bringing with them art, musical instruments, lamps, trade, and progress." As we've seen in the previous chapters our ancient ancestors were certainly more than "glorified chimpanzees". We'll also find that Diamond's view of Neandertal Man is quite different from reality. However, he was right about one thing! There is a decided fluorescence of culture about to occur.

Cro-Magnon Man came not from the south as before, but, as noted earlier, from the east - from Central Asia "... traveling either counterclockwise around the Black Sea basin through the Caucasus Mountains and the Ukrainian plains or clockwise through Turkey, Greece, and the Balkan Mountains. Almost all of the mitochondrial and Y-chromosome haplotypes found among Europeans today derive from haplotypes that still exist in the Middle East. Those first modern humans were not moving into uninhabited territory; the Neanderthal people had occupied the continent for hundreds of thousands of years. The Neanderthals had a culture well suited to the harsh climate of that time. They hunted big game using stone tools flaked from carefully prepared cores. They cared for those who were sick and crippled. At least on occasion, they buried their dead (Olson 2003)."

We know Cro-Magnon Man and Neandertals interacted with one another based on the findings of Svante Pääbo (Wong 2010) that showed that they interbred in the Middle East. He estimated the dates of this union to be between 50,000 and 80,000 years ago. However, as we've seen previously, proto-ancestors were gone from the Middle East by 80 *Kya* and Modern Man, Cro-Magnon Man, did not return until about 30 - 40 *Kya*. There's quite a bit of debate about whether Cro-Magnon Man/Woman and Neanderthal Man/Woman actually mated. Most of the discussion seems to center on the question "Would a good looking guy like Cro-Magnon Man have sex with such an ugly beast as Neandertal Woman?" As Steve Jones pointed out ..."most people would change seats if a Cro-Magnon sat next to them on the train", however, "they would change trains if a Neandertal did the same thing (in Stringer and McKie1996)." Paabo's research results should put an end to that discussion.

"Between about 45 *Kya* and 36.5 *Kya* the Chatelperronian culture appeared in France. The jury is out on whether the Chatelperronian culture was that of the Neanderthals exclusively or if they shared this culture with the new entrant on the scene - Cro-Magnon Man. Problems with dating these sites led to confusion. Determining the ownership of this culture will have a lot to say about how advanced Neanderthal culture was - for instance did they create art or borrow it from the more advanced Cro-Magnon? It may be that the Neanderthals and Cro-Magnons shared this culture. A later culture, the Aurignacian (32 -26 *Kya*) was originally determined to be the sole purview of Cro-Magnon Man. However, even this has come into question. So we really can't assign the Neanderthals or Cro-Magnons to the Chatelperronian or the Aurignacian cultures! "In any case there is clear evidence now that such cultures were appearing across Eurasia at about the same time. We have seen this for the Chatelperronian in France and the same is true for the transitional and early Upper Paleolithic cultures of central and Eastern Europe, the plains north of the Black Sea, and east across southern Siberia to the Altai Mountains. So, rather than the spread of the Ancestors what the cultural evidence shows is widespread experimentation and innovation across Eurasia and the Middle East at about the same time. It is no coincidence that this was the time of climatic downturn when conditions were reaching the height of unpredictability (Finlayson 2009a)."

The "...edge between woodland and the treeless habitats waxed and waned many times along a narrow belt between the plains and the mountains. This was the tension zone, in places like the Altai, the Carpathians, and the Pyrenees, where humans were being stretched to the limits of their ingenuity and it is right across here that we see cultures and technologies that we define as transitional or early Upper Paleolithic. This was the novelty sector where over and over again, for fifteen millennia, survival battles were won and lost (Finlayson 2009a)." The people in the Middle East to northern India faced the same problems except their habitats were shifting from dry-steppe to desert.

"What we can conclude about the period between 50 and 30 thousand years ago when we put together what we have seen so far in this chapter? Climate became progressively inhospitable. Repeated and sharp oscillations did not allow particular habitats to become well established. Changes were most abrupt where plains met mountains and different habitats were in close proximity. Most people were concentrated along the edge between the mountains and the plains, in places of high ecological diversity within small distances. These places offered a range of options for making a living.

"Few people had managed to get further north, away from the contact zone and into the open plains. Humans had been living to the south of the contact zone for a long time and had responded to periods of aridity, when the desert encroached, by developing projectile technology that enabled them to switch from ambush hunting to long-distance tracking and hunting of gazelles and other desert animals. This flexibility may have assisted their subsequent expansion. The human panorama across northern Eurasia between 50 and 30 thousand years ago was therefore one of colonization, extinction, and innovation (Finlayson 2009a)."

The Neanderthals held out in southwestern Spain in Gorham's Cave at the base of the Rock of Gibraltar until about 24 - 28 thousand years ago. Gorham Cave was well south and west and, at the furthest extent of Europe. The climate remained mild even under the worst of Ice Age conditions. The Neanderthals up to this time were eating the same range of foods as their predecessors had been 100 thousand years earlier. The last Neanderthals were making the same tools and weapons no different from those of their ancestors. They were still using the Mousterian technology and using it quite successfully. Their world had not changed so there was no need for them to change. "It was no coincidence then that the last Neanderthals survived in the extreme south-west of Eurasia. The Strait of Gibraltar seems to have prevented them from moving further south into North Africa - they had no get-out clause and they died out. Like the last populations of living pandas or tigers they became an endangered species with few chances of pulling through. In the end the last ones may have died out simply because so few of them were left that they were inbred, or a random fluctuation in numbers brought them to zero or perhaps a disease swept across the last population (Finlayson 2009a)."

The newcomers that settled into Gorham's Cave were Cro-Magnon. "In the west the descendants of the Gravettians, with a developed technology that emphasized small, portable, stone armatures, survived in the plains south of the ice sheets. Undoubtedly their sophisticated technology and social systems enabled them to survive this harshest of periods. They managed to do so from Italy in the west, right across the eastern European plains into areas north of the Black Sea. The Central Asian deserts further east probably marked their limits. It seems that, so long as there were grasslands with food, these people could survive the cold and the long winter nights. In Iberia and south-western France we find a different culture known as the Solutrean, from a site in France. It is thought to be the product of people descended from the Gravettians and its distinctiveness comes from the exquisitely shaped flint arrowheads. The Solutreans may have been among the first people to use the bow and arrow. They had wonderful artists among them who painted the animals of the cold world, especially the horse and the steppe bison, but they also penetrated deep into the south where their numbers swelled beyond recognition. The big modern human demographic explosion in the south of the Iberian Peninsula did not come with the Gravettians, who barely made an impact down here, but with the Solutreans (Finlayson 2009a)."

***

"Sometime around 21 thousand years ago, just as the world was freezing up, a group of Solutreans entered a cave in the very southern limits of Europe, in full view of Africa. They settled in the cave and started a fire in the deep part where they were protected from the wind and the constant blasting of the sand from the dunes outside. Here they were also safe from hungry hyenas and wolves and from here they made forays into the outside world where they hunted deer and ibex, and collected shellfish from the coast; they caught birds, rabbits, and seals and they scavenged beached dolphins. They took the seeds of the pine trees. The Solutreans were living in Gorham's Cave, Gibraltar, oblivious that other people had lived here for millennia before them. It was now maybe five thousand years since the last Neanderthals had lived here but the new group of people seem to have chosen a similar spot to start their fire: it was the optimal spot where the smoke rose and did not pollute the cave. And they hunted the same animals: what else would a human do if not exploit fully what was available? If it was not for their different build and tools you would not have been able to tell whether the people in the cave were Neanderthals or Ancestors.

"One other thing did separate them, but it was cultural, of the kind which also separated different populations of Ancestors. The Solutreans had kept the front teeth of a number of different red deer that they had hunted and one member of the group, a skilled craftsman perhaps, would stay behind in the cave carefully polishing the teeth and drilling perforations into them. These people had carried the Gravettian traditions with them. Here the cave substituted the mammoth bone tent but it was part of a base camp, a place where some people stayed behind and where others came back to exchange intelligence. Another of their group was tasked with a different skill. She was an expert at painting and the wall of a deep part of the cave was turned into a panel in which red deer, which had contributed to the necklace, was painted. As if wanting to leave an indelible mark of her work of art, she put her hand on the wall and sprayed paint over it. Taking her hand off, its imprint was left for an archaeologist to discover 20 thousand years later.

"The Solutreans thrived in Iberia at the height of the cold but so had Neanderthals on many other cold moments long before them - there was nothing to special about that. At the other end of the land mass, in eastern Asia, some hardy people had managed to eke out a living in the narrow pockets of steppe-tundra that survived between the Arctic and Himalayan ice sheets; and there was, of course, Beringia, where people managed to make a living out of reindeer and mammoth. Their chance would come with the thaw. People survived down the Pacific coast of Asia, in Japan where fishing economies became prominent, and in grasslands and woodlands along the Yellow and Yangtze rivers. These populations had connections with the tropical south, unlike the western Eurasians who were cut off from Africa by the Mediterranean Sea and the Sahara Desert (Olson 2002)." We'll hear more from the Solutreans and Beringians again with the settling of the Americas.

"A few thousand years after the death of the final Neandertal Europe entered a climatic crisis. Marine core records from the Mediterranean indicate that between 24 - 18.5 thousand years ago the climate turned very cold and very dry leaving Gorham Cave uninhabited for that period of time. The weather became even colder; glaciers pushed south until they were within a hundred miles of modern-day Stonehenge, Amsterdam, and Moscow. Average temperatures were as much as 20 degrees Fahrenheit lower than they are today. So much ocean water was locked up in the polar ice caps that the English Channel dried up and England became part of the continent. During the height of the Ice Age, between about 20,000 and 16,000 years ago, modern humans gave up on northern Europe, abandoning what are now Britain, northern France, the Low Countries, Germany, and most of Poland. Small groups may have wandered into those areas during the summers, but they left no traces of their visits. Europeans retreated into the warmer areas around the Pyrenees and the Balkans and north of the Black Sea. This period of intense social crowding was one of great innovation. Artwork flourished among the dense populations of southern Europe. New technologies were developed, such as spear-throwers that allowed hunters to launch projectiles toward their prey with great force. Groups seem to have heightened their cultural distinctions from each other, as if they were marking off separate territories for themselves. Then the glaciers slowly began to retreat. By 13,000 years ago, people had moved back into northern Europe, including England and northern Germany. At first they undoubtedly resumed their hunting of big game. But as the warming continued, the forests expanded, swallowing up the grasslands of the north. Forests have much less game than do the wide-open spaces of the subarctic plains, so they can support fewer hunter-gatherers. The huge herds of animals began to disappear, in part because of overhunting by humans. The polar ice sheets melted and sea levels rose, encroaching on the plains from the north. By 8,500 years ago Britain was again an island, cut off from the mainland by the rising waters of the channel (Olson 2002)."

"As the forests spread, people were forced to depend more on fishing, hunting birds and small animals, and gathering. Archeological investigations have shown that populations in northern Europe declined as the plains contracted, and people lived in smaller groups that were farther apart. The creation of artwork fell off drastically, replaced by geometric engravings and paintings on pebbles and bones. Even the stone tools became simpler, less varied, and less carefully made. By 10,000 years ago Europe had entered a period of cultural torpor. The conditions that had led to the flourishing of Stone Age culture in Europe were gone (Olson 2002)."

***

As mentioned earlier, Diamond (2006) labeled the period beginning about 40,000 years ago the "Great Leap Forward." This was the beginning of the arts! It is also known as the "cognitive explosion". "In 1908, during excavation works in the Willendorf area, related to the building of a railway line between Krems and Grein in lower Austria, a worker discovered a small statue that became world famous. It was the Venus of Willendorf, a magnificent example of the kind of art that the Ancestors were making on the European Plain around 26 thousand years ago. Right across the plains of the major eastern European rivers, such as the Danube and the Don, Ancestors with a new culture - the Gravettian - were settling in large camps that could best be described as hunters' villages. In the east, on the Don River, Russian scientist, I. S. Polyakov, had been excavating such villages since 1879, Kostenki Sites, - that have become the hallmark of this culture Dolni, Vestonice, Pavlov, Avdeevo, and Kostenki itself – which include the remains of huts, kilns, storage pits, tools, jewelry, and figurines.

The surge in art began with the Gravettians. "On the plains it took the form of portable art, small statuettes and sculptures made out of baked clay at high temperatures in kilns. The origin of pottery is usually attributed to the Middle East, around 8 thousand years ago, but the earliest clay pots date back to at least 13 thousand years ago in eastern Russia. Although they did not make pottery vessels, the Gravettians of the Eurasian Plains had mastered the fire-using technology needed for ceramics 15 thousand years earlier than the first potters of the Russian Far East and 20 thousand years before those of the Middle East. We pick up other evidence in their paintings in caves in south-west France, in sites like Pech-Merle and Les Garennes. These paintings are of perfect execution which suggests that the Gravettians had been painting for a long time prior to this, presumably in perishable media in the open, and that the preservation inside caves simply retained a biased sample, reflecting the numerous limestone caves in south-western France, for us to marvel at (Finlayson 2009a)."

The newcomers to Europe brought with them an entirely new tool kit. "The stone tools they made were specialized for particular tasks, such as scraping hides. They fashioned bone, ivory, and antler into projectile points, awls, punches, and needles. They developed regional tool making traditions, as if the modern humans of Europe were dividing into groups, each with its own distinctive style. They also created art, something the Neanderthals had never done. This artwork is so accomplished that it still has a profound emotional power. The cave paintings of Southwestern France, the animal sculptures of central Europe, the carved figurines of women found throughout the continent - all exhibit the visceral connectedness of great art, the ability to reveal to its viewers something about how the artist perceived the world. In the hunter-gatherers of Stone Age Europe we can recognize something of our own minds, which is not easy to do with the Neanderthals they replaced (Olson 2003)."

This picture of Neandertal Man being "primitive beings for whom art and progress scarcely existed" is changing. Discoveries in southeastern Spain at Cueva de los Avinones and Cueva Anton of perforated and painted cockle and scallop shells dating from 48 - 50,000 years ago - 10,000 years before Cro-Magnon Man entered Europe - are changing the way we think of Neandertal Man. In with these finds are those of an oyster shell that likely served as a paint mixing bowl and a sharpened horse bone used to mix, dab and apply paint to objects and to the face further this belief. Joao Zilhao (in Wong 2010a) notes that "The one thing these finds make clear is that Neandertals were behaviorally modern. They were not like early modern humans anatomically, but they were cognitively as advanced or more so. There are several possible conclusions one could draw from this observation. Either modern cognition and modern behavior emerged independently in two different lineages, or they existed in the common ancestor of Neandertals and anatomically modern humans: or the groups we call Neandertals and modern humans were not different species and therefore we should not be surprised that despite the anatomical differences there are no cognitive differences, which is the conclusion I favor (Joao Zilhao in Wong 2010a)." This ties in directly with the findings of Hauser (2009) in the last chapter.

So did Neandertal Man Just copy Cro-Magnon Man's art work or create it independently? Zilhao (in Wong 2010a) thinks they did not copy Cro-Magnon. He notes that "Prior to entering Europe, modern humans did not have pierced or grooved mammal teeth like the ones found in the Chatelperronian, nor did they have perforated bivalve shells like the ones found in Spain. But once they enter Europe, they have them. Where did the moderns get these ornaments?"

Neanderthals may have contributed to this cognitive explosion in another way. Cochran and Harpening (2009) propose that the fastest way of acquiring new and useful genes is through a mechanism known as introgression – the transfer of alleles from one species to another. They note that: "Natural selection may solve some problems differently in different populations, and what appears to be the most elegant solution at the time may not in fact turn out to be the one that works best in the long run." In fact they propose that the transfer of alleles from Neanderthals to modern humans allowed for the "Great Leap Forward" in humans 40,000 years ago. "We [Cochran and Harpening] argue that even limited gene flow from Neanderthals and perhaps other archaic humans would have allowed anatomically modern humans to acquire most of their favorable alleles. We believe that this sudden influx of adaptive alleles contributed to the growth of the capabilities that made up the 'human revolution,' and we believe that this introgression from archaic human populations will prove central to the story of modern human origins." They conclude that "... by 40,000 years ago, humans had become both anatomically and behaviorally modern (which is not to say they were exactly like people today). They had vastly greater powers of innovation than their ancestors, likely owing in part to genes stolen from their Neanderthal cousins."

Renfrew (2008) contends that the "human revolution" did not begin in France or Spain 40,000 years ago but in Africa between 150,000 and 70,000 years ago. He asks "Why there should have been this remarkable and localized creative explosion in Upper Paleolithic France and Spain and why such remarkable scenes of animals [referring to cave art] did not occur elsewhere until very much later, is not at present clear. It is one of the intriguing mysteries of prehistory." He goes on to say that "... to achieve a coherent view of human development, it is important to see that this localized yet long-lasting episode belongs firmly in the tectonic phase, after the great dispersal [out of Africa], and that it is special to its time and place. Such art was not a general feature of the human revolution or of early *Homo sapiens* in Africa, nor is it yet seen elsewhere during the Paleolithic period. To assert otherwise is to misunderstand the nature of that transition."

Diamond (2006) speculates that the driving force behind this "Great Leap Forward" was the development of language. We've touched on language in previous chapters noting the progression of language ability from *Homo habilis* through *Homo erectus/Homo ergaster* to the emergence of *Homo sapiens* in southern Africa. Cavalli-Sforza (2000) provides a tree of origin of human languages. He shows the initial development as occurring in *H. sapiens* in Africa between 100 -70 *Kya* (thousand years ago). This agrees with Marean (2010). Olson (2002) proposes a thesis on language set in the scene of our ancestors preparing to cross the Red Sea from their homeland in Africa. "If they built rafts to sail across the Red Sea, they were almost certainly using language to communicate. Felling logs and lashing them together to make a seaworthy vessel is a technological feat far more sophisticated than making stone tools. The migrants would have needed to provision themselves for the journey and rapidly adapt to new environments as they moved. Earlier societies of archaic and modern humans exhibit no clear signs of the technological creativity and social flexibility associated with language. The migration of modern humans from Africa to Australia is different - a journey of such epic scope seems inconceivable without language." He also creates a pidgin version of what words may have been spoken before the crossing: "Kun mena mana? Kun mena aqwa?" It means "Who thinks we should stay? Who thinks we should go across the water?" Many linguists believe there was an original language termed "Proto-World." The origins of "Proto-World" would likely have been with these ancestors standing on the shores of the Red Sea and contemplating the incredible journey ahead of them. What wonders lie ahead? What new dangers will we face?

***

Recall Darwin's hypothesis: *If we possessed a perfect pedigree of mankind, a genealogical arrangement of the races of man would afford the best classification of the various languages now spoken throughout the world; or all extinct languages, and all intermediate and slowly changing dialects, were to be included, such an arrangement would be the only possible one.* Are we able to test this hypothesis? No! We simply can't identify all the *various languages now spoken throughout the world; or all extinct languages, and all intermediate and slowly changing dialects.* However, we can make some sense of Darwin's proposed genealogy with the various languages we do know and understand. And, if we combine this information with that of genetics we can validate many of the assumptions we need to make. A pioneer in this integrated approach was Cavalli-Sforza.

If there is an association between language groups and groups of people and languages diversify the way groups of people do "then the worldwide patterns of language should match the worldwide patterns of human DNA. Where gaps appear in the genetic record, linguistic analysis could fill them. And both linguistics and genetics could help explain the archaeological record. In other words, by building on one another, the disciplines of archaeology, genetics, and linguistics could be on the verge of a 'new synthesis' ... that could shed light on vast spans of human prehistory (Olson 2002)."

Our first ancestors that crossed the Strait of Bab el Mandeb likely spoke the same language. Those that continued on and crossed into Australia probably spoke a different language that would have changed, maybe more than once, since the original group crossed the Red Sea. "For most of human history, languages have tended to be localized, spoken by particular groups in particular places. If a group split, the languages spoken by the secondary group tended to diverge, just as Darwin supposed. In this respect, the divergence of languages does bear some similarity to the divergence of DNA. As long as a group of people is intact, they share both their genetic mutations (through interbreeding) and their language (through conversation). But once a group divides, innovations in language have few ways of getting from one group to the other. In each group new pronunciations come into vogue, old words are replaced by new ones, and new ways of combining and inflecting words take hold. As a rule of thumb, say historical linguists, if two groups speaking the same language are separated for more than about a thousand years, their languages will change so much that they will be mutually incomprehensible (Olson 2002)."

Dawkins (2008) agrees. He states that "... I doubted that the details of language evolution are favored by any kind of natural selection. I guessed that language evolution is instead governed by random drift." He puts this into a cultural context in which "memes" are the cultural currency of genes. "Language evolves in a quasi-biological way and the direction its evolution takes looks undirected, pretty much like random drift. It is handed down by a cultural analogue of genetics, changing slowly over the centuries, until eventually various strands have diverged to the point of mutual unintelligibility." In addition, he notes that "It is just conceivable that certain vowels or consonants carry better than others through mountainous terrain, and therefore might become characteristic of, say, Swiss, Tibetan and Andean dialects, while other sounds are suitable for whispering in dense forests and are therefore characteristic of Pygmy and Amazonian languages."

Today there are 16 language families and more than 6,000 languages that all developed in the past 25,000 years, and, all of which arose through three basic mechanisms. "The first is colonization, in which modern humans entering a new area introduce their language. The second is divergence, in which a language spoken by different groups changes over time. The third is language replacement, in which a group of people adopts a new language introduced or imposed by another group (Olson 2002)." The top 100 languages are spoken by 90 percent of the people in the world today. Around 1500 AD there were an estimated 15,000 languages spoken - that's a loss of about 60 percent in just over 500 years.

"The classic example of genetic correspondences comes from western Asia and Europe. For more than two centuries, scholars have known that most of the languages spoken in a broad swath from Brittan to India are derived from a single original language. These languages belong to what is called the Indo-European family, and they are spoken by more people than are the languages of any other family in the world. They include Indic languages such as Hindi and Urdu, Iranian languages such as Farsi and Kurdish, Slavic languages such as Russian, and Serbo-Croatian, Romance languages such as Italian, and French, Germanic languages such as Norwegian and English, and Celtic languages such as Irish. Today many of these languages seem to bear only the slightest resemblance to each other. Yet most linguists believe they all are descended from a single ancestor language spoken by a small group of people - probably a few thousand - living in a relatively small area, perhaps just a few thousand square miles (Olson 2002)."

Geneticists use markers on the mtDNA and Y-chromosome to trace the expansion of human populations around the world. In the case of the Y-chromosome, they label these "M" followed by a number - i.e., M168. And, a Y-chromosome haplogroup may contain a string of markers denoted, for example by haplogroup C, M169 – M130. Similarly, the mtDNA haplogroups are assigned a letter such as haplogroup *A* which includes the string of letters *L1/L0 – L2 – L3 – N – A*.

Cavalli-Sforza has linked this language family tree to the genetic family tree. "There were two caveats made by Cavalli-Sforza and his colleagues in their study. The first is that the genetic markers they were studying did not *cause* the pattern of linguistic diversity - there was not Bantu gene that forced its hapless carriers to speak those languages. Rather, similar genetic markers reflected the common history of the speakers of that language, as markers of descent. The second caveat is that in many cases relationships suggested by genes and languages disagreed, showing that the correspondence wasn't absolute. The reasons for this might be language replacement, in which people learn to speak a new language without a corresponding influx of outside genes, or gene replacement, in which there was a significant influx of genes but the language stayed the same (Wells 2002)."

So, what do genetics and linguistics tell us about the expansions out of Africa and the settling of the world and the languages that emerged as a result? We need to start back in East Africa where the mtDNA haplogroups *L1*/L0 and the Y-chromosome haplogroups A, consisting solely of M91 are the most diverse lineages in the world. "One of the interesting things to come out of the Y-chromosome analysis is the pattern of diversity within Africa, seen in the distribution of deep genetic lineages within the continent. While all African populations contain deeper evolutionary lineages than those found outside the continent, some populations retain traces of very ancient lineages indeed. These groups are found today in Ethiopia, Sudan and part of eastern and southern Africa, and the genetic signal they contain is very good evidence that they are the remnants of the oldest human populations. The signals have been lost in other groups, but today these eastern and southern African groups still show a direct link back to the coalescence point – Adam (Wells 2007)."

Let's look at four Y-chromosome haplogroups A, B, and C and F. Figure 5-1 shows their relationship to each other in the form of an ancestral tree.

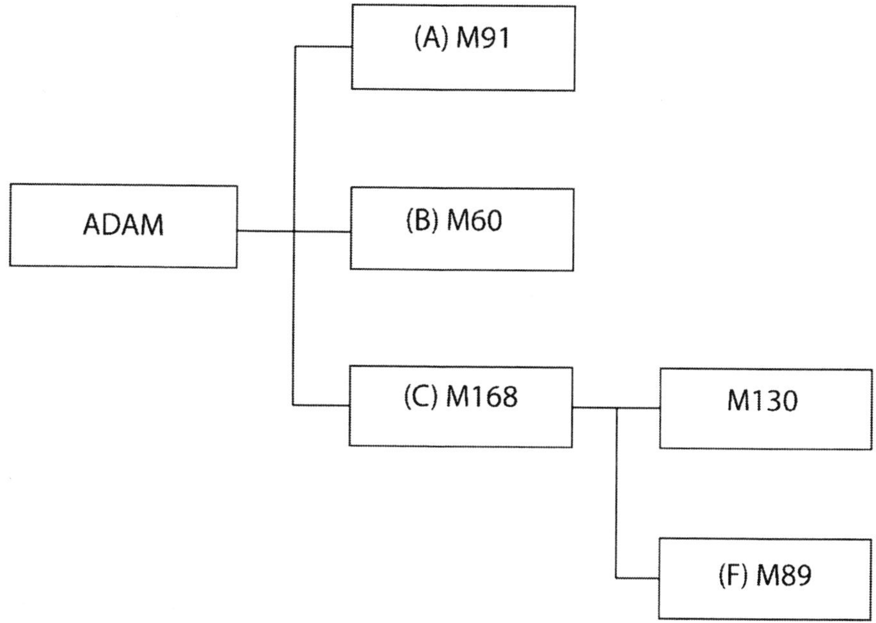

Figure 5-1. The ancestral relationships of Y-chromosome haplogroups A, B, C and F.

"The most diverse of all Y-chromosome lineages, haplogroup A dates back to roughly 60,000 years ago and is defined by the marker M91. Genetic diversity increases with age, so M91 provides a genetic link to the earliest common male ancestor of all humans,'Adam'

"Today, many individuals carrying the marker M91 live in Ethiopia, the Sudan, and southern regions of Africa. They often practice cultural traditions that are representative of the ways of life of their distant ancestors. For example, some live in traditional hunter-gatherer societies once common to all humans. They also may still speak ancient click languages, like those of the San Bushmen of the Kalahari and Hadzabe of Tanzania (Wells 2007)."

"M160 defines haplogroup B, an ancient African lineage that originated some 50,000 to 60,000 years ago. As with most very old lines of descent, it has a broad dispersal, is found today across the African continent, and is shared by many different African peoples. Often the most unique peoples and cultures, like that of the Biaka and Mbuti Pygmies, are themselves quite ancient (Wells 2007)."

The genetic marker M168 "...comes from one man in particular, who had a rather important, random mutation on his Y-chromosome between 31,000 and 79,000 years ago. He has been named, rather prosaically, M168. More evocatively he could be seen as the Eurasian Adam ... The journeys taken by his sons and grandsons defined the subsequent course of human history (Wells 2007)." We'll give him a date of 60,000 years ago and call him our 3,000th great grand-father. Members of his clan were the very ones who crossed the Red Sea to the Arabian Peninsula. Over the next 15,000 years they established themselves in the Middle East.

In Africa three distinct lineages of language families formed; the Khoisan language family and the Congo-Saharan group which split into the Niger-Kordofanian and Nilo-Saharan language families (Ruhlen 1994) (Figure 5-2)." "The populations involved encompass the African Rift Valley, extending into south-western Africa, where people known as the San - formerly called Bushmen - have a strong signal of the diversity that characterized the earliest human populations. They also speak one of the strangest languages on the planet, notable for its use of clicks as integrated parts of words - like the clicking sound we might make when we guide a horse, or imitate a dripping tap. No other language in the world uses clicks in regular word construction ... (Wells 2002)."

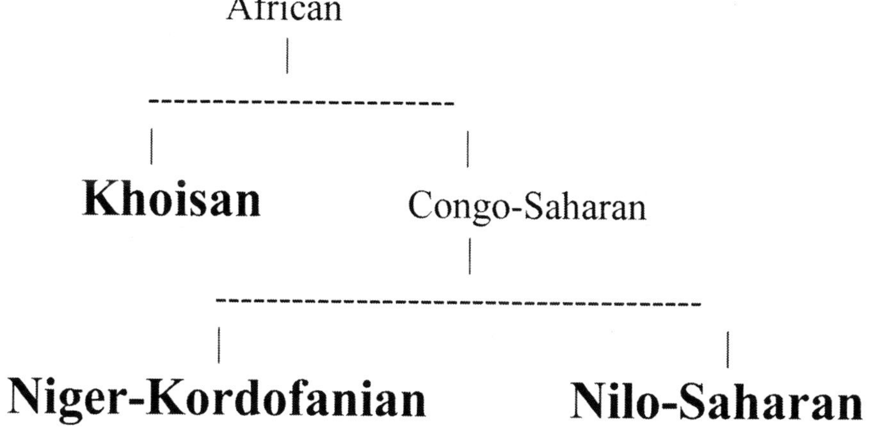

Figure 5-2. Genealogical structure of the language families of Africa (language families are in large-bold font)

"Approximately 50,000 years ago, probably in southern Asia, a man was born carrying the marker M130. His recent ancestors had just begun the first major wave of migration out of Africa. They followed the African coastline through the southern Arabian Peninsula, India, Sri Lanka, and Southeast Asia. Some members of the group would ultimately cross the Torres Strait to populate distant Australia. Because these early travelers were already quite adept at exploiting coastal resources, they didn't need to learn new skills as they followed the coastline, and migration to Australia occurred quickly, taking less than 5,000 years (Wells 2007)."

"The marker M89 first appeared around 45,000 years ago in northeastern Africa or the Middle East, and today is found in more than 90 percent of all non-African men. [The members of haplogroup F] "...appear to have followed the grasslands and plentiful game from eastern Africa to the Middle East and beyond. They were part of the second great wave of migration out of Africa (Wells 2007)."

We'll call the genetic markers M130 and M89 our Middle Eastern great grand-parents. One pair of these great grand-parents (M130) would be responsible for migrations along the coastal route to Southeast Asia and on to Australia. They would also provide one of the migrations to the Americas - more on this in the next chapter. This pair of great grand-parents would give rise to language families of the Southeast Asia/Pacific Group. There are two separate lineages in this group - one that leads to the Austric language family and a second that leads to the Pacific group. The Pacific group contains the Indo-Pacific and Australian language families (Figure 5-3). They would represent our 2,500th great grand-parents. The other group (M89) would go on to colonize the rest of the world. They represent our 2,250th great grand-parents. About 40,000 years ago climate change expanded the Sahara Desert and effectively trapping our 2,250th great grand-parents in the Levant. All of the language families represented in these two figures line up exactly with the genetic classification put together by Cavalli-Sforza and shown in Ruhlen (1994).

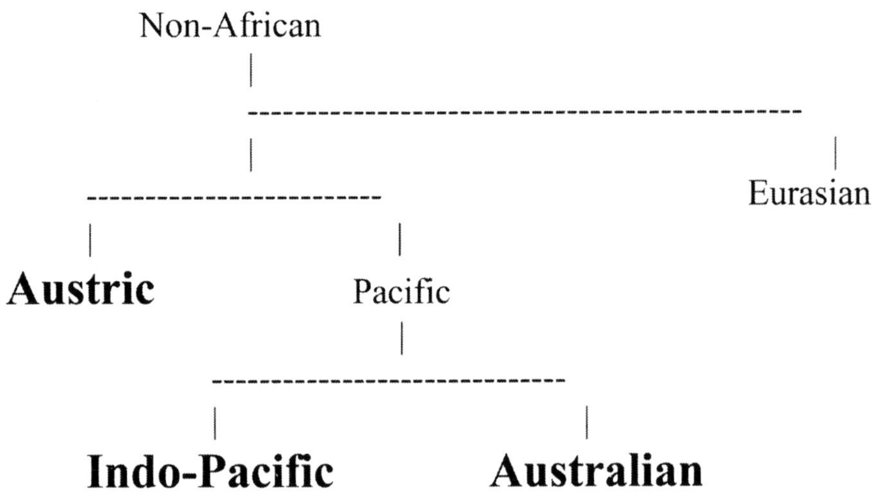

Figure 5-3. Genealogical structure of the language families of the Southeast Asia/Pacific group (language families are in large-bold font)

"The shutting of the Saharan gate after these M89-bearing populations were allowed through is suggested by the low frequency in north-eastern Africa of Eurasian markers that occurred later on the M89 lineage. If Africa and the Levant had been part of a continuous range occupied by humans throughout the Upper Paleolithic, we would expect to see a relatively homogenous distribution of markers throughout. In fact, it seems that the emigration of populations bearing M89, which we can call a Middle Eastern marker, signified the last substantial Upper Paleolithic exchange between sub-Saharan Africa and Eurasia. The world had been divided into African and Eurasian, and it was to be tens of thousands of years until significant exchange was to take place again. A second major branch of the Non-African language groups, the Eurasian, gave rise to all the other families. The Eurasian group is divided into the Dene-Caucasian language family and the Eurasia/Americas group (Figure 5-4). These languages line up fairly close to Cavalli-Sforza except that he does not address the Dene-Caucasian language family and he has a slightly different lineage from the Indo European family. He is more specific about the Indo-European in that he has it under a separate North Africa/ Eurasia group versus the Eurasiatic group of Ruhlen.

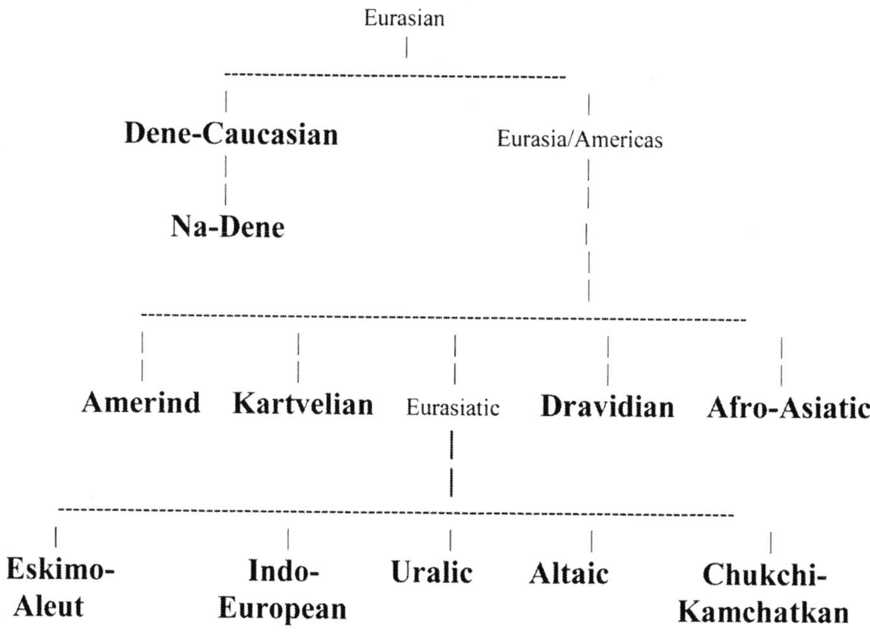

Figure 5-4. Genealogical structure of the language families of the Eurasian group (language families are in large-bold font)

"The presence of M89 in both north-eastern Africa and the Middle East, and the age of the Upper Paleolithic archeological sites in the Levant, helps us to answer the question of whether Eurasia was settled in a single southern coastal emigration from Africa. M130 chromosomes are not found in Africa, suggesting that this coastal marker arose on an M168 chromosome en route to Australia. Conversely, M89 Y-chromosomes are not found in Australia or south-east Asia - but they appear at fairly high frequency in north-eastern Asia. The implication is that M89 appeared slightly later than M130, in a population that stayed behind in Africa after the coastal migrants left for Australia. It was these people, sans M130 chromosomes, who first colonized the Middle East. There is archeological evidence for a modern human presence in the Levant from around 45,000 years ago, consistent with the arrival of modern humans from somewhere else. North-eastern Africa is the only nearby location with archeological sites dating from around the same time - and, crucially, the same genetic markers we see in the Levant. Thus, the genetic and archeological patterns tell us that there was a second migration from Africa into the Middle East. Once our Upper Paleolithic migrants had arrived in the Levant, the road into the heart of Eurasia was open. There was a continuous highway of steppe - not unlike African savannahs in terms of its species composition - that stretched from the Gulf of Aqaba to northern Iran, and beyond into central Asia and Mongolia. The hurdle of the Sahara having been overcome, the subsequent dispersal of these fully modern humans would have been limited only by their own wanderlust. They had all of the intellectual building blocks that would enable them to conquer the continent, and the process began with gradual migrations along this Steppe Highway, the continental equivalent of the southern Coastal Highway (Wells 2002)."

"Our canonical Levantine Upper Paleolithic lineage, M89, is found at frequencies of only a few percent in western Europe. It may have been these few Middle Eastern immigrants who introduced the earliest signs of the Upper Paleolithic to Europe, a culture known as the Chatelperronian [or as previously noted it could have been the

Neanderthals], but they did not leave a lasting trace. The true conquest of Europe, and the demise of the Mousterian, would have to wait for a later wave of immigration - people with a few more ingredients in their genetic soup (Wells 2007)." Figure 5-5 shows the distribution of Y-chromosome markers on the first waves out of Africa.

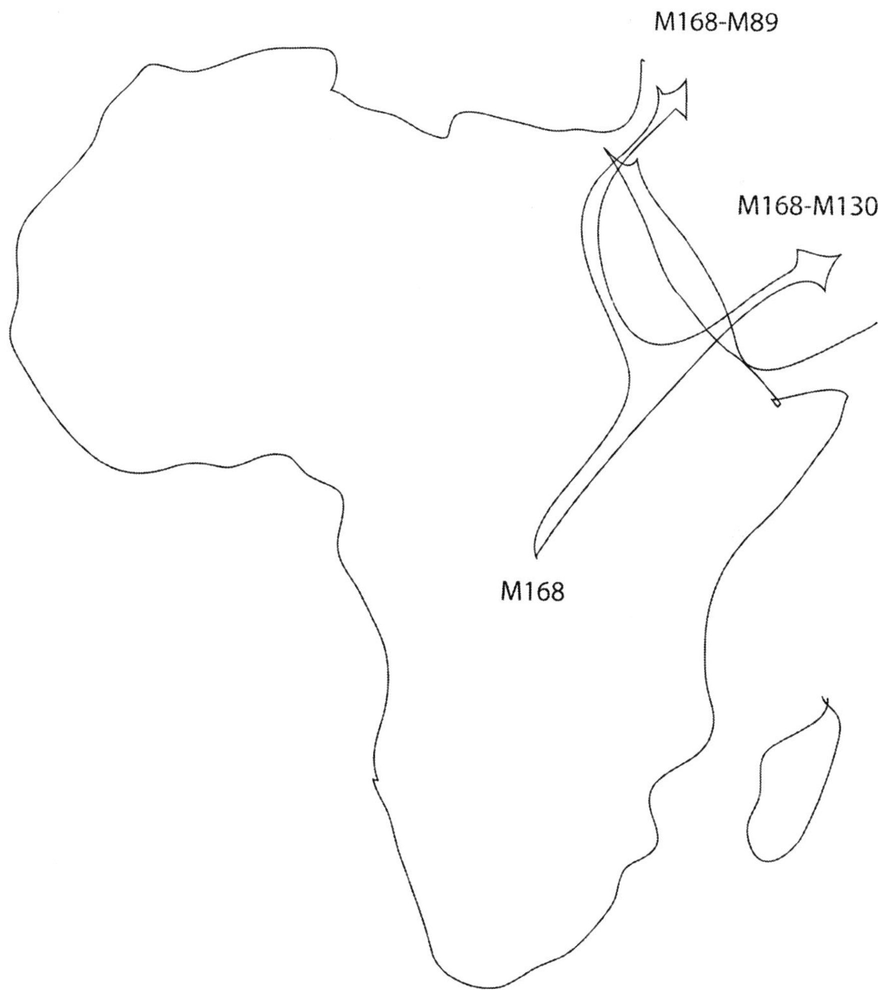

Figure 5-5. Origins and migration routes out of Africa for the Y-chromosome markers

The same general patterns of origin and migration can be seen in the mtDNA haplogroups. Figure 5-6 shows the ancestral relationships of these early haplogroups. "Mitochondrial Eve represents the earliest female root of the human family tree. Her descendants, moving around within Africa, eventually split into two distinct groups, characterized by a different set of mutations their members carry. The older group is referred to as *L0*. The individuals have the most divergent genetic sequences of anybody alive today, which means they represent the oldest branches of the mitochondrial tree. Importantly, current genetic data indicates that indigenous people belonging to these groups are found exclusively in Africa, which supports the notion that the earliest humans originated in Africa.

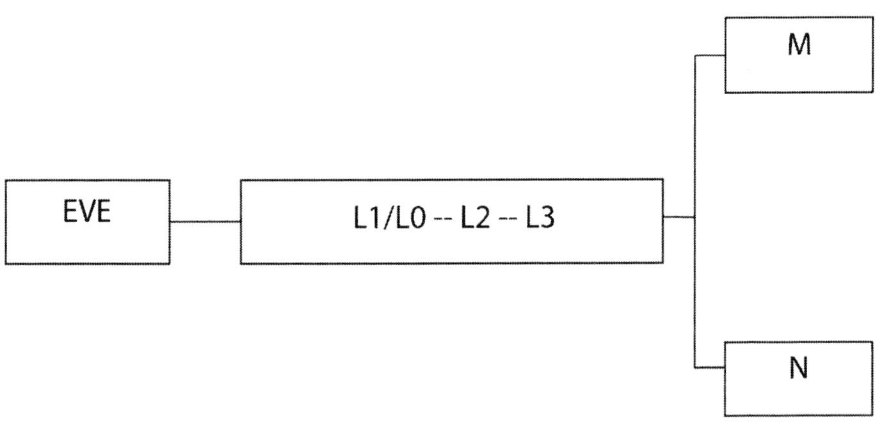

Figure 5-6. The ancestral relationships for mtDNA haplogroups out of Africa

"The descendants of Mitochondrial Eve eventually formed another group called *L1*, which coexisted with *L0* individuals during these early years. *L1* descendants would eventually leave Africa and populate the rest of the world, while the descendants of *L0* remained in Africa.

"Haplogroup *L0* likely originated in East Africa around 100,000 years ago. Over the course of tens of thousands of years, these early ancestors migrated throughout much of sub-Saharan Africa, and today are found at high frequencies among geographically diverse populations such as the central African Pygmies and the Khoisan of southern Africa. Today haplogroup *L0* is found in 20 to 25 percent of people in central, eastern, and southeastern Africa, but is found at lower frequencies in northern and western Africa (Wells 2007)."

"*L2* individuals are found in sub-Saharan Africa, and like their *L0/L1* predecessors, they also live in central Africa and as far away as South Africa. But whereas *L0/L1* individuals remained predominately in eastern and southern Africa, *L2* ancestors expanded into other areas. It is... likely the *L2* group, which arose around 70,000 years ago from a single female ancestor, first emerged in western and west-central Africa (Wells 2007)."

"The most recent common ancestor of haplogroup *L3* is a woman who lived around 80,000 years ago. While *L3* individuals are found all over Africa, *L3* is particularly known for its northerly movements. They were the first modern humans to have left Africa, representing the deepest branches of the tree found outside of that continent (Wells 2007).

"Today *L3* individuals are found at high frequencies in populations across North Africa. From there, members of this group went in a few different directions. Some lineages within *L3* in the mid-Holocene headed south and are predominant in many Bantu groups found all over Africa. One group of individuals headed west and is primarily restricted to Atlantic western Africa, including the islands of Cabo Verde.

"Other *L3* individuals kept moving northward, eventually leaving the African continent completely. Descendants of these people currently make up around 10 percent of the middle Eastern population, and gave rise to two important haplogroups that went on to populate the rest of the world (Wells 2007)."

"Haplogroup *M* is considered an Asian lineage, as it is found at high frequencies east of the Arabian Peninsula. Members of this group are virtually absent in the Levant (a coastal region in what is now Lebanon), though they are present at higher frequencies in the southern Arabian Peninsula at around 15 percent. Because the group's age is estimated at around 60,000 years old, members were likely among the first humans to leave Africa (Wells 2007)."

"Haplogroup *N*, like *M*, is one of two groups that descend directly from haplogroup *L3*. The first of these groups, *M*, made up the first wave of human migration to leave Africa. The second great wave, also of *L3* individuals, moved north rather than east and left the African continent across the Sinai Peninsula. Faced with the harsh desert conditions of the Sahara, these people likely followed the Nile Basin, which would have proved a reliable water and food supply in spite of the surrounding desert and its frequent sandstorms. Descendants of these migrants eventually formed haplogroup *N*.

"Some members bearing mutations specific to haplogroup *N* formed many groups of their own which went on to populate much of the rest of the globe. These descendants are found throughout Asia, Europe, India, and the Americas. However, because almost all of the mitochondrial lineages found in the Near East and Europe descend from *N*, it is considered a western Eurasian haplogroup (Wells 2007)" (Figure 5-7).

So, at this point we've got the folks out of Africa and along the coast to southeastern Asia and Australia. We can now begin to trace the lineages out of the Middle East and to the rest of the world.

Figure 5-7. Origins and migration routes out of Africa for mtDNA haplogroups.

The main body of Upper Palaolithic people began to disperse eastward. As with the other early human migrations, it almost certainly wasn't a conscious effort to move from one place to another. Rather, it seems that the continuous belt of steppe stretching across Eurasia provided an easy means of dispersal, gradually following game further and further afield. It was during this time that another marker appeared on the M89 lineage, given the name M9. It was the descendants of M9, a man born perhaps 40,000 years ago on the plains of Iran or south-central Asia, who were to expand their range to the ends of the earth over the next 30,000 years (Wells 2002)." M9 would represent our 2,000th great grand-parents. At about 35,000 years ago, we find during the continuous migration the development of a pair of our 1,750th great grand-parents M45 and M175. M45 went north into Central Asia. They would become the Gravettians. M175 would continue on south and east into Asia. About 30,000 years ago our 1,500th great grand-parents would spin off with M20 going south into India. At this same time, about 30,000 years ago, a set of great grand-parents would emerge from M45 on the trek to Europe. And about 20,000 years ago our 1,000th great grand-parents would emerge from M45 on their trek to Siberia, Beringia and finally to the Americas.

"As the steppe hunters migrated eastward, carrying Eurasian lineages into the interior of the continent, they encountered the most significant geographical bollards [this is English for obstacles] so far. These were the great mountain ranges that define the southern central Asian highlands - the Hindu Kush running west to east, the Himalayas running north-west to south-east and the Tien Shan running south-west to north-east. The three ranges meet in the center, at the so-called Pamir Knot in present-day Tajikistan, and each radiates off like a spoke on a wheel.

"The first humans to see them must have been absolutely awe-inspired. Although they had encountered the Zagros range in western Iran, it was a permeable barrier, with numerous valleys and low passes that would have allowed easy movement. The Zagros themselves actually would have been part of the geographic range of the prey species hunted by Upper Paleolithic people, with herds migrating into higher pastures during the summer and descending to the surrounding plains in the winter. The high mountains of central Asia were a different beast altogether. Each of the ranges has peaks that soar to 5,000 meters or higher (in the case of the Tien Shan and Himalayas, over 7,000 meters), and the radiating high-altitude ridges would have been formidable barriers to movement. Remember that the world was in the grip of the last ice age, and temperatures would have been even more extreme than today. It was because of these mountains that our Eurasian migrants would have been split into two groups - one moving to the north of the Hindu Kush, the other to the south, into Pakistan and the Indian subcontinent. How do we know this? The Y-chromosome again traces the route.

"Those who head north, toward central Asia, had additional mutations on their European lineage that we will trace below. The Upper Paleolithic people who headed south, though, had an unrelated mutation on their Y-chromosome known as M20. It is not found at appreciable frequencies outside of India – perhaps 1-2 percent in some Middle Eastern populations. In the subcontinent, though, around 50 percent of the men in southern India have M20. This suggests that it marks the earliest significant settlement of India, forming a uniquely Indian genetic substratum - which we can call the Indian clan - that pre-dates later migrations from the north. The ancestor of the Indian clan who moved into southern India around 30,000 years ago, would have encountered the earlier coastal migrants still living there (Wells 2002)."

"The genetic composition of the first Siberians was a mixture of both central Asian and ancestral Eurasian clan lineages. While M45 is the marker that we use to infer migration of the early central Asian steppe hunters, there were still many men alive who did not have Y-chromosome marked with M45 - they would have had unmarked Eurasian M9 Y-chromosomes. This is because new markers do not immediately increase in frequency to the point where all other markers - such as the ancestral M9 lineage - are lost. All of the Y-chromosome markers we study originated in a single man at some point in the past, so their original frequency was one (that individual) divided by the total number of men in the population - a very low frequency in all but the smallest groups. Over time, they become more common primarily due to the effect of genetic drift - the random changes in frequency that characterize all human populations. Thus the earliest people to colonize Siberia would have had members of both the central Asian M45 and the older Eurasian M9 clans, although drift appears to have caused them to lose most of their ancestral Middle Eastern chromosomes by this point.

"As with the Eurasians who entered India on the other side of the Hindu Kush, some of these Eurasian clan members would have migrated to the north and east, guided in their journey by the Tien Shan mountains. Some of them, perhaps taking advantage of the so-called 'Dzhungarian Gap' used thousands of years later by Genghis Khan to invade central Asia, made it into present-day China. It is likely that the majority were migrants along the Steppe Highway further to the north, avoiding the harsh deserts of western China by detouring through southern Siberia. However, make it they did. We know this because they left descendants from another Y-chromosome marker that is almost completely limited to East Asia, and is entirely absent from western Asia and Europe - M175.

"Today, M175, which arose on a Eurasian M9 chromosome, is found at highest frequency, around 30 percent, in Korean populations. Based on absolute dating methods, it appears to be roughly 35,000 years old, coinciding very closely with the appearance of the Upper Paleolithic in Korea and Japan. There are several more recently derived markers that have M175 as an ancestor (particularly M122 ...), and together these related lineages account for 60-90 percent of the Y-chromosomes in east Asia today ... M175 unites most Asian men living east of the Hindu Kush and Himalayas, defining an east Asian clan (Wells 2002)."

In addition to these genetic variants in Asia, we find our Coastal population (M130) also present. "The Coastal lineage is found at high frequency of 50 percent in Mongolia, and it is common throughout north-east Asia. How it reached this location remains a mystery, but it is likely that the early coastal migrants to south-east Asia gradually moved inland, migrating northward over thousands of years. The M130 chromosomes in the south are older than those in the north, consistent with such a migration. At some point, perhaps 35,000 years ago, they would have met the descendants of the other, mainline of migrants - our incoming Eurasians. The presence of both Eurasian and Coastal lineages in east Asian populations attests to the extensive admixture that occurred between them (Wells 2002)."

But let's return again to our 2,250th great grand-parents (M45) in Central Asia. We find that these Gravettians moved into Europe, and around 30,000 years ago our 1,500th great grand-parents emerged. The majority of all western Europeans alive today carry the M173 marker. About 20,000 years ago our 1,000th great grand-parents (M242) emerged from M45 on their trek to Siberia and beyond. About 10,000 years ago three pairs of our 500th great grand-parents emerged - M172 in the Middle East from M89 and migrated into Europe; M17 arose from the Ukraine and southern Russia into the Czech Republic across to the Altai Mountains in Siberia; and, M3 that arose from M242 and migrated into the Americas.

Cavalli-Sforza (in Wells 2002) using genes from current European and Middle East males found that about 20 percent were related to the migration of Middle Easterners (M172) into Europe about 10,000 years ago. In the next chapter we'll look at three separate migrations into the Americas - one from M130 and two from M3.

The two mtDNA haplogroups *M* and *N* were ancestral to all lineages out of Africa that ultimately populated the rest of the world. I won't cover all possible haplogroups but only those of interest to the colonization of the Americas. For a detailed account of all mtDNA and

Y-chromosome haplogroups and distribution maps see www.nationalgeographic.com/genographic/atlas.html.

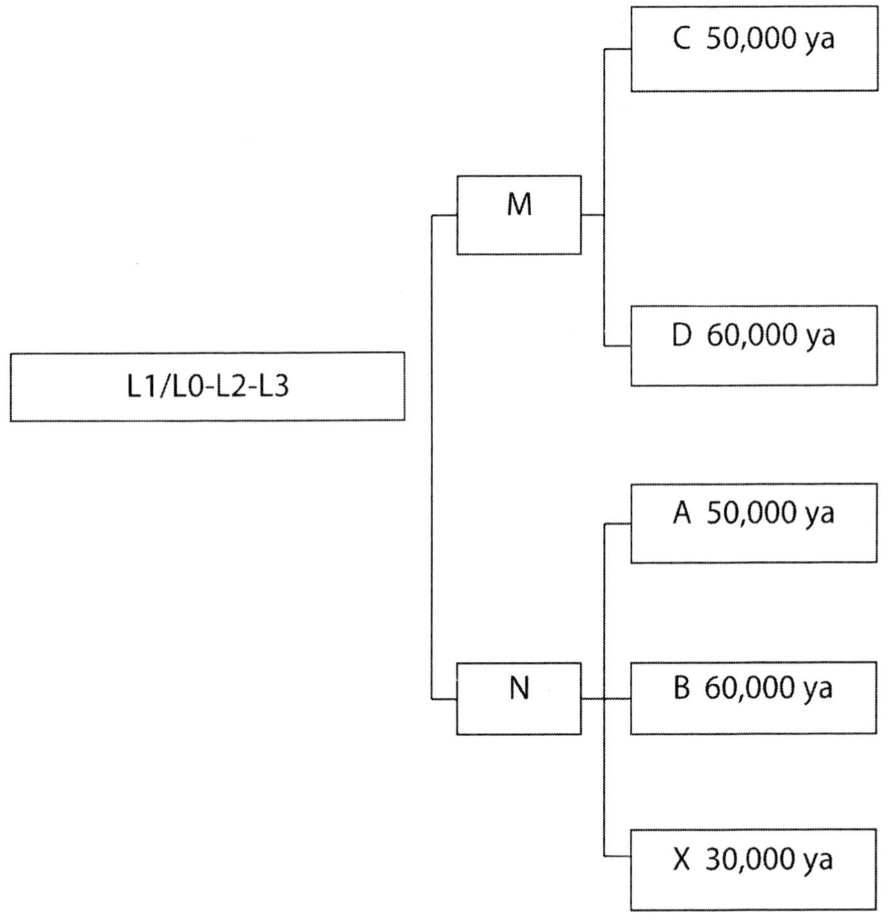

Figure 5-8. Ancestral relationships for the *M* and *N* mtDNA haplogroups

Around 50,000 years ago somewhere between the Caspian Sea and Lake Baikal, the first members of haplogroups *A, B, C,* and *D* began moving east across Siberia, the beginning of a journey that did not stop until reaching both continents of the Americas.

"Haplogroup *A* was first found among aboriginal American populations and has played an important role in allowing geneticists to use DNA mutations to date prehistoric migrations. With few exceptions, haplogroup *A* is the only lineage carried by Eskimos, an indigenous group native to Siberia, Alaska, and Canada (Wells 2007)."

"The first members of haplogroup *B* began moving into East Asia and ultimately reached both continents of the Americas and much of Polynesia. It is one of the founding East Asian lineages, and along with haplogroups *F* and *M*, comprises around three-quarters of all mitochondrial lineages found there today (Wells 2007)." "Radiating out from the central Asian homeland, haplogroup *B*-bearing individuals began migrating into the surrounding areas and quickly headed south, making their way throughout East Asia. Today haplogroup *B* is found in around 17 percent of the entire Chinese gene pool. It exhibits a very wide distribution along the Pacific coast, from Vietnam to Japan, as well as at lower frequencies (about 3 percent) among native Siberians (Wells 2007)."

Haplogroup C is considered a characteristic Siberian lineage, and today accounts for more than 20 percent of the entire mitochondrial gene pool found there. "Radiating out from the Siberian homeland, haplogroup *C*-bearing individuals began migrating into the surrounding areas and quickly headed south, making their way into northern and Central Asia. Haplogroup *C*'s frequency slowly declines the farther from Siberia one looks: It now comprises around 5 to 10 percent of the people of Central Asia, and around 3 percent of the people living in East Asia.

"Owing to their experience in the harsh Siberian climate, haplogroup *C* individuals would have been ideally suited for the arduous crossing of the Bering land bridge during the last ice age, 15,000 to 20,000 years ago. Today, haplogroup *C* is one of five mitochondrial lineages found in aboriginal Americans in both North and South America. While the age of *C* is very old (around 50,000 years), the reduced genetic diversity found in the Americas indicates that those lineages arrived only within the last 15,000 to 20,000 years and quickly spread once there (Wells 2007)."

"Fanning out from the Central Asian homeland, haplogroup *D*-bearing individuals began migrating into the surrounding areas and quickly headed south, making their way throughout East Asia. In northern Asia *D* is present in more than 20 percent of the population, and is found at around 17 percent in Southeast Asia. Haplogroup *D* has a gradual reduction in frequency moving west across Eurasia. It is found in 10 to 20 percent of central Asians, with several of its lineages there matching exactly with sequences found in the East, representing a much more recent mixture (Wells 2007)."

"Haplogroup *X* is primarily composed of two distinct subgroups, *X1* and *X2*, whose widespread and sporadic distribution have been the cause of much debate. *X1* is largely found in North and East Africa. The other subgroup, *X2*, is spread widely throughout western Eurasia. *X2* makes up around 2 percent of the European mtDNA lineages and is more strongly present in the Near East, Caucasus region, and Mediterranean Europe. In some western Eurasian groups, it is found at significant frequencies of 10 to 25 percent, though this is likely due to expansion events following the last ice age around 15,000 years ago.

"The real controversy surrounding haplogroup *X* is its place as one of the five haplogroups found in the indigenous peoples of the Americas, where it is found exclusively in North America at varying frequencies. In the Ojibwa from the Great Lakes region it is found at around 25 percent, in the Sioux at around 15 percent, the Nuu-Chah-Nulth at more than 10 percent, and in the Navajo at 7 percent. But *X2* is almost entirely absent from Siberia, the proposed land route of the first migrations into the New World.

"Unlike the four main Native American haplogroups (*A, B, C,* and *D*), haplogroup *X* is entirely absent in East Asian populations, indicating that it played no role in colonizing those regions. However, its age estimate in the Americas, around 15,000 years old, does indicate that members of this group were among the first modern humans there (Wells 2007)." While the distribution of this haplogroup seems to rule out a migration to the Americas via the Pacific, it surely does not rule out a migration across the Atlantic. In fact it's decreasing frequency in the Americas from east to west, points strongly to an entry on the east coast and expansion westward.

Grandfather's Tale

This tale was taken from the Moses Band of the Columbia River Indians

The Warm Wind Brothers vs. the Cold Wind Brothers

This is the story about two tribes that lived during the last Ice Age, many years ago. One of these tribes was called the Tribe of the Warm Wind. The people lived in Dry Falls-Vantage area. Wherever they camped, they were in warm country. The Chief of the Warm Wind people had five sons.

The second tribe was the Tribe of the Cold Wind. The Chief of this tribe also had five sons. Wherever the Cold Wind people settled, cold weather followed. All the lakes and rivers froze, and snow fell.

When the Tribe of the Cold Wind tried to move south, they were stopped by the Tribe of the Warm Wind. The Cold Wind people held council and decided that if they would kill the five brothers in the Warm Wind tribe, they could go south wherever they wished.

They asked Coyote to deliver a challenge for a dual between the five brothers of the Warm Wind tribe and the five brothers of the Cold Wind tribe. The challenge was accepted, and the date was set. Then Coyote traveled around to tell all the people in both tribes about the contest.

When the day arrived, both tribes gathered at the place for the duel. Two warriors fought at a time, one Warm Wind brother against one Cold Wind brother. The young warriors of the Cold Wind people were much stronger than their rivals. Soon all the Warm Wind brothers had been killed.

The Tribe of the Cold Wind now had the power to rule, and they ruled strongly and severely. The country became cold. The rivers and lakes froze solid and snow fell until the lodges were nearly covered. As far south as Dry Falls, the ice was piled as high as mountains.

Coyote was cruelly treated, and his work was never done. The Warm Wind people were miserable. They were made the slaves of the Cold Wind people. Any food they found was taken from them. They had to eat the scraps of food the Cold Wind people did not want.

Not long before the struggles, the youngest son of the Warm Wind chief had married a girl from a tribe farther south. She decided to go back to her people. Before she left, she told her husband's people," I am expecting a child. Pray that it will be a boy. If I have a son, I will train him to be the greatest warrior in the world. When he is grown, I will send him to you. Watch for him. He will avenge the defeat of his father and uncles."

A few moons later the woman gave birth to a son. When he was about three months old, his mother and her brothers had him follow a training course that would make him a strong warrior.

For years he trained. He became so strong that he could uproot trees and throw them over hills. He could throw large boulders many miles. At this time he believed himself the strongest man in the world.

Then his mother told him about the duel between the Warm Wind brothers and the Cold Wind brothers. The young man felt that he was ready to avenge the death of his father and uncles, and to set the people free. But his mother insisted that he train for one more year.

By the end of that year he could move small mountains. Then his mother told him that he was ready to go north to help his people. She told him just what he should do and what he should ask his grandparents to do to help him.

The young warrior started north, and a warm south wind went with him. As he neared the home of his grandparents, the ice on their lodge-poles began to melt for the first time since they became slaves. They were glad and asked each other, "Do you think that our grandson is coming?"

Before the sun set that day, the young man reached them. They saw that he was strong, and they believed him when he said that he had come to free them and their people from the Cold Wind tribe. He was sorry that they had been treated unkindly.

Coyote was sent to the camp of the chief of the Cold Wind tribe to deliver a challenge from the grandson of the chief of the Warm Wind tribe. It was accepted. The day and the place were decided upon.

In the camp of the defeated people, grandson asked them to follow his mother's instructions. "Boil some salmon, and put the broth in five containers."

On the morning of the duel, the people of both tribes gathered at the river at the chosen place. The grandson fought with the oldest brother from the Cold Wind tribe. The ice was slick. But the grandson's people threw down a bucket of hot salmon broth, and the ice became rough. So the young warrior defeated the first of the five brothers.

Then the second brother stepped forth, and the grandson fought him. The Cold Wind people threw water on the ice, hoping to make it slick. Then the Warm Wind people threw another bucket of hot broth on the ice, and it became rough. So the young warrior defeated the second brother.

The third, fourth, and fifth brothers he struggled with, each in turn. Each time he was helped by the hot salmon broth. When he had defeated the youngest brother, the Warm Wind people were free. They drove the rest of the Cold Wind people so far north that they could never find their way back. Soon the warm wind came in and melted all the ice.

When the young grandson travels north in the spring, warm weather follows. If he had not defeated the five brothers of the Cold Wind tribe, we still would be living in the Ice Age.

Chapter VI.
Settling the Americas
(20 thousand years ago to 10 thousand years ago)

A mere 35,000 years ago, Homo sapiens sapiens expanded rapidly into even more severe arctic landscapes and open steppe-tundra, where winters lasted three-quarters of the year, and where people were huddled in smoke-filled dwellings for months of total midwinter darkness. Eventually they came to the frontiers of a vast continent. Soon they were to wander into the heart of the Americas and explode into a new chapter of human history.

Brian Fagan, The Great Journey

We must view the settlement of the Americas as an integral part of a worldwide human social and cognitive explosion that first took off some 40,000 years ago and continues to this day. And within that broad movement early American prehistory is not a story of a single monolithic culture sweeping from pole to pole but rather an intricate, constantly shifting pattern of diverse local cultures, each with its unique history and lifeways.

Thomas D.
Dillehay, Settlement of the Americas: A New Prehistory

There are a number of plausible theories on the settling of the Americas. I'll discuss each of these in turn throughout this chapter. For many years the gold standard for theories on the settling of the Americas was the "Clovis First" theory developed in the 1960's to explain the rather rapid mega-fauna extinctions in North and South America. Bonnichsen (2005) provides a good overview of this theory. He notes that this model "...envisions that a band of Siberian big-game hunters numbering no more than 150 people came from northeastern Siberia to interior North America approximately 11,500 RCYBP [Radio Carbon Years Before Present]. Modelers envisioned that these people walked across the Bering Land Bridge near the end of the last glacial period when lower sea levels exposed the Bering Land Platform connecting Asia and North America. The band of hunters quickly found their way across eastern Beringia, the unglaciated portion of Alaskan and Yukon Territory, and then down the ice-free Corridor between the Cordilleran and the Laurentide Ice sheets onto what is now the Great Plains. These hunters allegedly found a virgin continent teaming with large and unsuspecting game animals. Using a sophisticated tool kit and a new weapon system, they were able to slaughter mammoths and other unsuspecting large animals." The premise was that early humans hunted down and killed these large animals over a period of about 1,000 years. In North America 33 genera of large animals went extinct. In South America over 55 genera went extinct. For a number of good reasons, this theory has come under question.

Haynes (2005), a major supporter of the "Clovis First" Theory, provides a scenario for the migration of people, destined to become the Clovis culture, out of Alaska through an ice free corridor in Canada to the Great Plains of Canada and the U.S. This scenario will set the stage for further discussions.

"One summer may have taken a dozen Nenana people from the upper Tanana Valley to the divide with the White River Valley, then via the Yukon Valley to the Pelly River in the Tintina Trench and a winter camp near Faro. Next spring the Pelly Valley would lead them around the northeast flanks of the Pelly Mountains and over the divide near Finlayson Lake to the upper Frances River. The Frances Valley would lead them to the Laird Plains near Watson Lake, from where with boats if a proglacial lake existed, they would follow the Laird Valley to Nelson Forks. As they exited the mountains, they would have their first view of the vast Interior Plains of Canada extending to the east as far as the eye could see. Laurentide ice would be at least 200 km [124 miles] to the east.

"At this point they might decide to return to the Yukon Plateau to inform others of their discovery. The return journey could be made in half the time of the outward journey because of travel over familiar ground and no side trips to explore tributaries. The next summer they returned to Nelson Forks with a much larger group, perhaps 50 people, the leader having talked several other bands into joining what amounted to a tribe for exploration southward.

"Early on, the question had to arise as to where the southward exploration might lead them. It is very likely that there were people whose curiosity and quest for knowledge of the beyond would urge them ever southward. Eventually a band leader would decide to go as far south as possible and not return to the base camp as winter approached. This decision could have taken them from winter camp in the Nelson Forks area of British Columbia to perhaps Lake Peace in Alberta by the end of summer. While camped on the north side, exploratory excursions south of the lake could be made by crossing on the winter ice. By spring time they would be ready to continue southward.

"As the band moved southeastward, paralleling the foothills of the Canadian Rockies, they probably avoided the Laurentide ice and recently deglaciated areas. Even so they had to cross many drainages carrying snowmelt northeastward out onto recently deglaciated plains, with many lakes and abandoned channels formed as drainage integration responded to isostatic adjustments and outburst floods as ice dams gave way. Some band members may have become victims of such catastrophes; but in spite of the geological hazards, the abundance of deer, elk, ducks, and geese made for good hunting. While dried meat and pemmican were basic staples, they probably consumed fresh meat and fruits as the opportunities arose.

"The first sighting of living mammoths may have been between Dawson Creek and Grand Prairie in Alberta, where poplar trees were growing around Boone Lake by 11,700 RCYBP. This site created intense jubilation among the hunters. Hunting was delayed as they observed the herd in order to develop a hunting strategy. The first attack was a

miserable failure, as they succeeded only in wounding a young female mammoth and infuriating the matriarch. Their atlatl spears were obviously too small and light to penetrate mammoth hide effectively. The more they attacked the mammoths, the more wary the herd became. This failure led to heavier spears with larger stone points; but in order to facilitate hafting in a longitudinally notched foreshaft, someone decided to remove longitudinal flakes from both sides of the nearly completed biface. With this new armament and better understanding of mammoth behavior, hunters were eventually able to bring down a young adult female mammoth, perhaps near Red Deer, Alberta.

"In the meantime, as small herds of bison were encountered, a new strategy had to be perfected. While not only concerned at first, bison, like mammoths, soon became wary of this new predator on the prairie. Whereas mammoths provided bone and ivory as well as meat, the hides were too thick and heavy for portable tepee construction. Bison, on the other hand, provided larger hides than deer or elk, requiring fewer of them for tepees. However, they did require thinning by scraping to make them more flexible for easy handling and lighter for transport by dog travois. Deer and elk hides remained the preferred material for garments.

"After wintering in the Red Deer area, perhaps one group decided to return to Tanana Valley to tell others of their discoveries. The two groups agreed to meet at the Red Deer camp in the third summer. The main group moved farther southeast, hunting, fishing, and gathering whatever plant foods were edible. From the Cypress Hills they moved to a new winter camp near Manyberries, Alberta. The following summer they followed the Milk River into Montana, and from near Havre they moved up the Big Sandy Valley to the upper Missouri. Following the White Cliffs, they turned south when they reached the mouth of the Judith River; and continued south to Judith Gap, from where they took a southwesterly course. The Potter Creek valley led them around the north end of the Crazy Mountains and into the Shields Valley, where they made their sixth winter camp near Wilsall, Montana. The tribal leader's infant son, having drowned, was buried with great ceremony at the base of a small rock cliff and with many offerings of well-worked stone and bone at what we know today as the Anzick site.

"By this scenario it took them six years to travel 5,700 km from the upper Tanana Valley to the Anzick site, estimated by multiplying valley lengths on maps by a factor of 1.5 to estimate actual kilometers traveled. Assuming summer travel is from 1 May to 30 September, there are 153 days each year for six years for a rate of 6.2 km/day. If they traveled at this rate during the winters instead of making winter camps, the journey would have taken 2.6 years. Even if they traveled at half this rate, the trip could have been made in 6 to 12 years (Haynes 2005)."

Does this scenario reflect reality? To be able to answer this question we need to examine information from a number of different disciplines including climatology, linguistics, genetics, and archaeology.

***

The previous scenario, the "Clovis First Theory", rests on the presence of a land bridge between Siberia and Alaska, and the presence of an ice-free corridor between eastern Alaska and western Yukon Territory and the Great Plains during glacial times prior to about 11,500 years ago. Did these two corridors exist, and if so when?

To get at the heart of this question, we need to develop an overview of the last Great Ice Age. This last Ice Age, often referred to as the Wisconsin Ice Age, or more properly the Wisconsin Glacial Episode, began about 130,000 years ago and continues into the present. There were a number of stadial (cold) and interstadial (warm) periods during this time. Elias (2002) notes that there were intervals of interstadial warming at 130,000 -120,000; 110,000 - 100,000; 87,000 - 75,000; and, 60,000 – 30,000 years ago; and, the Holocene (about 12,000 – 10,000 years ago to the present). There were stadial cooling periods, which corresponded to periods of glaciation, at 120,000 - 110,000; 75,000 - 60,000; and, 25,000 - 18,000 years ago. It's the period between 25,000 years ago and the present that is of most interest to us at this time.

From about 25,000 years ago to 18,000 years ago, most of North America was covered by two large ice sheets - the Laurentian and the

Cordilleran. The Laurentian ice sheet reached its maximum extent about 18,000 years ago, at a time when Cordilleran glaciers covered the mountain ranges of southern Alaska and British Columbia. As the Laurentian expanded, so did a Cordilleran glacier complex in the west. "This mountain glaciation mantled the northern Strait of Georgia after 25,000 years ago, developed around Vancouver after 17,500 and expanded vigorously to about 30 miles (48 km) south of Seattle by about 14,500 years ago. The ice expanded down the Pacific Coast, leaving only occasional ice-free pockets (Fagan 2003)." This Cordilleran ice reached its maximum extent later than the Laurentide at about 15,000 years ago.

At the height of the last glacial maximum so much of the world's water was locked up in continental ice sheets that the worldwide sea level was at least 85 meters (ca. 280 feet) below its present level, and perhaps as much as 130 meters (ca. 425 feet) below. Huge tracts of what are now parts of the submerged continental shelves of North America were then dry land. "From the paleoecological point of view, by far the most important of these tracts was the wide land bridge that linked present-day Alaska with present-day northeastern Siberia. It provided a corridor between Eurasia and North America for a large fraction of the sixty or seventy millennia of the Wisconsin glaciation and was finally broken about 15.5 ky B.P., when rising waters joined the Pacific and Arctic Oceans, thus creating the modern Bering Strait. The land bridge and lands immediately to east and west of it have acquired the name Beringia, which is an extremely important geographical entity in the ecological history of North America (Pielou 1991)." It is widely accepted that there was indeed a land bridge. The exact year for the creation of the Bering Strait is debatable. Many would have the date later than 15,000 years ago.

Climatic conditions improved significantly from 14,700 to 12,700 years ago during two separate warming periods known as the Bolling and Allerod warming periods. Brief cooling periods, the first around 14,000 years ago known as the Older Dryas and a second at about 13,250 to 13,100 years ago, known as the IACP (Inter-Allerod Cold Spell), interrupted the Bolling and Allerod periods. The Bolling and Allerod periods were immediately followed by an abrupt and substantial cooling period known as the Younger Dryas from about 12,900 to 11,500 years ago. We'll find this period is critical to our story both from the human standpoint and the animals they hunted. There are a couple of speculative causes for the Younger Dryas. The first relates to a change in the salinity of the North Atlantic due to an influx of fresh water from the breakup of Lake Agassiz, a large glacial lake in Canada, and the deglaciation in North America. This desalinization disrupted the normal circulation patterns in the North Atlantic bringing colder waters into the area. The second involves a shift in the jet stream northward in response to the changing topography from the melting North American ice sheet which increased rainfall and runoff to the North Atlantic and again caused desalinization.

We enter the present warming period in the Holocene at about 11,500 years ago. The Holocene is divided into two periods. The Hipsithermal was the warmest time period in our current interglacial and lasted some 6,500 years (about 12,000 to 5,500 years ago). The Hipsithermal is followed by the Neoglaciation (new glaciation). We've been in the Neoglaciation period, the first stage of the coming glaciation, for about the last 5,500 years. During this current Neoglaciation period there were two periods of warming and cooling in historical times. We'll discuss the Holocene in more depth in the next chapter.

As the climate began to warm sometime after some 15,000 years ago, sea levels began to rise. "The melting ice sheets and the rising sea interacted with each other to produce some interesting positive feedbacks. Where the margins of the ice sheets reached the ocean and formed floating ice shelves, the rising sea hastened the calving of icebergs. Besides being diminished by the calving off of bergs, the ice sheets were affected in yet another way. The melting bergs cooled the sea and led to reduced evaporation of water vapor from the surface. This in turn led to reduced snowfall, which eventually starved adjacent ice sheets. At the same time, a large quantity of fresh meltwater must have been flowing off the ice sheets into the sea. It would have spread out to form a thin layer of fresh water on top of the saltier, and therefore denser, sea water and, being fresh, would have frozen readily in winter. The resultant thin layer of ice would have formed a lid over the sea, reducing the amount of water vapor available to form snow. The effects of melting icebergs in summer, and an ice layer over the sea in winter, reinforced each other in hastening starvation of the ice sheets (Pielou 1991)."

Elias credits Bartlein (1991) with developing the important factors which controlled Beringian climate in the Late Pleistocene. These are "1) cooling effects from the continental ice sheets to the east; 2) a split in the atmospheric jet stream, in which one part flowed north of the ice sheets, and the other flowed south of them; 3) changes in insolation at high latitudes in the Northern Hemisphere through the various Milankovich cycles; 4) reduced CO_2 concentrations during glacial intervals and increased CO_2 concentrations during interglacial intervals; 5) sea ice, sea surface temperature and salinity changes; and 6) cooling effects of late-lying glacial ice on adjacent land areas." Though how these factors influenced Beringian climate are not well understood, Elias notes that "Some of these factors had unique impacts on Beringia. For instance, the split flow of the jet stream probably brought a stronger flow of southerly air into Alaska and the Yukon Territory, making summers cooler than otherwise, but perhaps making winters warmer. The combination of sea ice, sea surface temperatures, and salinity of the ocean waters off the southern coast of Beringia may have brought warmer autumns and winters than exist today. Increased CO_2 concentrations during interglacial periods served to amplify the warming brought on by increased insolation. Conversely, decreased CO_2 concentrations during glacial periods would have decreased the degree of greenhouse gas warming of the atmosphere." As a consequence of the climatic factors, most of Beringia is believed to have been ice-free. The Bering Land Bridge itself blocked warm Pacific waters from entering the interior making the climate too dry to allow for substantial snowfall accumulation. In addition significant concentrations of loess, windblown sand and other fine particles, attest to the aridity of the climate.

As a consequence of the climatic changes occurring in Beringia, the area became a major refuge for large animals living at high latitudes. This, of course, includes man. The only comparable ecosystem supporting such a large compliment of animals is the East African Savannah. Elias develops some interesting parallels between the Pleistocene steppe-tundra fauna and those of modern Africa. "In the Beringian ecosystems, the woolly mammoth filled the elephant niche. The woolly rhinoceros, probably a rare species in Beringia, filled the rhinoceros niche, and saiga antelope filled the antelope niche. The modern range of saiga antelope is presently restricted to Central Asia, where arid grassland habitats persist today. It died out in Alaska and the Yukon Territory sometime after 13,000 yr B.P., as changing vegetation patterns eliminated its steppe-tundra habitat. In place of African zebras, the steppe-tundra had Pleistocene horses. Instead of water buffalo and wildebeest, the steppe-tundra had large-horned bison and two species of musk-ox. The dominant grazers on this cold grassland were the woolly mammoths, Pleistocene horses, and large-horned-bison. Caribou were also present in the Late Pleistocene. Caribou feed on grasses and other herbs, shrubs, and lichens. Thus, their diet represents a mixture of browsing and grazing."

So, with people and animals safely ensconced in Beringia, was there a route for them to move out of Beringia and into the Great Plains – an ice-free corridor? "Nat Rutter of the University of Alberta divides the corridor route into three parts. The corridor begins on the eastern slopes of the Mackenzie and Richardson mountains and in the Mackenzie valley. Here the Laurentian ice penetrated to the mountains, but much of the range was unglaciated. The Laurentide and Cordilleran ice sheets did not coalesce there. Even at the height of the Late Wisconsin, there were ice-free areas in the mountains where people could have traveled, but the terrain was very rough. Between the British Columbia and Alberta borders and southward to the Edmonton-Jasper area, the two ice sheets did coalesce. Continuous sheets of glacial rubble linked ice flowing down the Rockies to the Laurentide sheet. There are signs that the linked sheets diverted ice flow southward. This area of the corridor is closest to the massive Cordilleran ice sheet west of the Rocky Mountain Trench. The ice spilled over the Rockies during the part of the Late Wisconsin, blocking the ice-free corridor, presumably during the glacial maximum, but the date is uncertain. Between Edmonton and the US-Canadian border, extensive Cordilleran and Laurentide glacial deposits never approach one another. Throughout the Late Wisconsin, an ice-free passage extended in places from the western edge of the Laurentide ice-sheet to the mouths of the major river valleys coming from the Rockies (in Fagan 2003)."

So what's the best estimate for when the corridor was physically open? Stalker (in Fagan 2003) believes that "...the best time for humans to move through the corridor was either around 19,000 years ago, at a time of minor glacial retreat, or after 14,500 years ago, post glacial times." These dates are highly unlikely since they are not supported by other evidence. Burns (1996) found that "Quaternary vertebrate fossils from Alberta, accompanied by numerous radiocarbon dates with a clear gap between 21 kya BP and at least 11.6 kya BP, now suggest that Alberta was overridden by ice during an extensive late Wisconsin glaciation." Other evidence suggests that "this route was blocked by ice from approximately 28,000 to 11,600 RCYBP (Stanford et al. 2005)." Recall that the Bolling and Allerod warming periods didn't begin until about 14,700 years ago.

"It wasn't until sometime around 11,500 – 13,000 B.P. that the ice had melted sufficiently to enable animals to reoccupy the region. By 11,000 B.P., the glaciers had melted enough to form what might be called a deglaciation corridor between the ice sheets. This would have enabled people to move from eastern Beringia southward into the more southern areas of North America. Conversely, if humans were already south of the continental glaciers, the melting of the ice would have enabled humans to move northward into the region that is now northwestern Canada and Alaska. Scientists agree that as the ice melted, plants, animals, and humans soon colonized the new land. Many archeologists believe that humans first entered the southern areas of North America by gradually moving into and settling the new environment created by the melting ice. This process would have enabled people to move gradually southward to places south of the continental ice.

"If this scenario were correct, the North American continent south of eastern Beringia would not have been colonized by humans until sometime after the glaciers had melted enough to permit people and animals to pass between the melting ice sheets, sometime about 11,000 years ago. The interior migration route requires an economy based on hunting terrestrial animals, gathering limited plant products, and probably freshwater fishing. This type of subsistence adaptation, characteristic of high-latitude environments, would have been an essential factor for colonizing recently deglaciated lands (Dixon 1999)."

Pielou (1991) suggests that a geological corridor is not necessarily an ecological corridor. "The ecological questions are: What were conditions like in the corridor when it was not under ice? And when, and for how long, was the corridor ecologically usable in the sense that it was habitable?" He further notes that "For much of the time when the corridor became glaciologically open, it must have been a forbidding place. Incessant, strong, bitterly cold winds would have blown off the nearby ice sheets. The land would have consisted of a mixture of bogs, marshes, islands of stagnate ice, and rock-strewn barrens, with dangerous, torrential rivers and icy lakes. Such vegetation as managed to grow would have been sparse. Food, clothing, and firewood would have been impossible to come by. While it was likely that the corridor could not have served as a home for humans, and a home rather than a migration route was what was needed. It is unlikely that people migrated in the sense of marching steadily forward. They did not know what sort of destination lay ahead. Instead of journeying southward deliberately, the growing population simply expanded its hunting grounds, in whatever direction expansion was possible."

Haynes' 6 to 12 year journey seems to be pushing the timeline too tightly. Recall from our previous discussion that a fairly quick migration out of Africa to Southeast Asia moved at a pace of about 0.50 to 0.75 miles per year. Haynes would have you accept that the migration through the corridor occurred at a pace of 6.2 km (3.8 miles) per day or about 575 miles per year. If the corridor needed to serve as a "home" for pre-Clovis people, their colonization would be substantially longer than the proposed 6-12 years. You also have to ask why pre-Clovis people would leave a fairly comfortable environment, one they knew well, in Beringia, to live in a relatively inhospitable environment, of which they knew very little, of the ice-free corridor, especially with no knowledge of where this journey might lead them?

So, if the ice-free corridor was first opened by 11.6 kya it was not likely an ecological corridor for some time to come. The times do not necessarily rule out a migration through the ice-free corridor, but they do place serious doubt that it was likely to happen at least in the time frame required by Clovis. At a later date, say 10,000 years ago in the time of the Folsom people, it may have been a viable corridor for entry into the Americas – or entry into Alaska from the Americas. What other routes were available during the late Pleistocene and early Holocene that would allow for early colonization of the Americas?

Support is growing for two theories that people using watercraft, possibly skin boats, may have moved southward from Beringia along the Gulf of Alaska and then southward along the Northwest Coast of North America, and/or along the Atlantic ice pack, possibly as early as 15,000 – 20,000 years ago. These routes would have enabled humans to enter the Americas prior to the melting of the continental glaciers.

Erlandson (2002) notes that "...glacial refugia existed along the northern Northwest Coast even during the last glacial, that most of the outer coast was deglaciated by at least 13,000 to 14,000 years ago, and that the ice-free corridor of the interior was far less hospitable to human migration than previously thought." Pielou describes the refugia along the Pacific Rim. He notes that "The ice-free part of the Beringian south coast was separated from the west coast of unglaciated North America by the long stretch of shore where the Cordilleran ice sheet reached the sea. Much of this coast must have been of high ice cliffs, but not all of it. Numerous ice-free refugia are believed to have been scattered along the shores and on the offshore islands of what are now Alaska and British Columbia." The Cordilleran ice front was not static, but "...made numerous local advances and retreats as ice lobes surged seaward and then wasted away by calving off bergs. Probably many of the lobes flowed as valley glaciers between the high coastal mountains; separating these glaciers were bare, ice-free ridges."

"While all this was going on, sea level was varying in a complicated manner. How it varied at any point was the result of three separate causes. First, there was a gradual eustatic rise as water flowed from the shrinking ice sheets into the world ocean [eustatic change is a rise or fall of the sea relative to the land]. Second, there were isostatic changes caused by the sagging of the earth's crust under the overlying load of ice [isostatic change is a rise or fall of the land relative to the sea]. Third, there was another kind of isostatic variation due to *diastrophism* (the deformation of the crust through folding, faulting, buckling, and the like) caused by events in the earth's interior, especially by drifting of the crustal plates. This third cause of sea level change was probably less important than the other two but appreciable nevertheless.

"The Cordilleran ice sheet thinned rapidly toward its seaward margin. At its westernmost edge, in the region of the Queen Charlotte Islands and the outer coast of Vancouver Island, it was quite thin and therefore light in weight. Consequently, at glacial maximum, the eustatic fall in sea level was greater than the eustatic rise, and land now submerged was above the sea, forming temporary sea level refugia. There were also mountain refugia (nunataks) where mountain summits rose above the comparatively thin ice sheet. A short distance to the east, where the ice sheet was thicker and heavier, the isostatic rise in sea level was greater than the eustatic fall, so that sea level along what is now the mainland coast was higher than it is now.

"There must have been a great variety of landscapes. For any given point in low-lying parts of the coastal region, at any given time, we can envisage four possible conditions: landfast ice, a floating ice shelf, ice-free land, or open sea. Perhaps many places experienced all these states several times over (Pielou 1996)."

The Atlantic Coast had similar refugia. Pielou (1996) notes that "The most important tracts of 'extra' land, that is, land that was dry at glacial maximum but is now submerged beneath the sea, were at fairly low latitudes, from Newfoundland southward. They are now part of the wide continental shelf off eastern North America, but were then part of the coastal plain.

"South of Cape Cod the exposed coastal plain formed an uninterrupted strip; north of Cape Cod it was in the form of big islands. The cause of the contrast is that the northern part of the continental shelf (as it is now) was gouged and scoured by the ice sheets of earlier glaciations. Only the high ground between the glacial valleys was above sea level at the glacial maximum of 18k BP ... Now that these one-time islands are submerged, they form the undersea 'banks' so well known to east coast fisherman: Georges Bank, Emerald Bank, Banquereau Bank, the Grand Banks of Newfoundland, and many others (Pielou 1996)."

In addition to the coastal plain, a variety of glacial refugia existed on and near the east coast. This coastal plain and various refugia plain were home to both mastodon and mammoth as well as a large proportion of plants and animals now living in the Atlantic Provinces of Canada and New England. The farthest north refugium was probably located around the mouth of Hamilton Inlet and probably existed as early as 21,000 years ago.

Glacial conditions in South America were substantially different than in North America. Dillehay (2000) notes that there were "...a sequence of six different glacial advances and retreats in the central cordillera of Columbia. The cold periods occurred before 48,000 B.P., between 48,000 and 33,000 B.P., between 28,000 and 21,000 B.P., and from 16,000 to 14,000, 13,000 to 12,400, and 11,000 to 10,000 B.P. Glaciers in this region apparently reached their greatest extent when the climate was cold and moist, but shrank considerably after 21,000 B.P. because of greatly reduced rainfall. The dryness and cold lowered the upper forest limit to perhaps 1,800 meters, in contrast to its modern altitude of 4,000 meters.

"Slightly to the east in Venezuela, ...the tropical lowlands, now covered by dense forest, were arid or semi-arid during the Late Glacial Period. After 13,000 B.P., the climate of eastern Venezuela fluctuated between colder and warmer periods, while the adjacent tropical lowlands grew warmer and drier. Between 12,500 and 11,400 B.P., there is evidence of a brief cold climate, indicated by a lowering of snow lines in the northern Andes. Pollen and glacial records suggest that just as this cold spell was ending in Venezuela, the regions to the south in Ecuador and northern Peru were cooling down.

"In Ecuador [there were] ... two late glacial advances in the eastern cordillera. These advances were similar in age to those in northern Peru, suggesting that a warming trend affected a wide area of the northern and central Andes south of the Equator, in the regions between 4,000 and 4,500 meters. Radiocarbon dating of organic material associated with the first advance indicates that a large ice cap began to shrink about 13,200 B.P. and that the area probably became ice free before 11,900 B.P. A smaller ice cap formed again, starting around 11,000 B.P., and advanced over the paramo vegetation of Ecuador before disappearing by 10,000 B.P.

"Many high, subtropical desert basins in the Andes of southern Peru, northern Chile, and western Bolivia, whose floors consist of dried, salt-encrusted playas, contained freshwater lakes in late glacial times. The nature of the Ice Age climate in these regions of the Andes has generated considerable debate among geologists. Whereas some maintain that expansion of the lakes came from an increase in rainfall, others believe that it reflects a decrease in rainfall accompanied by a decrease in evaporation that resulted from lower temperatures. Regardless of the specific causes, the lakes represent important environmental changes.

"Sediment records from high altitude lakes at 5,000 meters in Peru and Chile also show glaciers retreating rapidly between 10,700 and 9,700 B.P. In Bolivia, the records from Lake Titicaca are the most informative. The water level of the lake reached its maximum between 12,000 and 9,500 B.P. Deglaciation produced large quantities of melt-water, most of which quickly found its way to the ocean. During full-glacial times, glaciers carried heavy loads of sediment to their margins, creating moraines. These sediments were then reworked and deposited as outwash by swollen braided rivers.

"Retreating glaciers, in contrast, formed terraces, as streams cut downward into existing landscapes. Meltwater at the margins of the retreating ice was trapped by ice dams, which then gave way, producing enormous floods. The best evidence of these today can be found along the western slopes of the Andes in Peru and, especially, Chile, where impounded meltwater formed glacial lakes. When these lakes were suddenly released, they washed away surface sediments along with any archaeological remains that were in their path.

"In present-day fjords of southern Chile, the ice sheet extended beyond the present coastline, in many places all the way to the full-glacial shoreline. Relatively few stretches of ice-free coast are believed to have existed along the extreme southwest shore; nevertheless, unglaciated refuges may have been scattered between broad ice sheets, permitting people to move along the shoreline. Along most of the coast, the exact form of the glacial margin is not known, for the data lie largely submerged on the continental shelf. The massive glaciers of southern Chile, some fronting the ocean today, would have presented formidable barriers to people's movement.

"Even further south, in the far southern Andes of Chile, a late glacial advance ended by 14,000 to 13,000 B.P. There also is evidence of a glacial readvance in Peru and southern Chile at 15,000 B.P., a warmer climate between 13,000 and 12,500 B.P., and another glacial readvance between 12,500 and 10,000 B.P. In southern Patagonia, there is evidence of a rapid withdrawal of glaciers around 16,500 B.P. The Patagonia ice fields had withdrawn by 13,000 B.P, when they reached their present-day borders."

***

Assume for the moment that a migration through the ice-free corridor did indeed happen. What were some of the constraints on this population and how fast did they cover the ground between Beringia and the Great Plains and, in about a 1,000 year period, expand throughout North and South America? Was this colonization likely to occur given the time constraints imposed on it?

Meltzer (2002) identified two components critical to the colonization process - demography and landscape learning. He examined the colonization process in terms of prey selection, habitat use, niche expansion, and patch choice. He first considers the relatively straightforward migration of the Paleo-Eskimo groups followed by the Thule.

"The Paleo-Eskimo groups (badly misnamed: As McGee notes there is nothing to demonstrate they are ancestral Eskimo, or even spoke an Eskimo language), moved across the high Arctic from Alaska to Greenland, in just a very few centuries around 4700 - 4500 years ago. Their origins are uncertain, but they carried tool assemblages and left behind sites that have affinities to those of the Siberian Neolithic. These were groups who exploited seasonally freezing coasts and especially the adjacent interior highlands of the Arctic. They traveled far, and fast, but left little behind: the sites are few, widely spaced (implying vast stretches of ground were left unoccupied and unsettled), and sporadically occupied by what must have been small populations. How they pulled themselves through this territory is uncertain, given their lack of time to accumulate knowledge of its complex environments and resources. McGee raises the tantalizing hypothesis that they exploited a keystone species - musk ox - which had evolved the unfortunate defensive strategy of responding to threats of predators by bunching in a circle. This may have proven effective against wolves (their major predator until humans came onto the scene), but it would have enabled humans to come in for dangerously close shots.

"The descendants of these groups ultimately disappear from archeological site, replaced on the landscape by the Thule (Who *does* appear to be ancestral Eskimo). Like the Paleo-Eskimo, the Thule moved quickly across the High Arctic, but later. Around 900 A.D. [The Little Climatic Optimum 950 AD – 1250 AD] they exploded out of western Alaska, moved eastward across the Arctic Coast, and arrived in Greenland within just a few centuries. Like Paleo-Eskimo, they were tracking a keystone species, primarily bowhead whales, a species they'd hunted for millennia, mostly around the Bering and Beaufort Seas. There they may have stayed, save for a warming trend around this time that melted summer pack ice across the high Arctic, enabling bowhead whales to expand east through now-unfrozen seas. The Thule hunters went along for the ride, metaphorically speaking, leaving behind a tool kit remarkably similar from Alaska to Greenland (Meltzer 2002)."

"The Thule culture, ancestral to recent Inuit or Eskimos, developed on the west coast of Alaska and spread by a swift coastal migration from northern Alaska to Greenland and Labrador. Thule people traversed this distance of ca. 1,500 to 2,000 miles in less than 150 years, between A.D. 900 and 1050 [10 to 13 miles per year]. In the process, they replaced a long-resident precursor population of the Dorset culture, which had occupied the Arctic coast for 1,500 years. Whether there was face-to-face competitive contact between Dorset and Thule, or whether Dorset collapsed suddenly due to environmental change and Thule moved into the vacuum, remains uncertain (Fiedel 2005)."

"Yet, unlike the Paleo-Eskimo groups who had an easily exploited keystone prey in a relatively stable landscape, or the Thule, who were radiating outward in a familiar, homogenous, and expanding niche, Clovis groups had no such advantages. They were pioneering a trackless, increasingly exotic, and highly diverse landscape. Not only did the landscape vary across space, it *varied through time*, as longstanding biotic communities dissolved, and plants and animals responded to the climatic changes that accompanied the end of the Ice Age. How those changes took place, and whether rapid enough to trigger drastic and unpredictable fluctuations in prey populations on a temporal scale (annual or decadal) that would have been detectable to and directly impacted human foragers, we cannot yet say. Keep in mind that human life spans are (even today) less than a century - during Clovis times they were undoubtedly shorter. Biological or ecological processes that played out over several centuries may not have dramatically impacted human foragers, though certainly groups at the latter end of those Late Pleistocene changes faced a different world than those who first came to the land.

"Clovis groups also expanded across a much larger area than these Arctic groups - and just about everyone else we know of archaeologically. And unlike the landscape crossed by those fast-moving Arctic peoples, where the sparseness of resources made it advantageous to move quickly, Late Pleistocene North America was a very different place. Once groups were south of the ice sheets and beyond the range of near ice, periglacial conditions, they were in complex temperate and sometimes resource-rich habitats. The speed and frequency and distance that hunter-gathers move daily, seasonally, or annually, is a function of many factors, not least of which is the spatial/temporal availability and abundance and distribution of critical resources (food, as well as water and stone), the density and return rates of preferred prey populations (and how quickly those changed under predation), the structure and relative richness and patchiness of the resources on the landscape, and so on. Unfortunately, we cannot measure these variables for the Late Pleistocene of North America; what we can say, however, is that this was decidedly not the high Arctic, and need not and may not have been a setting in which rapid movement was necessary or even advantageous.

"A continent the size of North America, if colonized rapidly, must have stretched Clovis populations very thinly across the landscape. There are *demographic* costs to moving that far and fast, for it almost certainly involved moving away from other people whether kin or unrelated groups. There can be little doubt population density was lower at this time than at any subsequent point in American prehistory. And yet in the face of that, colonizers had to maintain a 'critical mass' of population and an accessible source of potential mates - in order to avoid inbreeding or, worse, extinction. These demographic demands would have been more or less severe depending on the local group's size, growth rates, age and sex composition, degree of environmental uncertainty (and the inverse: knowledge of the landscape's resources), as well as how rapidly it was moving from its geographic homeland and/or other groups and the local environmental constraints on group size and population densities. On a diverse and unfamiliar landscape, the risk of extinctions is greatest soon after dispersal, when population numbers and growth rates tend to be low.

"At the same time, groups moving into increasingly unfamiliar habitats had to *learn the landscape*, and identify and/or locate vital resources, since over the course of time in residence in an area the preferred or high-ranked resources in that area would decline due to foraging pressure. They had to learn what else would feed them, and what would clothe them, cure them, or kill them. They had to learn when and where and what other resources were available. Each new habitat colonizers entered would, for a time, have been *unpredictable* for the simple reason that the availability and abundance of plant and animal resources varies.

"With fixed, permanent and predictable resources - the outcrops of stone suitable for making tools - landscape learning could be rapid. Learning about more ephemeral resources in unfamiliar settings - plants and animals - took longer, requiring as it did that resources be identified, and their properties or behaviors, habitat, location, short and long term patterns in abundance and distribution over time and space be learned, which in some cases might have required substantial observation or experimentation. Just how much of this behavior went on, no one knows, but bear in mind there are over 30,000 species and varieties of plants in North America *alone*. It would have been all but impossible for each new generation to sample anew their properties. Once observation and experimentation was complete, there must have been a considerable selective advantage to maintaining knowledge of a plant's properties - as the similarity of medicinal plants across different areas attests. But still knowledge had to be gained anew, as groups moved progressively further south, and into areas with unfamiliar and exotic floras.

"Would that suggest groups may have then initially focused on big game, as a class of prey that would have enabled easy crossing of ecological boundaries or into unknown territories (much as musk-ox possibly enabled groups to penetrate the high Arctic)? Perhaps, but keep in mind not all animals, let alone all elephants (mammoths and mastodon, in our North American Pleistocene case) are alike in their behavior. Foragers harvesting different and perhaps unfamiliar species in the same size class (large game, in this example), might have been able to bring generalized weaponry, hunting skills, and tactics to bear on all such prey, but how those different species may have responded would have varied considerably by species, ... Even within a species, different animals may behave very differently across the landscape and at different times. Unless the environment is extraordinarily rich and animal encounter rates were very high, locating a specific prey species required knowledge of the *spatial and temporal and landscape behavior*, which varies by the animals' age, sex, group size and composition, the season, weather, time of day, competitive relationships, available water, physiological stage (breeding, pregnancy, lactation), species composition and heterogeneity of the vegetation, topography, and exposure to predation, among other factors. All of which, or course, presupposes some knowledge of the landscape.

"Obviously, one's knowledge of the environment and ability to predict resource availability and abundance increases with the time spent observing resources, and their cycles and patterns on the landscape. Colonizers new to a habitat haven't the knowledge base - and cannot ask the locals since there aren't any locals to ask - to successfully anticipate much beyond a seasonal or annual basis when and where resources would be available. Hence, patterns of hunting and gathering in an initial foray into a new landscape and habitat may have been more experimental and less systematic, more encounter-based and less intensive, than such patterns seen among groups after settling into an area.

"In this regard, we see why the ethnographic record of hunter-gathers is not an especially useful analogue for colonizers of empty continents. Virtually all the historically known hunter-gathers have *neighbors*. Neighbors, who are not only convenient sources of mates, but also handy sources of information about landscapes and resources and who, in some cases, are also competitors, which can fundamentally alter the calculus of landscape use. Our models of hunter-gather foraging assume a forager has relatively complete information about resource distribution and yields, by which decisions can be made about how long to stay and when to leave a patch. They acquire the information by constantly monitoring the landscape and, because one group can only cover so much ground, by talking to their neighbors and tapping their knowledge about resources available in other areas. Clearly, this was not an option for Clovis groups (or, for that matter any new colonizer).

"It has been argued, and I [Meltzer] think correctly, that the greatest effort in information acquisition should occur in patchy and unpredictable environments - the very ecological situation in which Clovis groups initially found themselves. Indeed, there are reasons to think that on an unknown landscape, natural selection *would favor rapid and extensive exploration* to see what's over the next hill. Under the circumstances, one would *expect* rapid dispersal.

"Demography and landscape learning are tightly linked, for several reasons, not least that the decision to stay in a patch or move onto the next is in part based on the suitability of a new patch relative to the current one, factoring in the costs of moving. Those foragers who can better calculate those costs (by having more knowledge) will increase their odds of survival. By gaining information about a landscape one reduces mortality, and thus can increase population growth rates.

"Arguably, then, the colonization process on a new landscape involved a trade-off between a series of competing demands, including: *maintaining resource returns*, or keeping food on the table, particularly as preferred or high-ranked resources declined, and in the face of limited knowledge of the landscape; *minimizing group size* in order to buffer environmental uncertainty or risk on an unknown landscape; *maximizing mobility*, in order to learn as much as possible, as quickly as possible about the landscape and its resources (in order to reduce environmental uncertainty in space and time), while, *staying as long as possible in resource-rich habitats*, in order to enhance knowledge of specific changes in resource abundance and distribution (you learn more by staying longer); and, finally and perhaps critically, *maintaining contact between dispersed groups*, in order to sustain information flow, social relations and, most especially, demographic viability.

"Under the circumstances, we expect to see among colonizers *large scale exploration*, in order to map the landscape; periodic *aggregations* of widely dispersed groups in order to exchange mates, resources, and information; and *extensive mating networks*, in which spouses can be drawn from distant groups. And central to making all this work are two processes that might actually be detectable archaeologically: *high settlement mobility* to maintain contacts with distant groups, and monitor resources and environmental conditions beyond the social and geographic boundaries of the local group; and *open social networks* which enable individuals to move easily between and be readily integrated within distant groups. Under the circumstances, strongly territorial behavior would be decidedly disadvantageous (Meltzer 2002)."

Our Clovis-first scenario proposed that Clovis people became hunters of mammoths once they reached the Great Plains. This seems to fair well with the archaeological evidence. Dixon (1999) notes "There is some, albeit equivocal, evidence for mammoth predation by humans in Beringia from Berelyoth, Broken Mammoth, Mead, and Swan Point sites, and the analysis of residue from some stone projectile points. However, mammoth hunting does not appear to be the primary, or possibly even an important, economic focus in Beringia. It has been suggested that in some cases mammoth remains may have been scavenged more than a thousand years after the animals have died. Perhaps some enigmatic sites south of the ice sheet, such as Lamb Spring, Dutton, Selby, and La Sena, may be ephemeral examples of a foraging tradition that incorporated scavenging mammoth bone from non-cultural bone accumulations or individual skeletons.

"There is little evidence to support the traditional paradigm of mammoth hunters expanding from the Asian steppe into Beringia and southward through what is now interior Canada into more southern regions of the Americas. It is equally plausible that New World proboscidean hunting represents an independent invention historically not related to Old World mammoth hunting (Dixon 1999)."

"So why did Clovis groups move so far and so fast - assuming they did? I [Meltzer] don't think it was because they were chasing mammoths - or because they were being chased by mammoths - or because the extinctions of these animals left Clovis groups with little to eat, so they had to press on. Rather, I think it is because they found themselves on a vast empty landscape that was unknown and unpredictable, in which there was a selective advantage to widespread movement. During the critical early centuries of their dispersal, they still maintained vital long-distance links to other groups that could provide mates, resources, and information. As their knowledge of the landscape and its resources became ever-more detailed, their foraging systems became more stable. This, in turn, increased their population size, lessened their odds of going extinct, and enabled them to extend their overall settlement range, and venture further from one another. In the end, Clovis and related groups were, with little doubt, highly successful colonizers. Within a thousand years, much of the North American continent was populated with their descendants (Meltzer 2002)."

However, Adovasio (2002) looking at the date of 12,500 B.P. for Monte Verde in Chile is a little more expressive. "This meant that serious archeologists had to swallow the idea of proliferating bands of hunters and their families trotting south ten or twenty miles a day, charging through deserts and tangled forests, pausing for fleeting moments to mow down the big animals that got in the way, bolting down hundreds of pounds of meat, breast-feeding babies on the run, leapfrogging whole regions, desperately seeking ... what? Some mystical, mythical destination that they knew lay south?" Maybe a bit overstated, but it gets the point home.

Given this discussion by Meltzer (2002), were the Clovis people the same people who expanded down the ice-free corridor? If it took 150-200 years for Paleo-Eskimos and the later Thule people to migrate from Alaska to Greenland, in a relatively simple though harsh environment compared to the ice-free corridor, would you expect that pre-Clovis people could have migrated from Alaska to the Great Plains in 6-12 years? This seems highly unlikely! At best they could not have moved faster than the Paleo-Eskimos or the Thule peoples. They would have required 200 plus years to make this journey. This means they couldn't have reached the Great Plains until about 11,400 years ago at the earliest!

***

Does linguistics provide any clues on the origin of the pre-Clovis people, or for that matter, any other founding population? Was there a single migration into the Americas and does the total diversity we see in the Native American populations rely on population isolation, genetic drift, natural selection, or other factors? Or were there multiple migrations that account for this diversity?

There are various estimates on the number of the language families in the Americas. "Current linguistic consensus allows for lumping of languages into no fewer than 21 families, at least 3 of which may require further splitting [Hokan may be 6 separate families and 5 isolates; Penutian may be only 4 isolates; Oregon Penutian may be only 4 languages]. There are 32 language isolates that defy classification into any of the 21 families (Coe et al. 1989)." Nichols, (2002), states that there are about 6,000 languages on Earth falling into about 300 separate genealogical families with a wide range of structural types. About 150 of these families, and the full range of structural types, are native to the Americas. Greenberg's (1987) analysis provides us with only 3 language families. So, we are not off to a good start linguistically speaking!

A key feature of the American languages that allows linguists to separate them from all other languages is the *n: m* pronoun system. Nichols (2002) considers the *n: m* pronoun system the result of a single unknown event. She notes that "...the first *n: m* system must have arisen in a language or language population on its way into the New World, and the spread of the system by diffusion and inheritance took place mostly in the New World." The *n: m* pronoun system is an invention of the early founders within the Americas. Ruhlen (1994) notes that "...the distinctive American pronomial system na/ma 'I/you', is in fact almost as old as the recognition of the Eurasiatic system." He further states that the widespread distribution "...could only be explained genetically, that is, by assuming that the numerous n - 'I' and m - 'you' pronouns found over and over in Native American languages must be modern traces of a single ancestral language. It is with this *n: m* pronoun system and a wealth of other cognate words that Greenberg (1987) was able to use to sort out the three main language families in the Americas - Amerind, Na-Dene, and Eskimo-Aleut. (Figure 6-1).

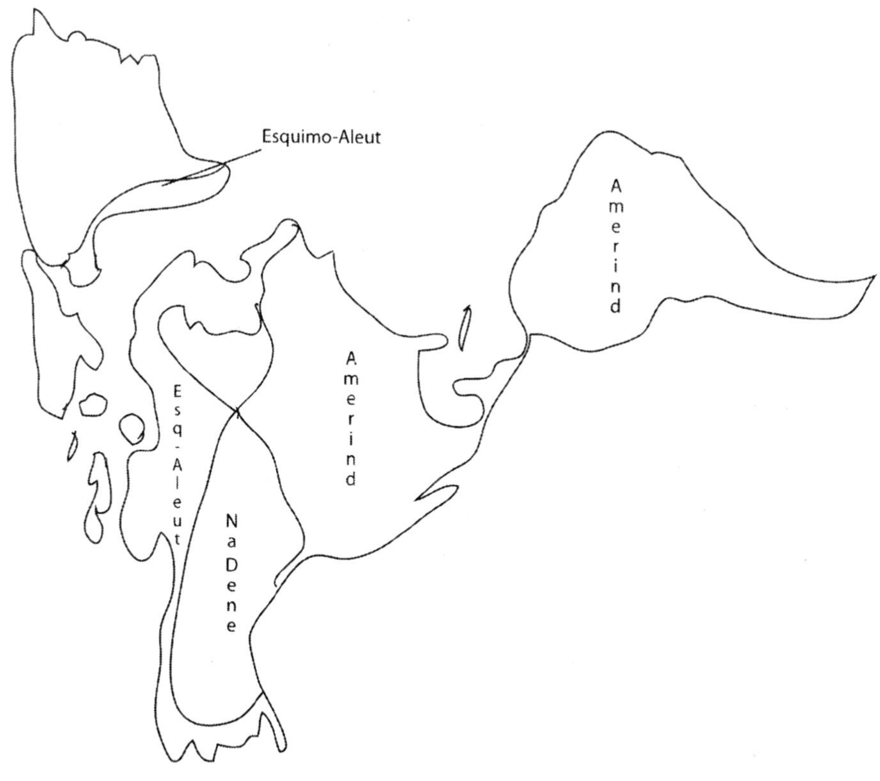

Figure 6-1. Distribution of the three main language families in the Americas

Greenberg proposed that these three language families were derived from three separate migrations from Asia: Eskimo-Aleut is a group of 10 languages spoken by Eskimos and the inhabitants of the Aleutian in Alaska, the Na-Dene is a family of 32 languages spoken by tribes in western Canada and by the Apache and Navajo in the southwest US, Amerind is a family of 583 languages spoken in both North and South America. "The broad distribution of Amerind languages and its considerable internal diversity both suggest that the first migration brought people speaking Proto-Amerind to the Americas. These people apparently spread very quickly throughout North and South America, which were at the time unoccupied by humans. The second migration, which brought the ancestors of the Na-Dene first to a homeland on the southern coast of Alaska, later saw the Athabaskan expansion into the interior. The final migration brought the ancestors of the Eskimo-Aleut, first to Southwestern Alaska, and eventually across the entire northern perimeter of North America as far east as Greenland (Ruhlen 1994)." A later migration within the last millennium would extend the Na-Dene into the Southwest and create the Navajo and Apachean languages.

Ruhlen (1994) confirms that "Eskimo-Aleut is the easternmost member of the Eurasiatic family, and that Na-Dene is the easternmost member of the Dene-Caucasian family; hence their immediate Asian relatives are already known. But what about the Amerind? Who are its Asian relatives? This question is perhaps more difficult than the other two, but my [Ruhlen] preliminary conclusion, based on a comparison of all the relevant families of Asia, is that, among all Old World families, Amerind is closest to Eurasiatic. But whereas Eskimo-Aleut is a branch of the Eurasiatic family, Amerind is related to Eurasiatic as a whole, and thus has no special ties to Eskimo-Aleut or any other branch of Eurasiatic. At the same time, it must be admitted that Amerind also shares certain roots with the Dene-Caucasian ..."

Nichols (2002) noted that there were two distinct linguistic populations in the Americas. Because of similarities to populations in Asia and Australasia, she calls the younger population the Pacific Rim population. There is an older population distributed throughout the hemisphere and overlapping geographically with the Pacific Rim population. "Here the older population will be called the *eastern* population, since it is the only occupant of the eastern parts of both North and South America (though it does not occupy only the east but is also found in the west). Evidence that the Pacific Rim population is the younger one comes from its more restricted distribution, its localization in the vicinity of the entry point, its continuity with populations which in Asia or Australasia spread after glaciation, and its high structural profile with several distinguishing grammatical markers. Especially good evidence is the clumpy distribution of those distinguishing markers in the Pacific Rim population. Grammatical features such as verb-initial word order and numerical classifiers, which have no inherent linguistic affinity and are not linked outside of the Americas, show a statistically significant tendency to co-occur in the same languages in the American Pacific Rim population. This clumpiness suggests that accidental combinations of properties that happen to occur in the ancestral languages still co-occur in their descendants. With enough time, language change will dissociate some of these clumps.

"The older population has none of these characteristics: its range is the entire hemisphere, and it has no continuity with any nearby Asian or Australasian population, no affinity to any population resulting from a postglacial spread, no distinguishing markers not also found in the Pacific Rim languages, and no clumpiness of grammatical features." Nichols (2002) estimates that the time of entry for the Pacific Rim population ranges between 12,000 and 24,000 years ago, suggesting that entry occurred from late in the last glacial period to the beginning of the Holocene.

Nichols (2002) provides a good summary of the movement of the populations within the Americas after initial colonization. "After entering North America, the Pacific Rim population evidently spread coastally, eventually extending to southern South America and hugging the coast with almost no eastward spread. In the last few millennia, a north-to-south trajectory is the most common one for language families in North America on or near the coast. The coastal Athabaskan languages and most of the California-Plateau Penutian groups have all spread mostly southward from their points of origin. The Athabaskan family originated in the subarctic (southern Alaska to northern British Columbia...), and its Pacific Coast languages extend like beads on a string from Washington to northern California. The Penutian languages probably originated on the Columbia Plateau and spread, in a series of migrations, into Oregon and California. This history is accepted for the Wintun, Maidun, Miwok-Costanoan, and Yokutsan branches and is a plausible trajectory for the possibly Penutian languages of western Oregon regardless of whether they are actual genetic kin.

"If the pre-Pacific Rim entries were also coastal - and it seems increasingly unlikely that any interior corridor was open to immigration and inviting to immigration during glaciation - then it is likely that once in North America they too spread coastally southward. This would mean that they spread from inhabited to uninhabited lands and that they kept to the coastal ecologies with which they were most familiar. Systematic settlement of the interior must then have begun after the end of glaciation. The spread of the Clovis culture marks one of the first expansions into the interior, and perhaps the very first.

"Colonization of the interior is likely to have come from the south. One piece of evidence for northward spread into the interior is the fact that in historical and reconstructable times the language families of the mid-latitude interior of North America have mostly spread from south to north. Prominent examples are the Uto-Aztecan family, which spread from approximately northern Mexico to southern California and the Great Basin; the Siouan-Catawba family, which spread probably from the vicinity of the lower Mississippi to form enclaves in various parts of the southeastern U.S. and well north along the Missouri, Mississippi, and Ohio River valleys (the northernmost members are the Stoney and the Assiniboine, both spoken in southern Canada); and Caddoan, which probably also originated in the vicinity of the lower Mississippi and spread far north into the Great Plains (its northernmost language, Arikara, is spoken in North and South Dakota). (Of these, the Uto-Aztecan spread in the Great Basin was a spread of non-farming people; the others were spreads of agriculturists.) Even the Algonquian family, which spread primarily eastward probably from the vicinity of the Columbia Plateau, once east of the Rocky Mountains underwent considerable northward expansion so that in historical times it occupied much of central Canada. [I'll later challenge this eastward assumption for the Algonquian family]. If Leer is right about the west central Canadian origin of the Na-Dene family, then it's mostly interior Athabaskan branch has spread far north of its origin. The only important exception to the mostly northward drift in the interior is offered by Navajo and the Apachean languages spoken around the periphery of the pueblo area in the southwestern U.S; they are the result of southward migrations from southern Canada within the last millennium.

"These examples of northward drift in the interior are based on language-family spreads, and pertain only to the last several millennia. Uto-Aztecan and Siouan-Catawba, two old language families of the interior, are both of about Indo-European-like age, *i.e.*, about 6,000 years old at most. Movements in this same time frame do not directly testify to directions of movement shortly after the end of glaciation, and furthermore the largest northward spread, that of the Siouan language family, involved farming people. Still, it is more parsimonious to assume that this northward drift continues a long-standing tendency rather than a recent development. Then the standing northward drift among language families in interior North America may provide indirect evidence that humans claimed interior North America after glaciation in northward movements. The source of these movements have been in the vicinity of the Gulf of Mexico."

"There has been some west-to-east movement bringing representatives of coastal and near-coastal families into the interior. The Athabaskan-Eyak-Tlingit family originated near the coast or on the eastern slope of the coast range somewhere from southern Alaska to central British Columbia, and its Athabaskan branch spread mostly eastward to cover most of interior central and northern Canada and all of interior Alaska. The Salishan family originated on or near the coast in northern Washington to southern British Columbia, and its interior representatives extend to Idaho and Montana. The Algic family probably originated on or near the Columbia Plateau, and its descendants now comprise Yurok and Wiyot of coastal northern California and the geographically vast but temporally shallow Algonquian family, which extends from the western Great Plains (Plains Cree, Blackfoot, Cheyenne, Arapaho) to the eastern seaboard (Montognais, Micmac, Delaware, etc.). Thus the entire non-Pacific part of North America - the Great Basin, the intermontane area, the Great Plains, the eastern woodlands and seaboard - have received some of their language population from the far west, while the converse does not hold (Nichols 2002).".

So Nichols (2002) has Clovis descended from an early "eastern" language group that entered the Americas about 20,000 years ago and migrated from the south to the north from the vicinity of the Gulf of Mexico. The problem with her assumption here is that she uses historical migration patterns of agriculturists to infer a pre-historic population movement of pre-agricultural, hunter-gatherer peoples, the Clovis. There isn't any special reason why this should be so! It may be more convenient than true!

Nichols (2002) notes, with respect to language groups, "Diversity is highest at lower latitudes, along coastlines, and in simpler societies; diversity is low at high latitudes, in continental interiors in areas with dry or cold climate, and in complex societies." By far the highest diversity of language occurs along the Pacific Coast. "In North America, two thirds of the language stock are found along the western edge of the continent, between the coast and the coast range." The largest, most widely distributed language group in North America is the Algonquians. This language group is distributed from North Carolina in the south, northward along the Atlantic coast to the maritime provinces of Canada, and west and north to the Rocky Mountains with two small groups in coastal California.

A study of natural history terms shows that the area between Georgian Bay and Lake Ontario in Canada is the most probable homeland of the Proto-Algonquian language (Goddard 1978) – and not the Columbia Plateau postulated by Nichols (2002). The Algonquian language group is made up of 41 different languages. The diversity of languages decreases from east to west. Eastern Algonquian has 29 separate languages, Central Algonquian has 7 languages, Plains Algonquian has 3 languages, and there are 2 languages on the California coast. There is a separate group that occupied the island of Newfoundland that may also be related to the Algonquian language group known as the Beothuk. This distribution of Algonquian languages follows the thinking of Nichols (2002) on diversity distribution; however, it implies a migration from east to south as well as east to west. If we use Nichols reasoning that the migration of the current peoples reflects earlier migrations, we would have to assume that some much earlier group of people originated in the east, possibly on the Canadian coast, or inland along the St. Lawrence River, and moved across the country to the west and south.

Nichols (2002) ran a number of different scenarios using 10-12 entries beginning between 20,000 and 30,000 years ago, a single scenario at 50,000 years ago and, a few entries before the height of the last glaciation that may have yielded 150 language stocks. The results strongly indicated that human settlement of the Americas began over 20,000 years ago. She noted that "...the large number of separate language families attested in the Americas seems far too high for colonization models that regard the Clovis Culture or something in its time frame as the initial colonizer."

Linguistic studies imply multiple migrations to the Americas at a relatively early date – certainly pre-Clovis. The oldest migration may have its origin in the Eastern U.S. or Canada – Nichols (2002) *eastern* Population. However, Nichols believes this population entered on the west coast and moved north and east. This seems like a bit of a dichotomy! There probably was an early, as well as a later, migration along the Pacific Coast – which may have given rise to the Na-Dene - and a separate early migration along the Atlantic Coast. A fourth, and much later migration, involved the proto-Eskimos and later the Thule or Eskimo-Aleut already mentioned.

***

We can pose the same set of questions for genetics that we did for linguistics. Do we get the same answers, or do we just confound the questions even more?

Genetics research has provided an invaluable amount of information related to early Americans. Schurr (2005) provides a good overview of the genetics of New World origins. His review covers current state-of-the-art for classical protein polymorphisms, HLA genes,

Immunoglobin genes, Alu polymorphisms, Mitochondrial DNA, and Y-chromosome DNA. I'll briefly summarize his findings with regards to origins and distribution within the New World.

Looking first at classical protein polymorphisms, Schurr (2005) notes that current research shows that there is a somewhat greater heterogeneity among South Amerindians than North Amerindians with Central Amerindians being less diverse than either one of them. Central Amerindians were more similar to Northern Amerindians than South Amerindians. This research also indicated that the South American populations can be divided into three major groups: Tropical Forest, Paleo-American, and Andean language groups. There were also differences between groups living north and south of the Amazon River.

Based on protein polymorphisms the current thinking is that Amerindians originated about 19,000 years ago at which time they migrated to the Americas; Na-Dene and Bering Sea Mongoloids from which Eskimo-Aleut populations evolved, diverged around 15,000 years ago with Bering Sea Mongoloids emerging as a distinct entity around 10,200 years ago. In addition, Eskimo-Aleuts diverged about 9,000 years ago with Inupiaq- and Yupik-speaking Eskimos splitting around 5,000 years ago.

The current research on HLA genes shows genetic links between North-East Asians and Native Americans with differences among Native American groups (Eskimo-Aleut, Na-Dene, and Amerindians). Northwest Coast Indians and Na-Dene Indians are grouped together and outside of the Amerindian branch and South American Indian groups generally cluster together away from North and Central Amerindians.

Immunoglobin gene studies generally support the tripartate model - Amerindian, Na-Dene, and Eskimo-Aleut -with one modification. They break out Amerindians into two separate migrations - a

South Amerindian migration and a north and central Amerindian migration.

Alu polymorphism studies indicate an Asian origin for Native American populations with only a single migration followed by population isolation and genetic drift accounting for the differences among groups.

Mitochondrial DNA studies have identified five separate haplogroups in the Americas - A, B, C, D, and X. Haplogroup A appears at high frequencies in North and Central America but diminishes in frequency in South America. Haplogroups C and D occur in high frequencies in South America and diminish in frequency in North and Central America. Haplogroup B occurs at nearly the same frequency throughout the Americas. Haplogroup X is found only in North America. Most Siberian populations contain A, C and D only with B confined to Southern Siberia along the Mongolian and Chinese borders and East Asia. Haplogroup B might represent a separate migration to the Americas. Siberian and East Asian populations lack the X haplogroup. Haplogroup X is found only in Eurasia - Europe and western Asia. The dates of entry into the Americas for all five groups range from 20 kya to 35 kya. Siberian and Eskimo populations appear to be genetically similar to each other having high frequencies of haplogroup A, while Aleuts are different from Eskimos by having mostly haplogroup D. Both groups have low levels of haplogroup C suggesting that C represents a founding mtDNA in Eskimo-Aleut populations. The highest concentration of haplogroup X is associated with the Algonquian language group in the Northeast U.S. and Canada. The X haplogroup comprises about 25% of the mtDNA of the Algonquian speaking peoples. This percentage drops off as you move east to west.

Merriwether (2002) brings up a problem in sampling living populations of Native Americans. Current native populations are estimated to be about 12.5 million individuals - 2.5 million in North America and about 10 million in Central and South America. At the time of first contact with Europeans, native populations were estimated at 43-65 million individuals. By the 1500's an estimated reduction of about 89% occurred - due mainly to introduced diseases. "So we have mtDNA sampled a few tenths of a percent of the entire living Native American population, and less than a hundredth of a percent of the deceased Native population." We have no idea how representative this sample is of the actual ancestral population. The only way to get at the actual distribution of genetic material in the ancestral population is to sample ancient DNA.

From a Y-chromosome perspective, there are clear linkages between Siberia and the Americas. The known linkages are summarized by Wells (2007).

M168 – M89 – M9 (Haplogroup K)

"Most people native to the Northern Hemisphere trace their roots to the Eurasian Clan. Nearly all North Americans and East Asians are descended from this man, as are most Europeans (Wells 2007)."

M168 – M89 – M9 – M45 (Haplogroup P)

"The M45 central Asian Clan ... was the common ancestor of most Europeans and nearly all Native American men (Wells 2007)."

M168 – M89 – M9 – M45 – M242 (Haplogroup Q)

"M242 arose some 15,000 to 20,000 years ago, with a man born in the savagely cold climate of Siberia. His descendants became the first explorers of North America. Despite frigid temperatures, some of the Siberian Clan generally crossed Siberia's ice-free tundra to eastern Siberia. Once they reached the northeastern edge of Asia, they were ideally poised to enter a new world.

"There has been some debate about whether humans reached North or South America much earlier, prior to 20,000 years ago. However, the genetic data coincide with archeological evidence for a Beringia crossing and lend support to the theory that this migration occurred about 15,000 years ago.

"While some of the Siberian Clan remained in Asia (marker M242 can be found in India and China as well as Siberia), almost all Native Americans are descendants of these people (Wells 2007)."

M168 – M89 – M9 – M45 – M242 – M3 (Haplogroup Q3)

"Soon after the first explorers reached the Americas a new marker, M3, arose. This ancestor, born in North America 10,000 to 15,000 years ago, is the patriarch of the most wide-spread lineage in the Americas. Nearly all native South Americans and most native North Americans are descended from this line.

"While descendants of the Siberian Clan, the first wave of migrants to reach North America, can be found in both Asia and America, Q3 descendants are found only in America. Because of this, geneticists believe that their common ancestor was born after the Bering Strait land bridge was once again submerged around 10,000 to 15,000 years ago (Wells 2007).

So what are the implications about the work of the geneticists? I'll use mtDNA work as the basis for this discussion. As noted earlier, there are 5 haplogroups found in the Americas - A, B, C, D, and X. Haplogroups A, B, C, and D are found only in Asia while haplogroup X is found only in western Eurasia. Within Asia haplogroups A, C, and D are found in Siberia - B is absent from Siberia. Haplogroups A, B, C, and D are all found in Eastern Asia.

Haplogroups A, C and D from Siberia are the likely founders of the Bering Sea Mongoloids which later became the Eskimo-Aleuts. And, that the Bering Sea Mongoloids split from the Na-Dene (Athapaskan) about 15,000 years ago and that the Eskimo-Aleut split off from the Bering Sea Mongoloids about 10,200 years ago (based on protein polymorphism results), and that the Eskimo and Aleut populations split off from each other about 9,000 years ago. The Eskimos have a high frequency of haplogroup A and a low frequency of haplogroup C. The Aleuts have a high frequency of haplogroup D and a low frequency of haplogroup C. The Eskimo populations then split into the Inupiaq and Yupik speaking peoples about 5,000 years ago.

What about North, Central and South America? First, let's look at haplogroup X. As we noted, haplogroup X is found only in western Eurasia. In fact, its' highest frequency is found around the eastern Mediterranean Sea and the Black Sea - its frequency distribution drops off with migration into Europe. The other location for Haplogroup X is in North America. Though it is fairly widely distributed, its highest frequency occurs in eastern North America among the Algonquian speaking peoples. About 25 percent of the Eastern Algonquian speaking peoples carry haplogroup X. Frequencies drop off as you move west through the Central Algonquians to the Plains Algonquians, and others. Since haplogroup X is found only in western Eurasia and North America and not in Asia, it is likely that the North American origin for this haplogroup is in Europe.

That brings us to the oldest of the language families - the Amerindians. Haplogroups A, B, C, and D are found throughout the Americas. Haplogroup B is found throughout the Americas at about equal frequency indicating a separate founding group. Haplogroup A is found at high frequency in North and Central America but in low frequency in South America. Haplogroups C and D are found at high frequencies in South America but at low frequencies in North and Central America. These differences are confirmed by classical protein polymorphism and Hla gene studies. Protein polymorphism studies show that there is greater diversity among South American populations than among either North or Central American populations. Hla studies show that northwest coast Indians and Na-Dene group together and away from Amerindians and that South American Amerindians group away from North American Amerindians. Adovasio (2002) notes that "Given the presumed mutation rate of mtDNA, the A through D haplogroups could have begun arriving as early as 25,000 to 35,000 years ago and the X people between 15,000 and 30,000 years ago."

Classical protein polymorphism, HLA genes, Alu genes and mtDNA studies all indicate an Asian Origin for Amerindians. Y-chromosome studies indicate that the Amerindians origin is in Siberia and not in East Asia. Based on protein polymorphism studies, the Amerindians migrated out of Siberia and into the Americas about 19,000 years ago. They likely went first to East Asia where somewhere around the border with China and Mongolia they acquired the mutation found in haplogroup B. They may or may not have mixed with existing populations in East Asia.

There may have been two separate migrations out of Asia into the Americas. The first around 19,000 years ago carrying a relatively high frequency of haplogroups B, C, and D and a low frequency of haplogroup A into South America and a second, at a later date, carrying a relatively high frequency of haplogroups A and B and a low frequency of haplogroups C and D into North and Central America.

***

"The Clovis-first model predicted that sites older than Clovis would not be found. Now at least 50 sites found in North and South America that have pre-Clovis radiocarbon assays seriously challenges the Clovis-first model (Sanford et al. 2005)."

Dixon (1999) reviewed 26 sites in North America that may be older than 11,500 B.P. He determined that "It appears that some, if not all, of these locales may have been visited by humans prior to 11,500 B.P." He cited six sites that were most likely to claim an age in excess of 11,500 B.P.: Manis (12,030±200 to 13,400±190 B.P.), La Sena (18,000 to 19,000 B.P.), Schaefer (12,220±80 to 12,310±60), Heboir (12,480±60 to 12,520±50), Cactus Hill (15,070±70 to 15,670±730 B.P.), and Little Salt Springs (12,030±200 to 13,450±190 B.P.). He also reviewed sites in South America and verified that Monte Verde has an average age of about 12,500 B.P. Many others would add Meadowcroft to this list as well as a number of other sites in South America (Stanford et al. 2005). Adovasio notes that "... the earliest inhabitants of the rock shelter [Meadowcroft in Pennsylvania] ... proved to be wholly consistent with new information about life at the glaciers edge." And that "... people were in western Pennsylvania as early as sometime between 13,955 and 14,555 radiocarbon years ago." Dixon (1999) suggests that "There is a growing consensus among American archaeologists that the 12,500 B.P. occupation at Monte Verde provides a secure minimum limiting age for human occupation and that colonization must have occurred earlier."

"The abundance and diversity of complexes with pre-Clovis dates in South America contrasts with their rarity in North America, where their ancestors should exist. They include rockshelters in Colombia, Peru, Chile, Argentina, and eastern Brazil, butchering sites on the pampus of Argentina and Patagonia, riverside camps in Amazonia, shell middens along the southern Pacific coast, and quarries and camps along powerline transects in several parts of Brazilian Amazonia. Several general factors are likely to be responsible for their relative abundance. First, much of South America escaped major remodeling by glaciations, inundation, erosion, and other physical processes during the Pleistocene and early Holocene, which repeatedly destroyed early North American archaeological sites. Second, caves and rockshelters suitable for habitation, which are widespread in South America, were occupied seasonally or continuously prior to the adoption of agriculture and have intact stratigraphic records. Third, major regions attractive to Paleoamericans, especially the Argentine pampus, Patagonia, the *cerrado* of northeastern Brazil, and the rain forest of Amazonia have remained little disturbed by modern settlement, roads, and agriculture; consequently sites are preserved and accessible (Stanford et. al. 2005)."

"North American archaeologists have generally discounted the antiquity of South American sites with initial dates exceeding 12,000 RCYBP because South American technologies do not conform to the

North American stereotypes. Nevertheless, increasing numbers of early sites are well documented and their abundance, continental distribution, diversity of subsistence adaptations, and distinctive lithic complexes provide conclusive evidence for their pre-Clovis age (Stanford et al. 2005)." "People lived at Monte Verde in Chile around 12,500 years ago and they practiced a generalized hunting and gathering lifeway, not just big-game hunting (Dillehay 2000)." "In short, there is no doubt that South America was populated by at least 15,000 RCYBP (Stanford et al. 2005)."

***

What do the anatomical features of known ancient Americans and living Native Americans tell us about early colonization of the New World?

"Unlike the Old World, the New World lacks human remains that are anatomically similar to very early human forms such as *Homo erectus*, Neanderthal, or even archaic *Homo sapiens*. Human remains found thus far in the New World appear to be completely modern humans *Homo sapiens* [*sapiens*]. There is some evidence possibly contradicting this, including the inconclusive identification of a human supraorbital ridge from the Chapala Basin, Mexico, and a curious report with illustrations of an archaic human calvarium (skullcap) from Brazil that has subsequently disappeared.

"Since the early 1900s, physical anthropologists have recognized that the earliest human skeletons found in the Americas tended to have longer and narrower (dolichocephalic) crania. Later human remains in the Americas and current Native Americans tend to have broader (brachycephalic) heads more closely resembling Asian Mongoloid populations (Dixon 1999)."

Steele and Powell (2002) provide evidence of these separate populations from the fossil record. Their findings closely agree with those of Nichols (2002) for the Pacific Rim and eastern language groups. Their research documents the following:

"1. The earliest remains recovered from the Americas have consistently been found to have a craniofacial shape that is distinguishable from the more recent prehistoric and extant populations;

2. The North American Paleoindians resemble more closely the features of living and Late Prehistoric Southeast Asians than they do Northeast Asians;

3. The South American Paleoindian remains resemble more closely the North American Paleoindian craniofacial form than they do any other American Indian groups (living or dead);

4. The South American Paleoindians, while resembling North American Paleoindians, differ from them by more closely resembling Australian and African samples than do the North American Indian populations;

5. The Paleoindians from both American continents are structurally more similar to Archaic (mid-Holocene) populations than they are to Late Prehistoric and living American Indians; and

6. There is as much variation in early and mid-Holocene populations as there is in Late Prehistoric populations."

In summary, Steele and Powell (2002) note that "...the oldest human skeletal remains found in North America share broad similarities with northern Asians, Southeast Asians, Pacific Islanders, and recent Native Americans. Among these populations North America's earliest remains resemble more closely South East Asians and some Pacific Island populations than they do northern Asians and Native Americans. The modest but consistently documented differences between early North American remains and recent North American Indians has challenged our traditional view that the Americas were colonized by Northeast Asians in a time too recent for evolutionary forces to have altered the features of their Native American descendants. While the differences between the earliest and more recent skeletal remains have become well documented, identifying the evolutionary force or forces which created these differences is more equivocal." Gill (2005) working with fossils in the Northwestern Great Plains suggests that "...the migration of new people into the Northwestern Plains and the changing natural-selection forces brought about by the new life-style they introduced, led to the genetic/morphological changes we have documented through our osteology analysis. Had it been migration alone with total replacement, the transition from the Archaic to the late prehistoric would have been even more abrupt, with no sharing of traits with late-Archaic individuals. Yet these shared traits do exist. Had the process been drift or selection without the entry of new people with additional Mongoloid traits, the process could not have moved as swiftly as it did."

"Early dates on human skeletal remains also support the existence of a pre-Clovis occupation of South America. Several dozen crania dating prior to 8,000 RCYBP encountered in rock shelters in Minas Gervais share with Kennewick Man a long narrow skull, low narrow face, subnasal prognathism, and low wide orbits and noses. Statistical comparison of four female and five male crania from Cerca Grande in the Lagoa Santa region employing principal components analysis and multidimensional scaling verified the strong relationship with Africans, Australians, and Polynesians obtained on the earliest North American individuals (Stanford et. al. 2005)."

Dental work by Turner and others suggest that an origin for early Americans is in Asia. This confirms the work of the geneticists for haplogroups A, B, C, and D but not for the haplogroup X. We'll find that most authors either bitterly object to haplogroup X having any part in early origins or they conveniently ignore the group altogether. Turner (2002) "...proposed that East Asia contained two geographic variants of the 'Mongoloid dental complex' ... These were a Southeast Asian pattern of retained and simplified features I [Turner] called Sundadonty, and one in Northeast Asia the morphological pattern was more specialized and complex. This latter pattern I call Sinodonty in pre-historic New World teeth." Turner notes that "... the dental relationship in the Americas lies with peoples of the American Arctic, Northeast Asia, and Neolithic Southeast Asia (a result of migration southward of northerly agricultural Sinodonts). There are no close relationships with modern Southeast Asia (Thailand), Europe, Oceania, or Africa. Like many other studies of teeth and crania, Australians and Africans are closely linked but show no relationship to Native Americans. Hence, on the basis of dental morphology comparisons, the homeland of Native Americans continues to be best viewed as having been in Northeast Asia. To test this hypothesis requires demonstrating that only Northeast Asia was the Pleistocene homeland of the Sinodont dental pattern. To date, Sinodonty has not been found in any non-Asian skeletal assemblage regardless of geological time." He also notes that "... the ancestors of Native Americans were Northeast Asian Sinodonts, who, in turn, had evolved from Sundadont or Proto-Sundadont stock that earlier began in Southeast Asia or Australia."

Turner summarizes this discussion noting that the "...ultimate regional homeland can reasonably be hypothesized as having been in north China, Mongolia, and Southern Siberia, a rich ecologically diverse region occupied long before humans set foot further north. An easy to remember focal point of this region is Lake Baikal, near the headwaters of the Anadar, Ob, Yenisei, Lena, and Amur rivers that together drain much of central and eastern Siberia." This agrees very well with the assumptions provided in the discussion on genetics.

Turner notes that "Sinodont variation in the New World, both past and present, seems to group into three divisions, with some widely separated North and South samples showing random convergence - useful evidence for thinking that genetic drift is an important factor in dental microevolution. The three New World dental divisions correspond rather well with the three linguistic families defined by Greenberg." Regarding the Na-Dene he qualifies this thinking by adding that "...some analysis suggests possible admixture event(s) responsible for the formation of the Na-Dene ("Greater Northwest Coast") dental group."

Kennewick Man, found near Kennewick, Washington, may represent an individual that differs from Turners Sinodonts. Dixon (1999) notes that "Dental characteristics fit Turners ... Sundadont pattern rather than the Sinodont pattern indicative of later North Americans. His teeth showed little wear and were all present at the time of death." Kennewick Man is dated at about 9,300 to 9,550 B.P.

"The most controversial proposal for the peopling of the New World comes from Dennis Stanford, who along with stone tool-making expert, Bruce Bradley, envisions a Late Pleistocene migration reaching the New World from Spain by traveling along the southern edge of the north Atlantic pack ice. Their north Atlantic pack ice route is grounded in their belief that a 'genetic' relationship exists between American Clovis and Iberian Solutrean points, as well as the apparent absence of this type of point manufacturing in Asia during Late Pleistocene [remember haplogroup X denial I mentioned earlier]. All known European Late Pleistocene dental evidence is unsupported of this idea, although, admittedly, I know of no dental studies on whatever Solutrean skeletal remains may have been recovered in Spain (Turner 2002)."

"It has long been argued that Clovis represents a single culture produced by a single biological population. The signature calling card of this complex is the Clovis toolkit. However, this proposed kit is not known in its entirety from a single site. Rather, it is a composite construct based on tools associated with Clovis points from multiple sites over the entire continent.

"The second and related argument is that there are different regional variants of Clovis. These include types such as East Wenatchee, Washington and Debert, Nova Scotia ...[t]he variability in point styles and toolkits has led to the suggestion that Clovis was not a culture, but represents the diffusion of a technological complex across populations already in the Americas during a time of severe environmental stress (Stanford et al. 2005)."

"The Clovis-first model predicts that the Clovis homeland is in Siberia, but after more than 40 years of research, no evidence has been found in Siberia to support the Siberian homeland proposition. The model also predicts that fluted points should be younger in South America, but fishtail points in Argentina and Chile are approximately the same age as those in North America (Stanford et. al. 2005)." "The oldest dated evidence from the Siberian side comes from the 11,000-year-old level at the Ushki site on the Kamchatka Peninsula, which is younger than many of the Nenana sites in Alaska. However, the widespread occurrence of microblade industries in Japan and in eastern Siberia in the Lena and Aldan river regions suggests strong ties in the far north between Asia and the eastern Beringia during late-Glacial times (Stanford et al. 2005)."

Dixon (1999) notes that "Three prehistoric cultural traditions have been documented in eastern Beringia that are important to understanding the colonization of the new World. These traditions are the Nenana complex, the American Palearctic tradition, and the Northern Paleoindian tradition. The earliest human occupation of eastern Beringia is ascribed to the Nenana complex, which is characterized by distinctive triangular bifacial projectile points and tapering ovate bifaces, probably knives. The Nenana complex persisted until circa 10,500 – 10,000 B.P. at which time it becomes difficult to distinguish it from the Northern Paleoindian tradition in interior Alaska. This suggests a possible blending of technological traits of the Nenana complex and the Northern Paleoindian tradition. This postulates that although both may have shared a common technological origin, the Northern Paleoindian tradition is derived from a northward movement sometime between 11,000 and 10,500 B.P. of the Paleoindian tradition from the Great Plains. This movement was probably associated with colonization of recently deglaciated terrain in central Canada. Distinctive microblade technologies are introduced into eastern Beringia sometime around 10,500 B.P. The microblade technologies are contemporaneous with both the Nenana complex and the Northern Paleoindian tradition. The Denali complex represents an inland adaptation by microblade using people. Because the American Paleoarctic and Northwest Coast Microblade traditions are found in ecologically very different coastal areas, they are recognized as distinct co-traditions."

"South America is very different archaeologically from North America because no single culture dominated the continent the way the Clovis culture, with its representative spearpoints, dominated North America. The first South Americans were not Clovis clones. In South America, the earliest technologies consisted of different kinds of stone tools, including a wide variety of spearpoints, unifacial tools made of flakes, and sling stones. Furthermore, many areas in South America witnessed the development of broad hunter and gather diets before 11,000 B.P., a pattern usually thought to be associated with later Holocene or early Archaic cultures dated after 10,000 B.P. And big-game hunting was simply one of many different economic practices; it never achieved the importance it did in North America (Dillehay 2000)."

"All of the South American sites ... testify to varied patterns of technology and subsistence in different environments, including big-game hunting in grassy plains and savannas, generalized foraging in forests and parklands between at least 12,500 and 10,500 B.P., and active plant manipulation in some areas of South America by 10,000 B.P. (suggested by the presence of domesticates possibly as early as 9,000 to 8,000 B.P.) (Dillehay 2000)."

"Furthermore, despite some rough similarities, the South American biface industry generally has few striking relationships to its northern counterparts. The distinction between the two continents is profound and complex. In North America, only bifacial technologies are seen as diagnostic of particular cultures and periods. Those technologies appear in clear stylistic and technological sequences that can be followed from the beginning of the Paleoindian period around 11,200 B.P. and from one elongated projectile point type to another (for example Folsom, Plainview, Dalton, Cumberland). These common stratigraphic sequences suggest an obvious technological and cultural progression. As a result, North American sites are dated and classified solely from their bifacial tools; unifaces are rarely diagnostic of anything. In South America by contrast, there are few linking traits to indicate technological evolution, even when there is a stratigraphic sequence of diagnostic projectile points. Where these points occur, they generally are regional types and appear with very low frequency, occupying 2 percent or less of the total stone tool collection in a site. In sum, the geographic distribution of these points in the southern hemisphere together with the presence of other bifaces such as knives and drills and of unifaces suggests complicated mosaics of technological and subsistence practices in which both bifacial and unifacial types occur independently (Dillehay 2000)." Dillehay believes that the unifacial tradition was invented in South America and that the bifacial tradition was imported from North America. In a comparative study of the Fishtail and Clovis points Dillehay noted that "...the two types represent distinct technologies and inventions and should not be equated with either a single technology or a single human migration throughout the New World."

"There are four primary reasons for the great regional variation of late Pleistocene archaeology in South America. The first reason concerns the relationship between ecology and subsistence. For example, given the range of available species between the high Andes and the lowland savannas, parklands, tropical forests, and deserts, the potential for technological variation is enormous. Thus we should expect great variation in specialized tools geared to specific local resources. Second, as the entry populations dispersed and increased and as regional variation accelerated, there was more and more likelihood of local cultural exchange from group to group. These borrowings would have brought changes in artifact form and in the inventory itself, as ideas diffused and brought about more change. Third, as people entered and dispersed throughout the continent, they settled into rich temporary environments and immediately adopted the generalized economy and technology of a proto-Archaic lifeway. The classic big-game hunting lifeway would have been restricted to specialized hunters in certain open environments. Finally, several archaeologists also would attribute some stone tool differences to the ethnic identities of the groups inhabiting different sites, although others would see the same variations simply as representing different activities taking place at different locations (Dillehay 2000)."

Adovasio (2002) notes that "... it was the technology – the idea of the fluted point – that traveled quickly, not the people. It now seems reasonable to suppose that Clovis points struck the fancy of people for some reason, perhaps religious, and the technique of making them spread across the country to people who were already there."

Meltzer (2002) noted that "...there are no obvious historical or technological links between the pre-Clovis age complexes of South America and the Clovis assemblages of North America. It looks as though Monte Verde and Clovis represent distinct archeological traditions and separate migratory pulses." In fact "There are more *Clovis* artifacts from here [Meadowcroft and Cactus Hill in the eastern U.S.] than any other area of North America, implying a relatively longer period of occupation – though whether it began earlier or lasted later we cannot say. Meadowcroft, Cactus Hill, and the great abundance of Clovis in eastern North America have led some to suggest that Clovis was invented here, and from whence it rapidly spread west."

Goodyear (2005) takes these eastern Clovis finds a little further. He reviewed four sites located in the eastern U.S. for pre-11,500 years ago occupation. He found "In each case, credible evidence of artifacts in deposits that pre-date that age had been recovered. Given the presence of bifacial projectiles at Meadowcroft and Cactus Hill and perhaps Saltville (SV-2) based on debitage [pieces flaked off as the point is being made], it is not difficult to see Clovis emerging from these technologies. Even the lithic raw materials at Meadowcroft are typically Paleoindian.

***

So what can we conclude, if anything, from these discussions? Did we address and answer all the questions posed throughout this discussion? Are there other questions that need to be asked?

As noted at the beginning of this discussion, the "Clovis First" model for the colonization of the Americas has been the gold standard about which all other theories have been evaluated. But, is the "Clovis First" model correct? There is growing evidence that it is not! Stanford et. al. 2005 note that "The Clovis model has collapsed under the weight of a substantial body of contrary evidence that refutes its basic premises. What is known about the spatial, temporal, and assemblage diversity does not support either the Clovis-first model or the revised models. They do not explain the co-traditions in North and South America and do not account for pre-Clovis-age sites." A new set of models is emerging emphasizing three principle routes to the Americas: The Bearing Land Bridge and ice-free corridor, a maritime route using the North Pacific Rim, and a maritime route using the North Atlantic.

Bonnichsen (2005) lays out the basic propositions for these new models. "First, the Americas were peopled before 11,500 RCYBP. Second, groups with different cultures and languages and from different genetic stocks peopled the Americas. Third, once these groups were in the Americas, a complex interaction occurred among peoples with different origins to create new peoples, cultures, and adaptive systems that were uniquely American. The new Perception does not exclude the possibility that the appearance of Clovis could have represented a distinct colonization event. It does, however, hold that Clovis colonists were not the first and only colonizers."

We'll discuss the pros and cons of each of these models. Let's first examine the key factors in any migration of this immense proportion, the evidence from linguistic studies, genetics, archaeology (tools and anatomical relationships) and other important information that will help sort out the current thinking.

***

"The field [of anthropology] is now at a very interesting juncture. This is a time when First Americans studies lack a prevailing model or theory. After more than 100 years of research, we have neither definitive answers as to when, where, or how the Americas were first colonized, nor indisputable evidence of how many colonization events were involved. We also do not have a clear understanding of the processes that led to the myriad of diverse cultural and biological groups found throughout the New World at the time of European contact. Although we have more scientific evidence now than at any time in the past, we lack carefully articulated, well-defined models for explaining the available evidence (Stanford et al. 2005)."

"A substantial body of evidence ... strongly supports the proposition that humans were in North and South America well before 11,500 RCYBP. Exactly how much earlier is a focus of ongoing research. An important new development has been the demise of the Clovis-first model; its failure created a void that has resulted in three different responses by the professional community. First, and not unexpectedly, there are a few residual Clovis-first advocates who still cling tenaciously to the hope that some of the propositions of the Clovis-first model can be reformulated to correct the model, or that definitive proof will be found to provide credence to this 'failed' model. Some 'hard core' Clovis-first proponents believe 'all pre-Clovis age sites' and biological evidence that supports a pre-11,500 RCYBP occupation must reflect flawed science. Therefore, all pre-Clovis sites and biological evidence are the focus of negative critiques. The second type of response to the collapse of the Clovis-first model is by former Clovis-first advocates who have accepted the 12,500 RCYBP dates from Monte Verde. They speculate that there must have been a peopling event that occurred slightly earlier than Clovis, 13,000-14,000 RCYBP. Scholars of this persuasion tend either to ignore or discount sites and evidence that support a pre-glacial maximum occupation of the Americas. The third response has been to create a series of new alternative models to explain the peopling of the Americas (Stanford et. al. 2005)."

"We suggest our efforts should focus on creating descent models. Rather than using terms such as colonization or migration, which imply a directed goal, descent models emphasize the temporal, spiritual, cultural, and biological relationships among different discrete archaeological and paleobiological data sets that are supported by evidence recovered from the fossil record (Stanford et. al. 2005)."

The Clovis-first model rests on a number of assumptions. First, they originated in northern Asia and migrated to eastern Beringia across a land bridge connecting Siberia to Alaska. Second, that there was an ice-free corridor between Beringia and the Northern Great Plains prior to 11,500 RCYBP. Third, that the habitat in the corridor would provide for subsistence for a human population to include the big game animals and that the corridor was indeed passable. Fourth, that when pre-Clovis people arrived at about 11,500 RCYBP there were no other individuals in the Americas. Fifth, they developed their Clovis technology once they entered the Great Plains. And sixth, that they were able to rapidly colonize all of North and South America in about a 1,000 year time period.

With the demise of Clovis-first two new models have been hypothesized: The North Pacific Maritime and the Atlantic Maritime. Note that disproving the Clovis-first model does not prove these new models are correct. The state-of-the-art in Science does not allow us to prove either at this time. They simply are the next best guess – which is what a hypothesis is after all. However, both should provide direction for new and innovative research.

For any maritime theory to work requires that the early people who dispersed to the Americas must have been mariners! And, therefore they must have had water craft capable of navigating, if not the open ocean, then certainly the coastal areas. So do we have any evidence that there were ancient mariners? The first obstacle these folks faced was a short stretch of water at Bab-el Mandeb. Wells (2007) notes that "These people most likely moved across the Horn of Africa, where a narrow span of water between the Red Sea and the Gulf of

Aden separates the East African coastline from the Arabian Peninsula at Bab-el-Mandeb. The short ten miles would have been easily navigable for humans possessing early maritime technologies. This crossing continued the start of a long coastal migration across the Middle East and southern Eurasia, eventually reaching all the way to Australia and Polynesia."

As mentioned in the previous chapter the Y haplogroup C consisting of markers "Adam" – M168 - M130 was the founding population of Australia. To get from Southeast Asia to Australia would surely have required the founding population to be skilled mariners – they surely had some form of seaworthy craft. "The best evidence for advanced maritime abilities in the late Pleistocene came from the southwestern Pacific. Greater Australia, which at lower sea level included the huge island of New Guinea, was colonized around 50,000 years ago. People migrating from Southeast Asia would have had to make a voyage, probably jumping off from Timor, of forty to fifty-five miles. From the combined Australia-New Guinea landmass, people next moved across deep (but not quite so wide) channels to very large islands of New Britain and New Ireland in the Bismarck Archipelago of northern Melanesia. The latter was reached by 32,000 or 33,000 years ago, and the required crossing to get there was a bit less than 30 miles (Koppel (2003)."

The next migrations in the southwestern Pacific, onward from New Guinea to New Britain and New Ireland, were almost certainly intentional. Obsidian, a dark, glassy rock formed from silicon at Hadean temperatures inside volcanoes, is ideal for making viciously sharp tools, and the specific volcanic source can be determined by laboratory analysis. Obsidian from New Britain was carried to New Ireland, where it turned up in sites some 15,000 to 20,000 years old. There would have been a water gap of around twenty miles to cross. Small marsupial animals were also taken from New Guinea to much smaller islands in the region. Presumably this was deliberate, to stock those islands with succulent beasties for future hunting. Still, New Britain and New Ireland are very large and mountainous islands. Reaching such visible target destinations did not require striking out into the unknown.

"Moving onward from New Ireland and colonizing the Solomon Islands required a much longer crossing. Archaeologists specializing in that region had long assumed that the Solomon's were colonized only a few thousand years ago. Then, in 1987 Stephen Wickler ... found well-stratified deposits more than six feet deep. The upper levels did, indeed, contain pottery, as expected. The lower levels had no pottery, but they were full of mammal, lizard, and fish bones, as well as shells, fire hearths, and flaked stone tools. Radiocarbon dating of the shells, which were in close association with the stone tools, indicated that the people had reached the site as early as 29,000 years ago. This was a shocker (Koppel (2003)."

Visibility is a key factor in open ocean navigation. Researchers have determined that there are "... limits of visibility relevant to getting from one island to the next, and to define a situation ... called 'two-way intervisibility,' where the distances and height of land were such that both the jumping off island and the destination island remained visible all the way. With longer distances and lower islands, though, there might be only 'one-way intervisibility.' This meant that the mariner had to set out without the target land in sight. If the jumping off island were mountainous, though (like New Ireland), a raft or canoe might be able to proceed quite far offshore before losing sight of it. Meanwhile, unknown land ahead might appear over the horizon before the high land behind faded from view. This could happen while a boat was on an intentional search for neighboring islands, but also accidentally, while simply out fishing.

"Finally, there was a situation of no intervisibility, where a mariner would have to lose all site of land and where, before compasses, it would have been a challenge to make it back to the jumping off point if no other land came into view. At first glance, it might seem that anyone voyaging over the earth's watery curve in these circumstances would be making a wild, life-threatening roll of the dice. If the sailor did not happen upon a new and unknown island within reasonable distance, he could try turning for home, but all he would see was a blank horizon. How would he know the right direction to steer, and how could he maintain a steady course even if he knew? But, in fact, (and here's where the landlubber's fear of the sea is based on lack of knowledge) he would not be totally blind or beyond hope. A lifetime of experience with the prevailing winds and wave patterns in the region would help. Polynesian islanders have navigated that way for millennia. Observing the flight of shore-based birds also provides clues to the direction of the nearest land, and might even have helped ancient navigators to find new, unsuspected islands. In addition, the tropical sun rises so high in the sky that it is useless in judging direction for much of the day, but at dawn and dusk It hovers low over the horizon as a guiding beacon (Koppel (2003)."

While land birds in general were useful navigational assistants, the albatross was revered by early mariners. The albatross brought the wind. No harm was to come to this creature. The consequences of killing an albatross often brought death to the sailors as in the *Rhyme of the Ancient Mariner.* After the sailor shot the albatross with his crossbow, the bird that brought the winds, the doldrums set in.

There passed a weary time, Each throat
Was parched, and glazed each eye.
A weary time! A weary time!
How glazed each weary eye –
When looking westward, I beheld
A something in the sky.

At first it seemed a little speck,
And then it seemed a mist;
It moved and moved, and took at last
A certain shape, I wist.

A speck, a mist, a shape, I wist!
And still it neared and neared:

As if it dodged a water-sprite,
It plunged and tacked and veered.

With throats unslaked, with black lips baked,
We could nor laugh nor wail;
Through utter drought all dumb we stood!
I bit my arm, I sucked the blood,
And cried, A sail! A sail!

With throats unslaked, with black lips baked,
Agape they heard me call:
Gramercy! They for joy did grin,
And all at once their breath drew in,
As they were drinking all.

See! See! (I cried) she tacks no more!
Hither to work us weal;
Without a breeze, without a tide,
She steadies with upright keel!

The western wave was all a-flame
The day was well-nigh done!
Almost upon the western wave
Rested the broad bright sun;
When that strange shape drove suddenly
Betwixt us and the sun.

And straight the sun was flecked with bars,
(Heaven's Mother send us grace!)
As if a dungeon-grate he peered
With broad and burning face.

The sails he saw were those of the Ancient Mariner – those of death!

Alas! (thought I, and my heart beat loud)
How fast he nears and nears!
Are those her sail that glance in the sun,
Like restless gossamers?

Are those her ribs through which the sun
Did peer, as through a grate?
And is that woman all her crew?
Is that a Death? And are there two?
Is death that woman's mate?

"Offshore voyaging did not require the coddling comforts of warm water and balmy breezes. On the much colder northeastern coast of Asia, advanced boating skills must also have existed by the late Pleistocene. By the early 1990's archaeology showed that obsidian from Kozushima Island, south of Tokyo, had reached huge Honsu Island by 32,000 years ago, which required a deepwater crossing of around twenty miles. The Island of Okinawa was also probably colonized about as early, and certainly by around 18,000 years ago, a movement of people that required multiple island hops.

"If people in both the southwestern and northwestern Pacific had capable watercraft well before the end of the last glaciation, there was little reason to doubt their ability to move in a long arc around the North Pacific Rim. Such a coastal migration would mainly involve relatively short hops, allowing for frequent stops to seek safety from the weather, to rest, and to resupply. Much of the coastline is mountainous, making visibility from a considerable distance offshore, at least when weather was clear. Even if migrating people sometimes lost sight of the shore, their situation would be far from dire. For major stretches, such as the coast of Kamchatka and most of the Gulf of Alaska, the shoreline is a long, unbroken expanse of land. The 'target,' as with the coast of Australia, is almost too large to miss. Even the roughest sense of direction, determined from prevailing winds, swells or winging birds, or from the sun in the sky, would allow mariners to head for shore and safety (Koppel (2003)."

Would this same technology or an improved version have been available for the colonization of the Americas some 30,000 to 40,000 years later? We don't have any petrified boats in the archeological record for this era! But, we do have an excellent history of past seamanship to guide us. And, given man's and woman's ingenuity, it seems like a no brainer! But then science doesn't work with "no brainers". Let's go with highly likely.

"The greatest diversity of boat types is found in Southeast Asia which at least one authority on boats has called the 'Indonesian center of boat diversity'. This is where the invention of various types of watercraft first shaped human destiny. Whether they developed boats from scratch or picked up some rudiments from Javan Erects is immaterial to the fact that it was Moderns that first made water vessels the 'open sesame' of vast new resources. Such a revolution could not happen all at once. To have a boat was one thing, but what could be harvested from it and how far it could go depended upon a series of subsidiary and later inventions, such as fishing devices and sails (Kingdon 1993)."

"... [T]he boundary between a raft and a boat can be blurred when the craft is of crude construction. Nonetheless, several essential differences would have emerged very early on. Canoes, whether of bark or wood, *contain* the boatman, his equipment and catch. Unless surrounded by a border, rafts *support* him; the rougher the water, the more important this difference could have become, because of the need to tie things down to prevent them from slipping off. A more significant difference was the greater ease with which a canoe could be maneuvered and a considerable difference in the speed of the two types of craft (Kingdon 1993)."

"There have been tentative suggestions by specialists in the history of boat technology that single outriggers might have been one of the more primitive types of seagoing vessels. On the basis of distribution and design details, Sulawesi has been suggested as their center of diffusion.

"If an unadorned dugout log is the primary form from which all subsequent elaborations derive, the single outrigger offers the easiest and most economical modification it is possible to devise. The addition of a float attached by a few booms can be made with virtually no specialized tools and the design has several special advantages for a one- or two- person fishing team. Righting a capsize is much easier, quicker and safer for a canoe with one outrigger than with two. Fishing, spearing, boarding and loading are all easier and less encumbered with one side open. The making and trimming of a single outrigger is easier and repairs are less complicated and less frequent than in a craft with more components to absorb stresses. All in all, the single outrigger canoe remains the main candidate of mankind's earliest ventures in seafaring and possibly provided the means of colonizing Australia (Kingdon 1993)."

"The expansion of shoreline foraging to open-water fishing has many implications, among them, sailing. The use of sails is not a necessary adjunct to the catching of fish, but sailing would have revolutionized the ease with which prime fishing grounds could be reached and would have substantially enlarged the diet and catchment zones of fishing communities.

"In any event, the invention of sailing is unlikely to have been long delayed once people had begun to rely upon harvesting sea fish. It can be supposed that fisher-folk, continually exposed to the vagaries of currents and ocean weather, would eventually turn the flow of wind and water to their advantage. They would have known that it takes no more than a standing figure to drive a boat along at quite a clip when the wind happens to be right.

"The most likely sail structure would have been a rectangular mat of woven palm leaf incorporating one or more fronds of *mkuti* and possibly using the rib itself as a boom that was attached high on a pole or foremast. Such a primitive structure could still be called a 'boom lugsail' in modern sailing parlance (Kingdon 1993)."

"At the most immediate level, the possession of boats undoubtedly changed the definition of seashores from being the outer edge of habitable territory to being the central core of existence. Thanks to their boats a marginal population expanded very greatly, both in numbers and territory, to become a highly distinctive and specialized branch of the Modern human family. Developed in the relatively limited context of large rivers, boats took the waterside economy out of the estuaries, along the seashore and then out to the islands. The first incentive was an abundant and diverse source of food, previously unexploited. Combined with the monopoly of potentially vast new island territories, this gave huge demographic advantages to the boat people and there would have been a rapid expansion of numbers throughout island Southeast Asia (Kingdon 1993)."

With this background information in hand, let's first examine the proposed North Pacific Maritime model.

"The coastal hypothesis has gained increased support in recent years because the remains of large land animals, such as caribou and brown bears, have been found in southeastern Alaska dating between 10,000 and 12,500 years ago. This is the time period that most scientists formerly believed the area to be inhospitable for humans. It has been suggested that if the environment were capable of supporting breeding populations of bears, there would have been enough food resources to support humans. [Some] ... believe that the first human colonization of America occurred by boat along the Northwest Coast during the very late Ice Age, possibly as early as 14,000 years ago. The most recent geological evidence indicates that it may have been possible for people to colonize ice-free regions along the continental shelf that were still exposed by the lower sea level between 13,000 - 14,000 years ago (Dixon 1999)."

As noted earlier, there were a series of refugia leading from Alaska down the Pacific Coast to the Americas. Pielou (1996) notes that these refugia would form "stepping stones for human migrants." In addition he states rather clearly that "Travel by this route [the ice-free corridor] would have been difficult and dangerous, however, and the coastal stepping stone route would have been much easier and more attractive."

Pielou notes that the coast of the Aleutian Islands to probably Southeast Alaska "...like the inland part of Beringia, must therefore have been dry and cold." The climate of this coastal area "...was dominated by a persistent arctic anticyclone giving clear skies and bitter east winds off the Cordilleran ice sheet, conditions inhospitable to marine mammals that bear their young on shore or on ice flows."

"The climate of the coastal route was probably fairly mild, at least along the more southerly stretches, because the sea was warmed by the Japan current, undiluted by arctic cold currents, which were blocked by the presence of Beringia. And it would have been foggy because the cold air off the ice sheet flowed over the warm sea. It is not known where along the shoreline the clear arctic cold of south coast Beringia gave way to the foggy warmth of what is now the British Columbia coast. South of the changeover, however, the refugia were probably covered by luxuriant vegetation and well supplied with meat: seals and sea lions, fish and shellfish, and migratory birds that would have used the chain of refugia as stopping points along a flyway. Caribou, a subspecies of which occurred in the Queen Charlotte Islands until becoming extinct in 1910, would also have been found there. The human invaders from Beringia probably found it easy to travel from one refugium to another, by boat in the summer and over sea ice in winter. A coastal population of Paleo-Indians with a maritime culture may have become established independently of, and earlier than, the inland Paleo-Indian population of the Great Plains, which subsisted on herds of grazing animals, especially bison."

Dixon (1999) provides some clarity to the coastal alternative model to the Clovis-first/Ice-free Corridor model. "An alternative model for human colonization of the Americas is coastal migration with later inland movement and settlement within broad environmental zones (biomes or 'megapatches') that extend from north to south throughout the Americas. It is easier for people to exploit their current environmental zone, using their existing subsistence knowledge, than to move to adjacent foreign environmental zones. Thus, people were likely to settle the coasts before moving very far inland. Migration probably occurred in many directions at the same time. For example, some people may have been moving more rapidly southward along the Pacific Coast of the Americas while others were colonizing more slowly eastward from the coast to the interior of the continents.

"This model is drastically different from the traditional interior Beringian migration and subsequent unidirectional movement form north to south cross-cutting environmental zones and a wide array of physical obstacles. Colonization along large environmental zones is more consistent with New World archeological data and enables seemingly conflicting evidence to be reconciled into a single rational model for colonization of the Americas.

"The coastal route provided the environmental avenue essential for the initial human entry to the Americas. The coast formed part of a continuous marine coastal-intertidal ecosystem extending between northeastern Asia and northwestern North America and farther south to the southern hemisphere. It would have facilitated coastal navigation and provided similar subsistence resources in a continuous ecological zone linking Asia and North America. Old World adaptations could have facilitated rapid colonization without developing new technologies or subsistence strategies.

"This alternative model proposes that initial human colonization of the Americas began around 13,000 B.P. along the southern margin of the Bering Land Bridge and then continued southward along the Pacific Coast of the Americas. With the use of watercraft, possibly skin boats, the human population moved rapidly southward along the coastal-intertidal Pacific biome, or megapatch. Although evidence of this early migration may have been obscured by rising sea level at the end of the last Ice Age, evidence might be expected to be found in adjacent areas of the interior, such as Monte Verde, which is located along a river drainage only 15 km (9 mi) northeast of the Pacific Ocean. If this model is correct, the Pacific Coast of the Americas could have been occupied thousands of years before the continental ice in North America melted.

"Coastal environments provide many ecological advantages for generalized foragers, an economic adaptation best suited for colonizing populations. For example, intertidal resources, such as shellfish may be harvested by children and the elderly, and simply eaten raw. On the other hand, the traditional interior Bering Land Bridge – Ice-free Corridor model requires human groups to be dependent on hunters specializing in large terrestrial mammals. This concept requires groups to be dependent on a few strong adults to bring down large mammals, including mammoths. This model also requires greater territorial movement and presents greater difficulty for human groups that realistically include the elderly, the very young, pregnant women, and the infirm. Current data from the earliest sites in the Americas do not indicate subsistence traditions based primarily on specialized large mammal hunting but on foraging instead.

"The coastal route offers many advantages for colonization. For example, latent heat stored in the ocean results in a more equitable environment lacking both the low temperature extremes of the north and the low precipitation of the arid west. Local abundance of marine and intertidal resources and predictable runs of anadromous fish concentrated human populations in specific locales, such as sheltered bays, inlets, estuaries, and along streams in which fish spawn.

"Temperate coastal technological adaptations rely heavily on readily available materials such as drift wood, marine mammal products, beach cobbles, and shell, which in many cases may have been already partially modified by non-cultural processes. In such an environment, reliance on sophisticated lithic technologies was probably not as important as in other environments. For example, pre-shaped and pre-polished sling and bola stones, the only lithic material required for two effective, deadly weapons, can be easily and efficiently collected from non-cultural beach deposits. Monte Verde provides a rare glimpse into this type of technological adaptation because it de-emphasized the production and use of bifacially flaked lithic tools and placed greater reliance on simple flakes and organic materials.

"From an original and theoretical maritime subsistence strategy, several adaptive trajectories were possible as humans expanded across the landscape. Survival may have been best assured by continuing a pattern of generalized foraging, which could be adjusted or modified based on availability of resources and increasing knowledge of local geography and biological patterns. For example, along the West Coast, people may have continued their ancient adaptation to shellfish gathering, fishing, and marine mammal hunting. In interior regions of southern California, Arizona, and Mexico, the pattern of general foraging may have led to an increasing emphasis on harvesting and processing plant products and seed grinding. In the Great Plains, general foraging persisted throughout the Paleoindian period, but people emphasized and refined large mammal hunting, particularly communal mass bison kills.

"Although the initial colonization along the continental margins of the Americas may have occurred rather quickly, subsequent colonization of interior environments may have occurred more slowly. People probably moved inland from the coast along rivers. As populations increased and people gradually adapted to interior environments, they possibly advanced inland along river systems; at the same time, colonization probably continued to progress along environmental zones.

"Given this scenario, the western plains of North America may have been among the last places to be settled as well as one of the least hospitable environmental regions of the continent. Adaptation to the interior plains may have occurred later because they are geographically separated from the Pacific Coast by the rugged mountainous Cordillera. Classic Clovis sites such as Blackwater Draw and Murray Springs containing evidence of spectacular mammoth predation, may be representative of a rather unique cultural, technological and ecological adaptation during the late Pleistocene. In other words, the spectacular and well-publicized Clovis kill sites may be the least typical and the least useful sites for interpreting the colonization of the Americas and early New World adaptations.

"Although the Clovis culture is often associated with mammoth hunting, more current data demonstrate a greater emphasis on generalized gathering by Clovis people. In fact, only twelve sites have been documented in North America where Clovis points have been found in clear association with mammoth remains. A more realistic portrayal of Clovis economics suggests mammoth kill sites occur in marginal habitats that may have been some of the last to be colonized. Although they may provide the earliest evidence of human occupation in the western interior of North America, this region may have been among the last to be colonized (Dixon 1999)."

So let's take a look at the proposed Atlantic Paleo-Maritime model. This is not a model that replaces the North Pacific Maritime model but one that complements it. Many of the same arguments for a coastal migration of the Pacific Coast made by Dixon (1999) can be made for an Atlantic Coast migration. In fact, if you drop the assumption that Clovis originated in the west and moved north and east and replaced it with a new assumption that Clovis originated in the east and moved north and west, you have the makings of a new theory.

Stanford and Bradley (2002) have developed a model for an Atlantic Coast entry into the New World. "Paleooceanographic studies of the North Atlantic suggest that during the LGM [last glacial maximum], a permanent arch of sea ice connected the Southwest coast of Ireland to the Grand Banks, forming a North Atlantic ice rim. Winter sea ice would have formed each year, extending the ice further south between 45 degrees and 50 degrees north latitude depending on annual weather conditions. The Gulf Current that now extends across the North Atlantic shifted southwards, making a clockwise circulation pattern moving toward the coast of Portugal and returned westward from North Africa. The Gulf Current warmed a weak counter clockwise current that circulated in the North Atlantic. This northern current would have moved northwestward from the Irish coast toward the Grand Banks and returned to the Bay of Biscay (Stanford and Bradley 2002)."

"The Solutrean hypothesis, in simple outline form, is that during the Last Glacial Maximum, sometime between 25,000 and 13,000 years ago, members of the Solutrean culture in the southwest coastal regions of Europe were led by subsistence behavior appropriate to their time and place to exploit the ice-edge environment of the polar front across the North Atlantic and colonize North America to become – after several millennia – what we know as the Clovis peoples, who eventually spread far and wide across the Americas. This does not necessarily mean that Paleolithic northwest Asians did not also colonize the Americas. It does mean, in concert with other strands of evidence, that Clovis is part of the rich, complex, and wonderful story of the ebb and flow of people whose descendants are what we call Native Americans (Stanford and Bradley 2012)."

"Even though the LGM North Atlantic has been characterized as cold and stormy, these hostile conditions were not continuous, being interrupted by periods of short term and perhaps even longer term intervals in which sailing conditions on open water would have been possible even for less experienced navigators. Regardless of weather conditions, it would not take too many generations of coastal dwellers to understand the signs of impending storms and the timing of these events. Nor would it have taken an observant hunter long to discover that floating ice islands and ice leads provided excellent havens during storms. These ice features reduced wave formation and provide pullout locations for setting up camps during inclement weather. An overturned, well-secured skin boat makes an excellent temporary shelter during such events. Further, sea mammals provide fuel oil for heating and cooking, and if necessary to assist in drying out skin boats (Stanford and Bradley 2002)."

Stanford and Bradley (2012) … "suggest that ice age people had cold-weather techniques at least as sophisticated as those required for survival during the warmer conditions of the Holocene … At the very least, tailored, water-proof clothing was a necessity, along with substantial all-weather shelters. With the skills to create these items, they no doubt also had the ability to construct sewn-skin boats. One has only to ask whether the peoples who live in Arctic conditions today could survive as they do without adequate clothing, shelter, or boats to cross and travel rivers and lakes. Lacking good archeological evidence to the contrary, we assume that this knowledge was integral to successful Artic colonization."

"We propose that in the two thousand or so years that Solutrean folks lived next to and observed these environmental parameters, they no doubt adapted to and capitalized on these resources. Further, as they became more skilled in seamanship, they would have ventured further and further out along the ice margin. Perhaps first as simple extended hunting trips and finally to spend entire summers hunting along the ice front. Eventually someone made it the mere 2500 km [ca. 1550 miles] between the LGM coastlines of the Grand Banks and Ireland. When arriving in the area of the Grand Banks they found a fishery that would have been unimaginable to a New England fisherman. This would have been big news back in the old country. Soon we suspect that more and more expeditions were undertaken, spreading the X haplotype into the Americas. Once established in North America, they spread southward along the now submerged continental shelf to eventually find new terrestrial hunting opportunities and perhaps new people with which to intermingle (Stanford and Bradley 2002)."

Stanford and Bradley (2012) note that "In North America, Clovis artifacts are relatively rare in the northeastern and north-central states, and some scholars believe this is because these areas were then recently deglaciated and not settled until late in Clovis times. Clovis sites are most abundant in the Mid-Atlantic and southeastern states and exhibit a wide diversity of activities, whereas diversity decreases to the north and west, where activities are focused on hunting. Western sites are either small campsites or kills that are sometimes associated with small temporary hunting camps. To the east and southeast, by contrast, they include not only kills but also stone processing areas, usually associated with large campsites; large overlook and upland surface sites; and at least one cave/rock shelter. Some of these differences in settlement patterns and site types may be the result of preservation or just plain luck in finding sites, but consistent patterns are emerging. The archeological evidence suggests that the contrast between the east and the west can be explained by the late ice-age prairie-forest boarder. The abundance of resources in the woodlands allowed at least semi-permanent settlements, while the prairies, plains, and deserts offered large game but little else at the time. Because plant and animal habitats were responding to abrupt and fluctuating climatic changes at the end of the ice age, this margin cannot be identified with great precision, but the general distribution of different Clovis site types gives a general indication of the location of this change." They also suggest that "Western expansions may have occurred in different ways, including exploration along major rivers and the Gulf Coast. The river orientation of Clovis people continues in the prairies, with major concentrations of artifacts along rivers and especially at the junctions of major rivers and streams." This is exactly the same concept of dispersal offered by Dixon (1999) for the west coast.

"We feel that the diverse amount of similarities noted between Solutrean and Clovis suggests a more complex situation than simple convergence. It appears possible that the dating and technology represented by Cactus Hill and Meadowcroft sites provide the link between Solutrean and Clovis. It may be bold of us, in the face of a dearth of proof to coin the term 'Atlantic Paleo-maritime' for these hypothesized ice-edge adapted people, however, we do so to spur future research on these issues and broaden the thinking of new generations of archaeologists (Stanford and Bradley 2002)."

***

So, what does all this stuff mean? Who were the first and successive Americans, when did they get here, and how did they get here? Who are the ancestors of modern Native Americans? It's at this point that I should say "I don't know!" Because, in truth, I don't know! However, since I don't have a dog in this fight, nor am I an up-and-coming young scientist who needs to protect a reputation, I'll not leave you hanging at this point. Here's what I think!

First, I believe that the Clovis-first model has been dismissed by the data. The first Americans to enter the west coast came from Siberia through eastern Asia and southern Beringia. They entered the Americas traveling along the Pacific Coast of both North and South America. They began this entry sometime around 19,000 to 20,000 years ago and reached southern South America probably around 15,000 years ago. Many of the early North Americans worked their way east to the inland west along the various river systems. In the southern hemisphere, they likely worked their way inland across the Isthmus of Panama, or thereabouts, and along the northern and eastern coast of South America and inland along the eastern front of the Andes. They later moved along the Amazon and other river systems to finally populate all of South America. After crossing the Isthmus of Panama they may also have worked their way north and east along the coast of Mexico and into the southwestern U.S.

At about this same time, maybe a little earlier at 20,000 to 25,000 years ago, there was a migration along the Atlantic ice-pack into eastern Canada and the U.S. to around the Carolinas. These Solutrean people from Europe brought with them a technology and culture that we would call Clovis in the Americas. Remember the Solutrean people had their roots in northeastern Asia. Both the people and the technology migrated westward and southward across southern Canada and the northern U.S. primarily and likely mixed with the peoples from the Pacific Coast migration. Later, at about 10,000 to 11,000 years ago they would move north through an ice-free corridor and into the Yukon Territory and Alaska and mix with the Nenana people.

A third entry of people into the Americas once again came from Asia, but this time from Southeast Asia at around 10,000 – 12,000 years ago. These are the folks who are anatomically different from modern

Native Americans and are Turner's Sundodants. They likely moved inland up the major river systems and either died out as a group or mingled with the folks that were already in the area. There were not likely enough of them to make a difference in the genetic makeup of modern North American natives.

A fourth migration brought the Na-Dene from Alaska along the coast and interior of western Canada. We see a split with two groups, the Navajo and Apache, moving into the American Southwest.

The Na-Dene and Bering Sea Mongoloids, from which the Eskimo-Aleut emerged, diverged about 15,000 years ago. There is a fifth migration of these Bering Sea folks along the western and northern coast of Alaska throughout the northern-most part of Canada and into Greenland, probably around 10,000 years ago. The Eskimo-Aleuts diverged about 9,000 years ago. The Aleuts remained in Alaska while the Inuit (Eskimos) migrated (6th migration) again across northern Alaska and Canada to Greenland. About 5,000 years ago they split into the Inupiaq and Yupik speaking peoples. They would certainly have encountered all the other folks to the south of them. What the nature of their interaction would have been is unknown.

Grandfather's Tale

The Creation of the First Humans

Long, long ago, the Creator, the Great chief Above, made the world. Then he made the animals and the birds and gave them their names – Coyote, Grizzly Bear, Deer, Fox, Eagle, the four Wolf Brothers, Magpie, Blue jay, Hummingbird, and all the others.

When he had finished his work, the Creator called the animal people to him. "I am going to leave you," he said. "But I will come back. When I come again, I will make the human beings. They will be in charge of you."

The Great Chief returned to his home in the sky, and the animal people scattered to all parts of the world.

After twelve moons, the animal people gathered to meet the Creator as he had directed. Some of them had complaints. Blue jay, Meadowlark, and Coyote did not like their names. Each of them asked to be some other creature.

"No," said the Creator. "I have given you your names. There is no change. My word is law.

"Because you have tried to change my law, I will not make the human being this time. Because you have disobeyed me, you have soiled what I brought with me. I planned to change it into a human being. Instead, I will put it in the water to be washed for many moons and many snows, until it is clean again."

Then he took something from his right side and put it in the river. It swam, and the Creator named it Beaver.

"Now I will give you another law," said the Great Chief Above. "The one of you who keeps strong and good will take Beaver from the water some day and make it into a human being. I will tell you now what to do. Divide Beaver into twelve parts. Take each part to a different place and breathe into it your breath. Wake it up. It will be a human being with your breath. Give it half of your power and tell it what to do. Today I am giving my power to one of you. He will have it as long as he is good."

When the Creator had finished speaking, all the creatures started for their homes – all except Coyote. The Great Chief had a special word for Coyote.

"You are to be head of all creatures, Coyote. You are a power just like me now, and I will help you do your work. Soon the creatures and all other things I have made will become bad. They will fight and they will eat each other. It is your duty to keep them as peaceful as you can.

"When you have finished your work, we will meet again, in this land toward the east. If you have been good, if you tell the truth and obey me, you can make the human being from Beaver. If you have done wrong, someone else will make him."

Then the Creator went away.

It happened as the Creator had foretold. Everywhere the things he had created did wrong. The mountains swallowed the creatures. The winds blew them away. Coyote stopped the mountains, stopped the winds, and rescued the creatures. One winter, after North Wind had killed many people, Coyote made a law for him: "Hereafter you can kill only those who make fun of you."

Everywhere Coyote went, he made the world better for the animal people and better for the human beings yet to be created. When he had finished his work, he knew that it was time to meet the Creator again. Coyote thought that he had been good, that he would be the one to make the first human being.

But he was mistaken. He thought he had as much power as the Creator. So he tried, a second time, to change the laws of the Great Chief Above.

"Some other creature will make the human beings," the Creator told Coyote.

"I shall take you out into the ocean and give you a place to stay for all time."

So Coyote walked far out across the water to an island. There the Creator stood waiting for him, beside the house he had made. Inside the house on the west side stood a black suit of clothes. On the other side hung a white suit.

"Coyote, you are to wear this black suit for six months," said the Creator. "Then the weather will be cold and dreary. Take off the black suit and wear the white suit. Then there will be summer, and everything will grow.

"I will give you my power not to grow old. You will live here forever and forever."

Coyote stayed there, out in the ocean, and the four Wolf Brothers took his place as the head of all the animal people. Youngest Wolf Brother was strong and good and clever. Oldest Wolf Brother was worthless. So the Creator gave Youngest Wolf Brother the power to take Beaver from the water.

One morning Oldest Wolf Brother said to Youngest Wolf Brother, "I want you to kill Beaver. I want his tooth for a knife."

"Oh, no!" exclaimed Second and Third Brothers. "Beaver is too strong for Youngest Brother."

But Youngest Brother said to his brothers, "Make four spears. For Oldest Brother, make a spear with forks. For me, make a spear with one fork. Make a two-forked spear and a three-forked spear for yourselves. I will try my best to get Beaver, so that we can kill him."

All the animal persons had seen Beaver and his home. They knew where he lived. They knew what a big creature he was. His family of young beavers lived with him.

The animal persons were afraid that Youngest Wolf Brother would fail to capture Beaver to make the human being. Second and Third Wolf Brothers also were afraid. "I fear we will lose Youngest Wolf Brother," they said to each other.

But they made the four spears he had asked for.

At dusk, the Wolf Brothers tore down the dam at the beavers' home, and all the little beavers ran out. About midnight, the larger beavers ran out. They were so many, and they made so much noise, that they sounded like thunder. Then Big Beaver ran out, the one Creator had put into the water to become clean.

"Let's quit!" said Oldest Brother, for he was afraid. "Let's not try to kill him.

"No!" said Youngest Brother. "I will not stop."

Oldest Brother fell down. Third Brother fell down. Second Brother fell down. Lightening flashed. The beavers still sounded like thunder. Youngest Brother took the four-forked spear and tried to strike Big Beaver with it. It broke. He used the three-forked spear. It broke. He used the two-forked spear. It broke. Then he took his own one-forked spear. It did not break.

It pierced the skin of Big Beaver and stayed there. Out of the lake, down the creek, and down Big River, Beaver swam, dragging Youngest Brother after it.

Youngest Wolf called to his brothers, "You stay here. If I do not return with Beaver in three days, you will know that I am dead."

Three days later, all the animal persons gathered on a level place at the foot of the mountain. Soon they saw Youngest Brother coming. He had killed Beaver and was carrying it. "You remember that the Creator told us to cut it into twelve pieces," said Youngest Brother to the animal people.

But he could divide it into only eleven pieces.

Then he gave directions. "Fox, you are a good runner. Hummingbird and Horsefly, you can fly fast. Take this piece of Beaver flesh over to that place and wake it up. Give it your breath."

Youngest Brother gave other pieces to other animal people and told them where to go. They took the liver to Clearwater River, and it became the Nez Perce Indians. They took the heart across the mountains, and it became the Methow Indians. Other parts became the Spokane people, the Lake people, and the Flathead people. Each of the eleven pieces became a different tribe.

"There have to be twelve tribes," said Youngest Brother. "Maybe the Creator thinks that we should use the blood for the last one. Take the blood across the Shinning Mountains and wake it up over there. It will become the Blackfeet. They will always look for blood."

When an animal person woke the piece of Beaver flesh and breathed into it, he told the new human being what to do and what to eat.

"Here are roots," and the animal people pointed to camas and kouse, and to bitterroot, "You will dig them, cook them, and save them to eat in winter.

"Here are the berries that will ripen in the summer. You will eat them, and you will dry them for use in winter."

The animal people pointed to chokecherry trees, to serviceberry bushes, and to huckleberry bushes."

"There are salmon in all the rivers. You will cook them and eat them when they come up the streams. And you will dry them to eat in the winter."

When all the tribes had been created, the animal people said to them, "Some of you new people should go up Lake Chelan. Go up to the middle of the lake and look at the cliff beside the water. There you will see pictures on the rock. From the pictures you will learn how to make the things you need."

The Creator had painted the pictures there, with red paint. From the beginning until long after the white people came, the Indians went to Lake Chelan and looked at the paintings. They saw pictures of bows and arrows and of salmon traps. From the paintings of the Creator they knew how to make the things they needed for getting their food.

Chapter VII.

The Great Extinctions, Agriculture, and the Rise of Cultural Diversity in the Americas

(10 thousand years ago to First Contact 400 years ago)

I am tired of fighting. Our chiefs are killed. Looking Glass is dead. Toohulhulsote is dead. The old men are all dead. It is the young men who say yes or no. He who led the young men is dead.

It is cold and we have no blankets. The little children are freezing to death. My people, some of them, have run away to the hills and have no blankets, no food. No one knows where they are – perhaps freezing to death. I want to have time to look for my children and see how many I can find. Maybe I shall find them among the dead.

Hear me, my chiefs. I am tired. My heart is sick and sad. From where the sun now stands, I will fight no more forever.

Chief Joseph of the Nez Perce, Surrender Speech

This final chapter takes place entirely within the Holocene period. There were three major events during this period which affected the lifeways of Native Americans: The extinction of a large number of large Pleistocene mammals (charismatic megafauna), the development of agriculture, and the arrival of Europeans. Once again it is important to understand climatic and ecological events and conditions in relation to the lifeways of the early Americans.

We enter the present interstadial period in the Holocene at about 11,500 years ago at the end of the Younger Dryas. Temperatures increased about 7 degrees C (12.5 degrees F) in about a 50 year period. This is an incredible temperature swing compared to the 1 degree C (1.8 degrees F) in 100 years that some people are trying to make you believe is catastrophic to the planet. Shame on them! Since this initial warming, the overall trend of the climate in the Holocene has been one of cooling. There are a number of different classifications that can be used to discuss Holocene climate. I'll touch on a couple of these and try to put together a scheme useful to our discussion.

The Holocene can be divided about equally into the Hypsithermal which ended about 5500 $\pm$ n years ago (this date is pretty soft) and the Neoglaciation which continues into present (Figure 7-1). The Hypsithermal was generally considered to be a warming period; however, as we'll see shortly, it was anything but uniform. During much of the Neoglaciation the climate was cooler than the Hypsithermal but, again, not uniform. As the name implies, the Neoglaciation, is the forerunner of the next glacial period – the Holocene being an interstadial period. During the Hypsithermal there was a period from about 9000 to 8000 years ago, referred to as the "Glacial Aftermath" characterized by cool temperatures. Mayewski et al. (2004) notes that it "… appears to have been generally cool over much of the Northern Hemisphere throughout this interval, as evidenced by major ice rafting [in the North Atlantic], strengthened atmospheric circulation over the North Atlantic and Siberia … [And] at low latitudes, this is a period of widespread aridity that occurs midway through a prolonged humid period that began in the early Holocene. This RCC [rapid climate change] interval occurs when the Northern Hemisphere was still significantly more glaciated than today and during the decline in summer insolation since its early Holocene Maximum. The 9000-8000 cal. yr B.P. [calendar years before present] interval may thus be interpreted as a partial return toward glacial conditions following an orbitally driven [recall the Milankovich cycles] delay in Northern Hemisphere deglaciation. At this time, changes in ice sheet extent and mass balance would still have played a major role in climate change. At least one pulse of glacier meltwater into the North Atlantic probably enhanced production of sea ice, providing additional positive feedback on climate cooling. This RCC interval represents the last major stage of deglacial climate affecting the Northern Hemisphere. The RCC interval 9000-8000 cal. yr B.P. is the only event that coincides with a significant increase in volcanic aerosol production and it occurred when bipolar ice sheet dynamics still had the potential for substantial effects of global climate."

Years X 1000												
12	11	10	9	8	7	6	5	4	3	2	1	present

Hypsithermal				Neoglaciation					
Anathermal			Altithermal						Medithermal
Megafauna Extinctions		Glacial After-math	Beginnings of Agriculture	H L T L L A		H L T L L A		H L T L L A	Near Current Climatic Conditions

Figure 7-1. Climate events over the past 12,000 years

Near the end of the Hypsithermal and continuing into the Neoglaciation there were three periods characterized by high latitude cooling and low latitude aridity (HLT LLA) – from 6000 to 5000, 4200 to 3800, and 3500 to 2500 years ago. "In the northern Hemisphere, the 6000 – 5000 and 3500-2500 cal yr B.P. intervals feature North Atlantic ice-rafting events, alpine glacier advances, and strengthened westerlies over the North Atlantic and Siberia (Mayewski et al. 2004)." These two RCCs were probably the direct result of a decline in solar output. There were two historical periods of interest in the Neoglaciation (Figure 7-2). The first was a warming period from AD 950 to AD 1250 often referred to as the "Little Climatic Optimum" or "Medieval Warm Period." It is this period in which the Vikings started to explore the North

Atlantic. The second is characterized by "cool poles, wet tropics". "High levels of volcanic aerosols occur at early stages in the event, perhaps contributing to its onset. In the Northern Hemisphere, glacier advance and proxy evidence for strengthened westerlies over the North Atlantic and Siberia suggest that climate changes in this interval have the fastest and strongest onset of any in the Holocene (Mayewski et al. 2004)." The period of global cooling is often referred to as the "Little Ice Age" and lasted from about AD 1300 to about AD 1850. The temperature changed about 7 degrees F in the transition from the Medieval Warm period to the Little Ice Age in about a 20 year period. It affected much of Europe and the Americas and caused widespread crop losses, famine, disease epidemics, and death by hypothermia, bread riots and more. In 1815, Mt. Tambora in Indonesia erupted. This eruption "... was the largest in historic times, and the aerosols from it reduced sunlight and global temperatures so much that 1816 became known as 'the year without summer' when New England experienced snow in July and August (Klein and Blake 2002)."

Years A.D.				
2000-1750	1750-1500	1500-1250	1250-1000	1000-0
Near Current Climate Conditions	Little Ice Age 2-4 degrees F cooler than current Collapse of Greenland's Norse Colonies		Medieval Warm Period 3-5 degrees F warmer than current Collapse of Mayan Civilization	Near Current Climate Conditions

Figure 7-2. Climate events over the past 2000 years

Mayewski et al. (2004), determined from their analysis that "... Holocene climate change can be quite abrupt, even in the absence of the large, unstable ice sheets that so dramatically disrupted Pleistocene climate. Further, Holocene RCCs have been large enough to have significant effects on ecosystems and humans. The short-lived

1200-1000 cal yr B.P. RCC event coincided with the drought-related collapse of Maya civilization and was accompanied by a loss of several million lives, while the collapse of Greenland's Norse colonies at ~600 cal yr B.P. coincides with a period of polar cooling that is minor by glacial standards."

A second classification system divides the Holocene into three separate divisions – The Anathermal, Altithermal, and Medithermal (Figure 7-1). "In at least the interior western portions of the continent climatic shifts toward warmer and drier conditions after the Pleistocene went beyond modern levels. Great Basin archeologists have made more use of hypothesized long-term fluctuations in temperature than have archeologists in most other regions of North America. The Anathermal, characterized by cool moist conditions, is supposed to have dominated the period 7000-5000 BC. The Altithermal maximum (5000-2000 BC) is recorded at many sites, and dates of around 4400 BC for the peak are known. The Medithermal (after 2000 BC) saw a reversal toward modern conditions. This seeming climatic overcorrection created strong adaptive pressure for Archaic populations. However the subsequent relaxation of heat and arid conditions toward modern norms left them well positioned to move from food collection to food production (Coe et al. 1986)."

Note that global warming and cooling events are not a new phenomenon as many in the popular press would have you believe. We've "Been there, done that!" before and will likely experience these trends again. Carbon dioxide, required by plants that produce oxygen required by animals, has been identified as the global warming culprit. As we noted earlier, carbon dioxide is released from glaciers as they melt and from the oceans as they warm and vice versa – it follows temperature increases. Increases in carbon dioxide have always lagged behind temperature rises and the lag involved an estimated 400 to 800 years. There has never been a time when carbon dioxide levels have led temperature increases (Duffy undated). "Much credence has been given to the 'hockey-stick effect' of temperature data (upward swing in mean temperature over just the last decade or so) proposed and adopted by the IPCC (intergovernmental Panel on Climate Change). Nations have grabbed this and are using this to base policies for actions on global warming effects, and the implementation of controls on carbon-based emissions by carbon taxing. The very computer program that gave IPCC those results was recently rigorously tested by inputting *random numbers*, and the computer-generated readout gave the same upward, hockey-stick, data trend with this meaningless input. This makes a mockery out of the early IPCC reports and subsequent actions. Of course IPCC cannot admit to that as their report has been regarded as 'gospel' by many nations (Duffy undated)."

"So what are the key players in [short-term] 'Climate Change'? The major driver is the sun. Warming depends on the sun. Cooling is due to the lack of sun's energy. Radiant energy enters the earth's atmosphere – air (on a dry basis) which mainly consists of nitrogen 78.08% and oxygen 20.94%. Of the 0.98% remaining, 0.934% (almost all) is the inert gas argon. Carbon dioxide is a trace. It is less than 400 ppm (parts per million) or 0.04%. Surprisingly, less than a fifth of that is man-made carbon dioxide (.008% of the total), and that is only since the beginning of the industrial era and the rapid increase in world population (Duffy undated)."

"The sun directly drives the El Nino-La Nina current motions that drive temperature changes worldwide. The sun sets up evaporative cycles, drives larger air and water currents or cycles, and changes weather patterns and therefore climate change. The varying degrees of lag and out-of-phase changes cause periodic oceanic oscillations. The El Nino Southern Oscillation (ENSO cycle) turns from warming to cooling depending on the net warming or cooling effect of the sun. This occurs quite rapidly. From about 1975 to 2000 there was a strong El Nino warming period (a positive Pacific Decadal Oscillation (PDO)). Now there is a La Nina period, and this has a cooling or decrease in warming (negative PDO). In essence ENSO and PDO switching is caused directly by the sun. Also there are similar periodic oscillations in other oceans (Atlantic and the Arctic Oceans) (Duffy undated)."

"Water vapor condenses to form clouds and it is by far the most abundant and significant of the greenhouse gases. Water accounts for about 95% of the greenhouse effect. The main atmospheric 'intermediary' between the sun and earth is water, and thus it dictates the *behavior* of the earth's climate. Without water vapor in particular and other greenhouse gases in the air in general, the surface air temperature worldwide would be well below freezing. The sun clearly must be a much bigger influence on global temperatures than any of the greenhouse gases, even water and carbon dioxide. Carbon dioxide is about 1/60 of water in air! It clearly is not the major player ... (Duffy undated)."

"The atmosphere may be heated by clouds by emitting latent heat of condensation as water vapor condenses. But clouds can both heat the atmosphere by reducing the amount of radiation transmitted, or cool the atmosphere by reflecting radiation. So of all the effects that can cause heating and cooling in the atmosphere and on earth, clearly water is the main greenhouse 'gas'. Other greenhouse gases (Carbon dioxide, methane, oxides of nitrogen) are 1/60 to 1/30 smaller in both quantity and effect. So with all the 'greenhouse gases' including water, human activity accounts for only minute amounts, just 0.28% of the total greenhouse gases. If we exclude the key one, water, then human activity would only account for about 5.53% of the total greenhouse effect (Duffy undated)." So where's the beef?

"A common misconception is that all or most carbon dioxide emissions from human activities accumulate steadily in the atmosphere with a proportional greenhouse effect. Yet, on average, the surface environment absorbs about half of those carbon dioxide emissions. In addition, each unit of carbon dioxide increase generally produces half the warming effect of the preceding one, and the atmosphere can become saturated to stop further effects (Bell 2011)."

I'm going to go beyond the basics on this issue because it is so important for people to understand what is really behind the "climate change" movement. Bell (2011) provides the clearest analysis I've found to date on the subject. He notes that "… the real impetus behind the cooked numbers and doomspeak of the global warmers has little to do with the state of the environment and much to do with shackling capitalism and transforming the American way of life in the interests of global wealth redistribution ('social justice')." It's the undeveloped countries against the developed countries. "Government restrictions upon carbon emissions are being promoted on the basis of three errant and deceptive premises: (1) that they will help protect our planet from dangerous climate change and pollution; (2) that they are needed to wean the United States and the world away from excessive energy consumption; and, (3) that they will incentivize energy technology and conservation innovations that will lead to independence from foreign oil.

"The initial premise is wrong on two accounts. First, there is no real evidence of any human-caused climate crisis. Second, there is no real evidence that any attempts to reduce atmospheric carbon dioxide emissions would have any significant climate influence. Simply because the EPA, parroted by media propaganda, condemns carbon dioxide as a 'pollutant,' that does not make it so. Such a declaration only misleads people and confuses this natural and essential molecule with real pollutants that truly should be restricted.

"The second premise, that carbon restrictions are necessary for energy consumption control, follows the ideological agendas of the UN and its IPCC (Intergovernmental Panel on Climate Change). Specifically targeted at the US and other affluent industrial countries, the restrictions are intended to artificially drive up energy costs to levels that curtail consumption-based capitalism. The burdens of this zero-sum-gain strategy will fall heaviest upon population segments that can least afford them.

"The third premise, that carbon penalties attached to fossil-fueled utilities will incentivize alternative technology innovations, is misleading in several respects. Heavily financed promotions fail to inform the public of the limited-capacity potentials afforded by 'renewable' energy sources, most particularly in regard to the urgent time frames required to substantially offset demands. Unfounded technology promises provide excuses for expanding government control and spending, unwarranted mandates, subsidies, and profit-taking fortunes for those who play the system [i.e. General Electric]. Free markets built upon delivery of competitive values are compromised when government is empowered to pick the winners and losers through politics that reward promises over performance [i.e. General Motors]. We the taxpayers and captive consumers, cover the costs (Bell 2011)." General Electric makes millions of dollars and pays no taxes. General Motors promised a cost effective electric vehicle and failed miserably – no one wants it! However, we are all stock-holders in GM. I'm still waiting for my stock certificate.

I feel we have much more to be concerned about with the coming global cooling period than the slight change in global warming that we are now experiencing. Global temperatures have been declining since 1998. It is not likely that humankind is having a marked effect on climate warming – since we can have no effect on the tilt of the earth or of its orbit about the sun, or on the continuous drift of the continents. And if we could affect these processes, it might be a "good" thing in holding off the inevitable global cooling that is leading us into the next long term stadial cooling – i.e. the next ice age.

***

There is no doubt that sometime between about 17,000 and 8,000 years ago 36 genera of "charismatic mega fauna" went extinct in North America. Many of the extinctions correspond with the presence of the Clovis people. Various causes have been suggested including a change in climate that changed the habitat of many of these large animals; human-caused extinction; and introduction, by man, of diseases that crossed over from man to animal. Martin (1995) champions the theory that man was the cause of the extinctions – known as the Pleistocene Overkill Theory (not well named since most of the large mammals died out in the Holocene).

As previously mentioned this was a period of rapid climatic change, particularly around 11,000 to 9,500 B.P. when most of the extinctions occurred and average annual temperatures at the end of the Younger Dryas increased an enormous 7 degrees C (12.5 degrees F) in a matter of about 50 years. Many of these mega fauna were cold adapted animals like mastodon (10,400 B.P.), mammoth (10,500 B.P.), horse (10,400 B.P.), camel (10,300 B.P.), stag-moose (10,200 B.P.), and giant beaver (10,200 B.P.) or predators that relied on these animals like the American lion (10,400 B.P.), and the saber-toothed tiger (9,400 B.P.). These extreme shifts in temperature would certainly have impacted the habitats of all these animals much more so than the limited ability of Clovis people to search out and hunt down all these animals in such a short period of time. The hunting ability of Clovis people has been severely over-estimated. Adovasio (2002) quotes an avid Clovis scholar, Jim Judge, regarding this theory: "...each Clovis generation probably killed one mammoth, then, spent the rest of their lives talking about it." Not a whole lot different than the stories told by modern day hunters and fishers – we've all had to suffer through stories about the "big one that got away".

Other researchers have suggested that an introduced disease carried by the first Americans "... jumped across species lines and caused the devastation. The large ones would have been most likely to suffer the most, since they not only reproduce more slowly but are always much fewer in number than smaller mammals such as mice. Thus it would have been the smaller ones that survived long enough to develop immunity. On the other hand, there is no known virus that could accomplish such mayhem, none known to leap from humans to other mammals, and no trace of such a virus has ever been located in any remains of mammoths or any other now-extinct Ice Age animal (Adovasio 2002)."

For all practical purposes, the Pleistocene Overkill Theory has been pretty well discredited by a substantial amount of data. The likely cause of these extinctions was primarily a change in climate followed by a significant change in the habitats of these large animals.

Humans may have hurried matters along but are not likely the villains in the extinctions. The extinctions, however, did change the focus of early hunters to smaller game. The bison became a focal point of their attention due to their large numbers and wide distribution. Before the Altithermal, they hunted the giant bison (*Bison latifrons*). During the Altithermal, higher temperatures and aridity had a negative effect on their habitat causing a decline in the giant bison population. The smaller bison (*Bison bison*), the ones we know today, were better adapted to these conditions and, therefore, were favored by the hunters. Of course, these same hunters would have hunted deer, moose, elk, caribou ... and bears in their individual habitats.

***

As we enter the Holocene we find that people are scattered across North America and occupying just about every niche in the environment. It was a time of transition from one lifeway to another – a story told mostly based on a change in tools. People in different environments responded with different tool boxes. In Alaska, the Nenana people persisted until about 10,500 - 10,000 B.P., at which time it became difficult to distinguish them from the Northern Paleoindians in interior Alaska. "This suggests a possible blending of technological traits of the Nenana Complex and the Northern Paleoindian tradition (Dixon 1999)." Dixon (1999) speculates that the Northern Paleoindian tradition "... is derived from a northward movement sometime between 11,000 – 10,500 B.P. of the Paleoindian tradition from the Great Plains. This movement was probably associated with colonization of recently deglaciated terrain in central Canada."

The microblade technology, known as the American Paleoarctic tradition, was introduced into Eastern Beringia around 10,500 B.P. and was contemporaneous with the Nenana and Northern Paleoindian technologies. It spread from interior Alaska where it is known as the Denali complex and then along the Pacific Coast where Dixon (1999) refers to it as the Northwest Coast Microblade tradition. These folks along the coast were well adapted to using marine resources.

"The bearers of the American Paleoarctic tradition may have first introduced the bow and arrow into North America from Asia possibly as early as circa 10,500 B.P. Although the American Paleoarctic tradition almost certainly used the atlatl, the introduction of the bow and arrow created a sharp contrast in the use of early weapon systems. The American Paleoarctic tradition added the bow and arrow to their weapons inventory, while the earlier Nenana and Paleoindian populations probably continued to rely on the atlatl as their primary weapon system. Apparently, bow and arrow technology was not shared with the Folsom complex or Plano tradition to the south, where there is no conclusive evidence for the use of the bow and arrow prior to 8,000 B.P. (Dixon 1999)."

It was at the beginning of this period when there was a distinctive shift from the Clovis Complex (11,500 – 10,900 B.P.) to the Folsom Complex (10,900 – 10,200 B.P.) on the Great Plains. This transition took place in perhaps less than 100 years. Dixon (1999) suggests that "... environmental change coinciding with the extinction of the mammoth quickly brought about an end to the Clovis lifeway. Although the Folsom people that followed continued a general pattern of foraging supplemented by big game hunting, they began to place greater emphasis on bison hunting, often killing many animals on a single hunt." On the Great Plains a second transition occurred from Folsom peoples to Plano peoples (10,200 – 8,000 B.P.). "The available archeological data suggest that on the Great Plains communal Paleoindian bison kills began in Folsom times and continued with increasing intensity. They appear to be seasonal occurrences, most commonly in the fall and winter when cold weather would preserve meat for long periods of time. Groups of people congregated to cooperate and participate in the kill. During these occasions, they processed the meat and hides and enjoyed a brief period of sedentary life. As meat supplies were depleted, or became tainted, these larger groups dispersed across the landscape. As smaller groups, possibly only a family or two, they survived based on more diverse resources and a broader foraging and hunting pattern. These communal undertakings were probably strongly reinforced by social and economic factors, including opportunities to trade, find marriage partners, and to share knowledge about changing environmental and social conditions. A prolonged period of drought beginning shortly after about 8,000 B.P. appears to have brought an end to this early way of life and ushered in a new era of economic and cultural change on the Great Plains (Dixon 1999)."

"Contemporary with Plano, Early Archaic developed in the eastern woodlands, characterized by distinctive side-notching of points for hafting (Dickason 1992)." About 8,000 B.P. eastern Early Archaic peoples migrated to the western Plains, where they came in contact with Plano peoples.

In the Great Basin the Western Fluted point tradition transitioned to the Western Stemmed Point tradition around 11,000 B.P. "The extensive use of manos and metates in the Western Stemmed Point tradition indicates an increased reliance on seed collecting and processing (Dixon 1999)." Dickason (1992) states that "... artifacts associated with seed grinding have restricted distribution in South America but are widespread in southern and western North America. And along the Pacific Coast there is a well-documented adaptation to the coastal environment by about 10,500 B.P."

"Overall the current view of the continent after 8000 BC [10,000 yrs. BP] is one that stresses an initially rapid environmental readjustment, which slowed gradually over time as conditions settled toward their modern states. For Indian bands the initial part of this readjustment must have entailed considerable uncertainty, and archeological remains reflect a series of broad unspecialized adaptations. As the environments became more stable Indian communities could risk more specialized adaptations to more predictable subsistence activities. Free-wandering patterns of movement settled into restricted wandering patterns, in which bands moved about according to regular habits within restricted territories. This long period, which archeologists have named the Archaic, follows the specialized big-game hunting of the Paleo-Indians, but precedes the emergence of full-scale plant cultivation. By definition Archaic Indians made optimal use of an increasingly wide range of plant and animal species (Coe et al. 1986)."

"The general problems of adaptation facing the Archaic Indians were similar in type to those faced by people throughout the world at this time. At a very general level the solutions were also similar. With the disappearance of Pleistocene game, the principle game animals became deer, moose, and caribou. As the trend toward smaller game continued, the Indians gradually replaced a relatively extensive exploitation of large species with an intensive exploitation of many small species. On the foraging side they intensified the exploitation of nuts, fish and shellfish.

"Increasing population density made it impossible for people to seek new resources by moving to unoccupied regions, for even marginal areas were filling up. As a consequence their patterns of movement settled into restricted wandering within defined territories, patterns that often took them to the same sites year after year in scheduled seasonal rounds. The archeological consequences of this shift include evidence of repeated occupations at carefully chosen sites, and the accumulation of midden debris at those sites. Simple portable tool kits were gradually replaced by larger more complex kits, and eventually heavy ground stone tools made their appearance. Heavy seed-grinding implements were made and used regularly, the transportation problem solved by the storage of separate implements at the sites visited during the course of a year. Food storage also increased, as did the keeping of material wealth.

"Earlier social organizations appear to have been relatively egalitarian. High mobility, large band ranges and impermanent habitations had facilitated the acquisition and exchange of exotic lithic materials.

"Archaic Indians had comparatively small band ranges, reduced mobility and less egalitarian organizations. Probably as a consequence of greater inter-band competition and reduced ranges, older patterns of extensive exchange of exotic raw materials were replaced by resort to local materials. Archaic sites often contain tools made of relatively low-grade but locally available raw materials. Better materials remained in the hands of bands lucky enough to have inherited them within their own reduced territories.

"By the end of the Archaic, many Indian bands were manipulating their environment. This involved the burning of tracts of forest to encourage deer grazing. It also included the occasional tending and perhaps unintentional propagation of plant species that were particularly desired. Also by the end of the Archaic new patterns of trade and exchange began to emerge. Superficially these appear similar to the pre-Archaic systems through which exotic lithic materials made their way from one egalitarian band to another. However the new systems appear to have been quite different in kind, judging by the use to which the exotic materials were put when they reached their archeological destinations. Many exotic goods have been found associated with burials, used in contexts that suggest that they were designed to mark differential social rank in increasingly competitive societies (Coe et al. 1986)."

"The development of grinding tools made a wider variety of seeds available for food, such as small seeds of grasses and amaranths. Peoples who depended on such resources tended to remain in one place, where their supplies were readily at hand. In other words, the mobile lifestyle in pursuit of different food resources cannot be assumed to have been universal at any period; and even when it was practiced, it followed a seasonal pattern within a known area. The vision of early humans as aimless wanderers in search of food does not equate with the evidence at hand; in fact, the contrary is strongly indicated, that they have always lived in communities that were as stable as food resources permitted (Dickerson 1992)."

"By the end of the Archaic, Indian cultures in several parts of North America had moved from intensive foraging to the actual cultivation of some indigenous plants. These plants by no means had the importance that true domesticates have for agriculturists, and their manipulation may not have even looked much like modern cultivation. Nonetheless the importance of sunflower, goosefoot, pigweed, knotweed, maygrass, marsh elder, gourd, squash and perhaps some early strains of maize was increasing, and the beginnings of horticulture were being practiced alongside traditional intensive hunting and gathering. The resulting new subsistence pattern has been called the 'cultivating ecosystem type' (Coe et al. 1986)."

The stage is set for the development of agriculture. In the west Dixon (1999) notes that "Two hypotheses have emerged, both of which suggest that the origins of American agriculture may have its roots in the early lifeways of people living in the Far West. One suggests that collection of seeds and a greater reliance on plant products that began in the Far West at the end of the last Ice Age led to plant domestication. The other hypothesis theorizes the abundant resources of the sea enabled coastal people to be sedentary. Sedentism enabled them to nurture and protect plants, thus making plant domestication possible." "In any event, people were firmly established throughout the hemisphere, and in some parts of Central and South America they began experimenting with domesticating plants perhaps as early as 9,000 years ago (Dickason 1992)."

***

"The Neolithic revolution of some ten thousand years ago was a profound change in human history, second only to the origin of culture itself. Although first described for the Middle East and Europe, similar changes are found to have taken place worldwide and largely independently in Africa, Asia, and the Americas. It was a social revolution – our domestication – a transformation as much conceptual and social as economic: Domesticated people are those who live (and mostly work) in houses grouped together in hamlets, villages and small towns as distinct from people of the past and the present who only use temporary dwellings (Wilson 1988)." "As mobile foragers, organized at the family level, humans maintained an open society based on personal intimacy, and enjoyed 'the only true form of liberty, liberty from the interference of others' (Wilson 1988)." "When humans settled down and built houses in villages, their cognitive and spacial worlds changed (Johnson and Earle 2000)." Several technological changes occurred in the Neolithic Revolution. Among these were the origins of plant cultivation primarily in a true agricultural context and related axe trade and forest clearing, the domestication of animals, and new forms of transport including carts.

"By the end of the Pleistocene ice age 10,000 years ago, the second human revolution was well in place, that of reengineering the mobile kin based, foraging band into a settled society, bindable by ties of altruism and religion into larger groups (Wade 2006)." "It is important to realize that farming was neither the production of food according to an economic rationale nor inevitably imposed on early Neolithic communities by large-scale events beyond their control. Instead, the adoption of agriculture was part of the profound transformation of the entire forager society and an adjustment to a wholly different set of societal values and meanings (Akkermans and Schwartz in Wade 2006)."

"What agriculture did was to create a massive population expansion. Ten thousand years ago, at the dawn of the agricultural era, the total world population numbered only a few million hunters-gatherers spread across the world's habitable continents. Today we number more than 6.5 billion, and this is on course to increase to around 10 billion by the middle of this [21st] century. This tremendous increase got off to a rapid start. Beginning in the Fertile Crescent, agriculture soon began to spread eastward toward central Asia's river valleys, into the Indian Subcontinent (although some scholars theorize a separate origin of agriculture in India), and westward toward Europe. If we look at the archeological evidence, we see quite clearly that the spread of agriculture changed the material culture of Europe. Carbon-14 dating shows that the first farming communities in southeastern Europe date from around 7,000 years ago, while those in northwestern Europe date from only the last 5,000 years. The pattern is entirely consistent with one in which farming spread from its central origin in the Middle East into Europe over the course of several thousand years.

"Two possible explanations for this spread were suggested in the 1970s. One was that agriculture, as a cultural phenomenon, was adopted by the indigenous population of Europe once they saw its advantages. At the time the very first person decided to plant seeds and settle down in one place (it was most likely a woman, since women traditionally did the gathering in hunter-gatherer groups and would have had ready access to the seeds needed to make this cultural leap) [They also would have the horticultural knowledge needed to develop agriculture in the first place], all of the surrounding groups would have been hunter-gatherers. The neighbors could have been inspired by the harvests and learned to grow crops as well, a process that could have been repeated all the way across Europe. It might have taken thousands of years for the inhabitants of the British Isles to learn about the fantastic new advance, explaining the gradient of agricultural dates we see across Europe.

"The other possibility for the spread of farming, suggested by geneticist Cavalli-Sforza and archeologist Albert Ammerman in the 1970s, is that it was the people who moved, not simply their ideas. In this scenario, the large number of descendants of the first farming communities swept aside their hunter-gatherer neighbors, spreading their own genes as they did so. If so, then we should see a genetic gradient across Europe that mimics the one for the spread of agriculture (Wells 2007)." And, that is just what we see!

Smith (1989) notes that in the 1970's research established that there were "... three localized centers of plant domestication (the Near East, North China, and Mesoamerica), along with three larger, dispersed non-central areas of domestication (Africa, Southeast Asia, and South America)." To this list, Smith (1989) adds eastern North America as a fourth independent center of plant domestication. Our focus here will be on the centers of Mesoamerica and Eastern North America and the dispersion of agriculture from these centers. My discussion will touch on Mesoamerica and provide a more in-depth discussion of North America. South America will be left to others.

The two determining factors for the practice of agriculture are a certain amount of sedentism and a need to store and preserve the fruits of your labor. Dickason (1992) notes that "What is essential for a sedentary way of living is an assured supply of food in one place, a situation not necessarily dependent on agriculture, at least in the distant period we are considering, when populations are usually small. Even though agriculture was closely associated with the development of permanent settlement, particularly as populations grew, it was not an essential requirement for the process to begin. Not only did permanent settlements in some cases long predate agriculture, in others the reverse occurred, and agricultural experimentation began long before the adoption of a sedentary lifestyle." She continues with "... the burial mound builders of Ohio's Adena culture (c. 600 B.C. – 100 B.C.) and those of the later but closely related and much more widespread Hopewell (c. 300 B.C. – A.D. 400) elaborated complex sedentary chiefdoms in rich self-renewing flood plain environments with comparatively minor assistance from agriculture. The silt deposited in the spring by the Mississippi and its tributaries ensured abundance, so it was not necessary to change village sites from time to time because of resource depletion." Sedentism required new ways of thought, new social relationships and a new kind of social organization, one in which people had to trade their prized freedom and equality for hierarchy, officials and chiefs and other encumbrances (Wade 2006)."

In addition, agriculture has commonly been associated with a number of social features: reduced territories, greater territoriality, more defined concepts of property, and formal definitions of property; more well-defined social boundaries, further closure of mating systems, and more complex social and political systems often leading to class differentiation. It is at this point that we see the development of the Regional Polity referenced in Chapter V. Johnson and Earle (2000) describe the Regional Polity as developing "... out of formerly fragmented local groups ..." and "... depending on scale of integration, these are either chiefdoms or states."

"Chiefdoms develop in societies in which warfare between groups is endemic but becomes directed toward conquest and incorporation rather than toward the exclusion of the defeated groups from their land. The subsistence economy is similar to that of a Big Man collectivity and requires similar management. Economic strategies, however, notably irrigation agriculture and external trade, provide opportunities for elite investment and control, which are used to extract surplus production from the subsistence economy to finance the chiefdom's operations. As the regional integration of the polity proceeds, clearly defined offices of leadership emerge at the local and regional levels and are occupied by members of a hereditary elite.

"Always in search of new sources of revenue, chiefs seek to expand their territorial control by conquest. Here a typical cyclical pattern is found, as local communities and thousands of people incorporate under the control of an effective chief only to fragment at his death into constituent communities. Competition is intense, both within a chiefdom for political office and between chiefdoms for the control of revenue-producing resources. Ceremonies legitimize the leadership and control of the ruling elite.

"The development of states and empires involves the extension of political domination, usually by conquest, to a still larger area. States formed by conquest may incorporate vast populations, often in the millions, that are ethnically and economically diverse. As in chiefdoms, elites carefully manage the economy in order to maximize surplus production that may be translated into power and political survival. Elite ownership of resources and technology is typically formalized in a system of legal property. National and regional institutions are developed – an army, a bureaucracy, a law-enforcement system – to handle the state's increasingly complex functions. Ceremonies mark significant phases in the annual economic round and legitimize unequal access to resources."

"The city-states in the Americas are best represented by the Incas of South America; the Olmec, Classic Maya, and Toltec in Mesoamerica; and the Adena-Hopewell cultures of the central and eastern United States. "The Inca Empire, Tawantinsuyu, was the largest and administratively most complex polity of the prehistoric New World. The empire, which extended from what is now Chile and Argentina through Peru and Bolivia to Ecuador and Columbia, incorporated about 350,000 square miles and perhaps eight to fourteen million persons (Johnson and Earle 2000)." The Olmec were the first to build temple-pyramids, to carve and move stone monuments on a grand scale, and to develop an involved religion and art focusing on deity complex (Coe et. al. 1986)."

There are a number of general themes related to the development and application of agriculture (adapted from Cohen undated):

1. Domestication of various crops was geographically a very widespread series of parallel events. In the Americas there appear to be three distinct centers where agriculture developed independently: South America, Mesoamerica, and Eastern United States.

2. Agriculture spread widely by diffusion and/or population movement. Maize was first cultivated in Mesoamerica and spread widely throughout North America.

3. There is often a substantial time-lag between domestication of a crop and actual dependence on it. The adoption of farming was a gradual process more than a rapid revolutionary adaptation.

4. A general increase in the quantitative dependence on domesticated crops with a corresponding quantitative shift from high quality to low quality resources over time – that is a reduction in nutrients, calories, protein, vitamins, minerals and fatty acids. Major staples – rice, maize, wheat, barley,

potatoes, sweat potatoes, manioc, and taro all cause dietary deficiencies when relied on too heavily as the sole basis of a diet.

5. An increasing use of more intensive agriculture on smaller units of land as land became less available and yields declined due to soil nutrient depletion. The three sisters (corn, beans and squash), grown in many areas, were part of an early switch to intensive farming where the beans grown with the corn replace the nitrogen used in great amounts by corn and squash provided shade to hold in soil moisture. The use of rest-rotation was another.

6. Technological changes were motivated by necessity or by demand, not by independent invention or technical advance. This differs from today's "gadget revolution" where gadgets are created and we search for a use later. The "zippo lighter" application on some hand held devices is cute but totally useless – you can't light a cigarette, or anything else for that matter, with it!

I'll touch on each of these themes.

In his review of Indian Agriculture of America, Hurt (1987) defines agriculture as "raising things on purpose." Agriculture was a slow transition from a gather-hunter lifestyle. In most gather-hunter societies the man is the hunter and the female is the gatherer. They were certainly keen observers of their environment for their lives depended on it. First, let's look at the men. The men were primarily hunters though the old, young and injured may have stayed back in camp and helped the women. Animal protein may not have supplied all, or even a majority of, the calories needed for daily existence. Recall the story of the man who once killed a mammoth and talked about it the rest of his life! The women on the other hand likely did supply the majority of the food needed for the clan to survive. They would have been excellent botanists in their own right. They surely had names for all the useful, as well as dangerous, food plants. They would have easily recognized a germinating plant, what the mature plant would look like and what value it had as a food item. And, it's likely that they would have names for the plants in all these stages of plant development as well.

Dickason (1992) noted that "... the New World domesticated plants that made such a contribution to world agriculture were all of undisputed American origin and were developed by Amerindian farmers. The two best-known of these are corn or maize (*Zea mays*) and potatoes (*Solanum tuberosum*), although such items as tomatoes (*Lycopersicon esculentum*), peanuts (*Arachis hypogaea*), pineapples (*Ananas comosus*), and cacao (*Theobroma cacao*, from which chocolate is made) are not far behind. More than 100 species of plants routinely farmed today were originally grown by Amerindians. Maize and potatoes are two of the world's four basic food crops: the other two are wheat (*Triticum vulgare*), developed in Mesopotamia and rice (*Oryza sativa*), domesticated in the Indus Valley. The most famous of all Amerindian crops, tobacco (*Nicotiana*), was grown for diplomatic, ritual, and some medical uses."

To get at the roots of Agriculture we must begin with the inadvertent discovery that useful plants grew from seeds "spilled" about their work areas. The first tillage would have been simply a matter of daily human activity clearing small patches of ground. Of course, "weeds" – plants we don't find pretty or useful – would invade these patches as well. It's not a big jump in imagination for individuals to have discovered that pulling the weeds up increased the productivity of the useful plants. Anyone who has worked with seeds knows that if they get wet they germinate quite readily. Our ancestors would have noted this as well and would have been fascinated by the process. They knew the roots and cotyledons were quite distinct parts of the plants and which should go in the ground and which should remain above ground. Would they have simply pocked a hole in the ground with their finger, put the root in the ground, packed dirt around the root, and possibly watered the plant and then stood back to see what happened next? Our sense of curiosity would demand it!

"Plant domestication could not have occurred without an extensive botanical knowledge already in place; it was no accident that agriculture developed first in warm, moderately rainy latitudes, where plant diversity was greatest and ecological conditions were such as to allow the necessary freedom for experimentation with a comparatively simple technology. Northerners were no less skillful and experimental in exploiting their resources, but the restrictions of their environment in combination with their Stone Age technology, ingenious as it was, meant that their options were more limited (Dickason 1992)."

Hurt (1987) notes that, in Mesoamerica, most evidence points to two centers of agricultural development: the Mexican State of Tamaulipas and the Tehuacan Valley, located in northeastern and south-central Mexico, respectively. "Indian farmers may have first raised bottle gourds (*Lagenaria siceraria*) as early as 7000 B.C. in the Tehuacan Valley and in Oaxaca; Tehuacan Valley farmers were perhaps the first to domesticate corn; and Indians may have originally cultivated pumpkins in Tamaulipas. Concurrent agricultural development is important, because it indicates that Indian farmers attempted to meet their own food needs according to the dictates of their environment."

"The Tehuacan Valley was well suited to be a center of agricultural development in the Western Hemisphere. Soil and moisture conditions enabled Indian farmers to practice agriculture. The semiarid climate of the valley was an asset, because it had prevented growth of vegetation. Indian farmers could plant without giving a great deal of attention to clearing the land and to fighting continuous battles with unwanted vegetation that would compete for soil nutrients and moisture. Tehuacan farmers probably planted their crops on the moist alluvial fans at the end of the water courses that spread out from the hillsides. Such sites had several advantages for early farmers. First, floodwaters periodically removed much of the unwanted vegetation. Second, torrents of water from heavy rains washed fertile soil down from the hillsides and replenished the richness of the alluvial fans. Third, when the runoff was minimal, the water irrigated the growing plants. Finally, subsurface moisture watered the crops even in the absence of surface runoff (Hurt 1987)."

"The Tehuacan people probably became committed to the agricultural process during the Coxcatlan cultural phase of Mesoamerican civilization (5200 to 3400 B.C.). During that period the Indians became more dependent on plants, seeds, and fruits. The Coxcatlan people grew corn; tepary, the black and perhaps the common bean; chili peppers; avocados; amaranth; bottle gourds; and two kinds of squash. Each spring, these agriculturalists probably came together into large groups, or macrobands, to plant their favorite seeds and to eat the fruits of their labors throughout the summer months until they exhausted the surplus. Then they broke up into small family groups, or microbands, for an autumn and winter hunting and gathering (Hurt 1987)."

"Independent agricultural development also occurred in Tamaulipas, on the northeastern periphery of Mesoamerica. During the Infiernillo cultural phase (7000 to 5500 B.C.), Indian farmers cultivated pumpkins and gourds. Pumpkins provided a very small part of their diet, but the Indians supplemented this food with wild runner beans and chili peppers. Between 2200 and 500 B.C., during the Flacco, Almagre, Guerra, and Mesa de Guaje phases, these Indians became more sedentary as agriculture became increasingly important. They raised corn, sunflowers, squash, and possibly manioc. Between 1500 B.C. and A.D. 900 the Tamaulipas farmers added new domesticates or cultigens – lima beans, squash (*Curcurbita mixta*), tobacco (*Nicotiana rustica*), and cotton (Hurt 1987)."

"Ultimately, these Mesoamerican farmers cultivated a wide variety of small grains, legumes, squashes, and fruits. The squash-bean-corn complex furnished the staple nutritional base, however, and plants in the squash family may have been domesticated first (Hurt 1987)." Various varieties of squashes were grown throughout Mesoamerica and were an indication that these farmers were very conservative since they raised the same squashes for many generations. "Beans were the second staple in the food complex. Mesoamerican Indians domesticated the common bean (*Phaseolus vulgaris*) about 5000 B.C. in the Tehuacan Valley and about 4000 B.C. in Tamaulipas, though they did not raise it abundantly until agriculture became well established in Mesoamerica (Hurt 1987)."

"Beans were an important agricultural crop for Mesoamerican farmers, one that contributed an important part to the nutrition of the people of that region. Indeed, these four species of beans [common beans, sieva beans, tepary beans, and runner beans] were closely associated with the cultivation of corn, and Indian farmers may have domesticated some varieties of beans and corn at the same time. Beans have an abundant supply of lysine, an amino acid necessary for good health. Corn lacks both lysine and the amino acid tryptophan. Beans complement corn by adding these two amino acids to zein, the amino acid in corn, thus creating a protein of high nutritional value. This union of amino acids occurs with any combination of corn and beans. While Mesoamericans lacked a scientific explanation, about seven thousand years ago they recognized that a combination of corn and beans provided a healthy diet, and thereafter, both crops remained basic to their agricultural process. Because of insufficient archeological evidence, no one can yet state precisely when Mesoamerican farmers first cultivated beans or when they began making careful selections for the development of specific varieties. Nor can anyone identify the origins of these cultigens. Beans, however, were important to Mesoamerican agriculture, and they spread northward to become just as significant to the Indian farmers within the continental United States (Coe et al. 1987)."

"By the sixteenth century, at least 150 varieties of corn, adapted to a wide array of different conditions, were being grown throughout agricultural America, except in the Andes above 3,900 meters), where it was replaced by quinoa (*Chenopodium quinoa*). The difficulty arises from the fact that if a wild corn once flourished, it has not been found; two wild grasses related to corn still grow in the highlands of Mexico – teosinte and tripsacum. Teosinte (the name is Mexican, meaning "mother of corn") is generally closely related to the domesticated plant and can be cross-bred with it. Teosinte is demanding in its requirements, wanting equal hours of daylight and darkness and warm temperatures, which effectively restricts its range; corn, in contrast, was being grown from Huronia in the Ontario midlands of Canada all the way through to southern Chile when Europeans arrived (Dickason 1992)."

"Multiple origins for the domestication of corn may exist, and the Tehuacan Valley probably was one center. Warm temperatures and a rainy season from April through October provided ideal circumstances for seed germination after the dry winter months. Archeologists have recovered prehistoric corn from five caves – Coxcatlan, Purron, San Marcos, Tecorral, and El Riego – in the Tehuacan Valley, which show the evolutionary development of the plant over a period of sixty-five hundred years.

"The earliest corn yet discovered has been dated at approximately 5000 B.C. from the Tehuacan Valley, but no clear evidence exists to indicate that the Indians practiced agriculture in this region at that time. In fact, no one is certain when or where Mesoamerican farmers began cultivating this plant. Between 3400 and 2300 B.C., during the Abejas cultural phase of Mesoamerican civilization, corn became part of the agriculture of the region, and Indian farmers were raising it along with bottle gourds, squashes, tepary, common, and jack beans, chili peppers, and avocados. By 900 B.C., corn provided the basis for irrigating projects, large villages, and more extensive agriculture than ever before. It was a dietary staple for Mesoamericans by that time (Coe et. al 1987)."

"By 1000 B.C., the Mesoamerican people were full-time farmers who were producing enough food so that large numbers of people could form permanent villages and develop distinctive cultures and civilizations based on agriculture. Although cultivated fields generally did not extend farther beyond the villages than the farmers could walk both ways in a day and still work, the variety of plants that the

Mesoamerican farmers domesticated and cultivated was very large; corn; common tepary, jack, and lima beans; pumpkins; squashes; chili peppers; amaranth; cacao; tomatoes; and avocados provided both staple foods and seasonings. The farmers understood and practiced the principles of selection, irrigation, and terracing – all with the aid of stone tools for clearing, planting, and cultivating. They also raised surplus crops and stored grain in underground pits or granaries (Coe et. al 1987)."

"... [I]n the prehistoric Southwest, agriculture had begun as early as 1300 B.C. By the time of Christ, the Indian farmers of the Southwest had made the seed selections and developed the plant varieties that were best suited for the climatic conditions in that region – from the cool, moist mountains to the hot, dry desert. The corn, for example, that the people of northern Mexico had first carried through the central Sierra Madres into the Southwest by 1000 B.C grew best in the wetter, high elevations. The corn was a Chapalote-like small-cob corn. Soon thereafter, the Mogollon farmers in the Bat Cave region of present day New Mexico developed a new type, or local 'race,' of corn. This variety was more variable and drought-resistant. It could be grown in the wetter high elevations as well as in the dry lower regions. Not only did this new variety produce larger ears with more rows of kernels, it also sprouted from deep planting. Dry, sandy soil necessitated this development. Planting depths of a foot or more prevented germination before late June or early July so that the summer rains would water the new plants. If germination occurred in May as a result of occasional spring rains, the ensuing high temperatures and dry conditions would ruin the crop. This corn became a staple of the Hohokam-Basketmaker (Anasazi) complex, and it provided the subsistence basis for Southwestern Indian civilization.

"These farmers also cultivated several varieties of squash and beans. The squash *Curcurbita pepo* was being raised along with the earliest corn at Bat Cave and as far north as Flagstaff, Arizona, by A.D. 1000. It was an important cultigen for the Hohokam. About A.D. 900 the southwestern farmers also began cultivating two other varieties of squash, *Curcurbita moschata* and *Curcurbita mixta*, using the seeds and flesh for food. Seeds provided twenty times the calories and twice the calcium of the flesh, as well as more vitamins. Because both parts could be dried for winter use, squash was an important food year-round. Until the Indians began cultivating beans, squash supplemented corn by providing the amino acid tryptophan, which corn lacked, for their dietary needs. Squash did not become an important crop until about 300 B.C, however, and was not cultivated on a widespread basis until about A.D. 900.

"Perhaps as early as 500 B.C. the people of Bat Cave were raising beans. The lack of a suitable technology for preparing beans for consumption may have influenced the relatively late adoption of this cultigen to their agricultural practices. Much later, sometime between 100 B.C. and A.D. 100, the Hohokam began to cultivate the common, or kidney, bean. Because beans usually have to be soaked and cooked for a relatively long time before the seeds are edible, they may not have been practical until the invention of pottery made such soaking and cooking possible. This is only a guess, however, because the Indians could have eaten beans in the pre-pottery age after parching and grinding them. The southwestern farmers primarily raised two varieties of beans: the common bean, in the plateau area, and the tepary bean, which the Hohokam favored.

"The tepary bean was particularly well suited for southwestern agriculture, because in the absence of irrigation, it had superior drought resistance, whereas high temperatures and evaporation rates tended to wilt the other varieties of beans. Indian farmers therefore preferred the tepary, since it tolerated hot, arid conditions and because it yielded more than the common bean by a ratio of four to one. Still, the adoption of teparies took time. Prehistoric southwestern farmers had relied on the common bean for as long as three thousand years before the drought-resistant tepary was developed, perhaps by the farmers in the Sonoran Desert or in southern Mexico. By A.D. 1000, however, the Hohokam farmers at Snaketown cultivated teparies, which had spread to the Mogollon in New Mexico by A.D. 1100.

"The southwestern farmers also raised two other bean varieties – lima and jack – on a small scale, adding both varieties relatively late in the pre-contact period. About A.D. 1000 the lima, or sieva, bean probably entered the Southwest from along the western slopes of the Sierra Madre by passing from farmer to farmer down the San Pedro Valley into the Gila River area. From there, it spread northward to the Anasazi. Sometime between A.D. 700 and 1200, the Hohokam also began to cultivate jack (*Canavalia ensiformis*) for food or for ceremonial purposes (Kingston 1993)."

"... [T]he Hohokam, the Anasazi, and the Mogollon were the major agricultural people in the prehistoric Southwest. They adopted plant varieties that were well suited to the severe growing conditions of that region. Over time, they improved their crops by using great skill in the selection of seeds. In the case of corn, for example, they chose the ears with the largest kernels, usually taken from the center of the ear, where the seeds were the most uniform. Dry-land farmers, such as the Anasazi, selected seeds from ears that matured first. Centuries later, the Zuni maintained this practice, which the Mormons also adopted. Moreover, the Indian farmers learned to plant their crops in locations that had different exposures to the sun and moisture to ensure at least a partial harvest during times of abnormally hot or cold temperatures and moist or dry years. For example, if a flood destroyed their crops along a stream or arroyo, they would still harvest a crop from the drier hillsides. Or in drought years, perhaps only the stream side fields would produce a crop. Thus, by using highly specialized crops, by carefully selecting field sites, and by depending upon soil fertility, elevation, and runoff, the Anasazi and the Mogollon successfully raised crops for centuries in a region that farmers still consider unsuitable for agriculture without irrigation.

"The Hohokam particularly left a distinguishing mark on the history of Indian agriculture. These 'canal builders' engineered an irrigation system on a twentieth-century scale. They not only established the only true irrigation culture in prehistoric North America, but they also made agriculture practical where farming would have been tenuous at best without irrigation. Certainly, irrigation requires farmers to have greater knowledge, such as an understanding of the behavior of water and the technical skill needed to control it, than dry-land farming requires. Indeed, the ability of the Hohokam to solve their farming problems, which the environment created, makes them an important agricultural people.

"The prehistoric farmers in the Southwest gained increasing control over their environment as they adopted and modified cultigens so that early maturing and drought-resistant crops would furnish a dependable source of food and fiber. By A.D. 500, agriculture was providing an important subsistence base for Anasazi, the Hohokam, and the Mogollon peoples. During the following nine centuries, the southwestern farmers continued to refine their agricultural techniques. By A.D. 1100, agriculture probably had superseded hunting and gathering in importance, and it provided approximately half of the Indians' dietary needs, thereby giving them greater security (Hunt 1987)."

"Corn arrived in the eastern woodlands by at least 400 A.D., if not earlier, again making no great difference in the life there. It was not until about 900 A.D. that it came into its own as a major part of the diet. It had a profound effect on the Plains, often thought of as the exclusive home of bison-hunting people and not the sort of place agriculture could flourish until Europeans brought a plow capable of turning the thick mat of prairie grasses over to expose the topsoil. But the Plains produced some of the continent's most important native farming communities long before European contact.

"By about the beginning of 1000 A.D., maize and beans were being grown in small, semi-sedentary villages on the central Plains, even as far north as North Dakota, which is about as far north as maize can be grown. The villagers also made pottery and buried their dead in small mounds, traits probably taken on from their eastern woodland neighbors. They also hunted bison, as did their neighbors on the Plains to the west. Over the centuries, village life became more widespread, with villages now permanent habitations, usually on terraces and bluffs above rivers. Corn and beans were the staples, and the farmers lived in substantial single-family dwellings with ample storage pits, the village or hamlet of fifty to seventy-five people typically being surrounded by stockades or moats. Seasonally these people gathered wild plants and trekked westward to hunt bison. Here, then, were people who lived between two quite different worlds – that of the new, elaborate farming societies of the woodlands and that of the ancient, still basically archaic nomads of the dry Plains – taking from each world what they deemed useful and ignoring the rest. They were the putative ancestors of the northern tribes (Hidatsas, Mandans, Arikaras), the central tribes (Pawnees and Wichitas), and the Caddos of Oklahoma and Texas (Page 2003)."

In the Eastern Woodlands a number of indigenous plants were being cultivated to include: sunflower (for both seeds and tubers), goosefoot, pigweed, knotweed, maygrass, marsh elder, gourds and squashes. These plants supplemented the diet of the local hunter-gatherers. "Seeds were important, but not all of the plants tended at this time were exploited for their seeds. Sunflowers included at least one species cultivated for its tubers, a species later named rather inappropriately the 'Jerusalem artichoke.' Tropical curcurbits (gourd and squash) came to North America through Mexico, and were used for their flesh as well as their seeds. Beginning around 2500 BC in a few places in the Mississippi drainage, and lasting to at least 400 AD in some, Indians exerted much more control over the propagation and production of these plants than earlier foragers could have imagined, if somewhat less than true horticulturalists can achieve. The low yield of the early cultigens relative to the later domesticates suggests that they served mainly to buffer temporary shortages in hunted and gathered wild foods. They were less than true staples, but they were storable and made a critical difference for people threatened mainly by the rare episode of starvation. This threat would have been sharpened by the gradual restriction of Indian communities to small, well-defined territories in which unpredictable failures in natural harvests could only be buffered by deliberate surplus production and careful storage (Coe et al. 1986)."

"Dependence on wild foods limited the size of permanent settlement rather than preventing it. The residents of Koster, on the upper Mississippi (c. 4000 B.C., although it was inhabited seasonally for several thousand years before that), never did find it necessary to turn from wild to domesticated food resources, except to a limited extent; the region was so rich in plant and animal life, both on land and in water, that there was little need to undertake cultivation. Similarly, the burial mound builders of Ohio's Adena culture (c. 600 B.C. – 100 B.C.) and those of the later but closely related and much more widespread Hopewell (c. 300 B.C. – A.D 400) elaborated complex sedentary chiefdoms in rich self-renewing flood plain environments with comparatively minor assistance from agriculture. The silt deposited in the spring by the Mississippi and its tributaries ensured abundance, so it was not necessary to change village sites from time to time because of resource depletion (Dickason 1992)."

"The new cultivating ecosystem type became the base upon which the Adena and later Hopewell cultural phenomena were built. Adena culture developed in the Ohio River Valley by at least 700 BC, perhaps as early as 1100 BC (Coe et al. 1986)." The Adena people were noted for their construction of burial mounds. "The burials under these monuments were placed sometimes in simple clay-lined basins, at other times in large log tombs big enough to contain up to three people. The dead found in simpler graves had often first been cremated. The more elaborate graves contain evidence that the dead were smeared with red ocher and graphite, and often contain more high-quality grave goods as well (Coe et al. 1986)." Adena women made pottery, which had been added to some Late Archaic household inventories, but it was not particularly well made and played no important role in burial ceremonialism. Adena grave goods included reel-shaped gorgets, carved tablets and tubular pipes. All of these were typically made of banded slate or some other fine-grained stone that would take a high polish. The tablets bear carved curvilinear designs, or abstract zoomorphic designs, usually a bird of prey. They may have been used to apply designs to fabrics, or perhaps to lay down designs on skin, either temporarily or in preparation for tattooing.

"Hopewell culture was waning by 400 AD. Although it may have continued for a time in a few localities, the trade network that had sustained it was breaking down, though the reasons for this are not yet fully known. Perhaps new strains of domesticated plants were being introduced from Mexico and the cultivating ecosystem that sustained Hopewell communities was under pressure from a more adaptive ecosystem. Competition for the rich river bottomlands that supported the new intensive horticultural system could have been incompatible with maintenance of the aging Hopewell trading network. Another possibility is that subtle climate shifts, perhaps to drier conditions, depressed food production overall and weakened their ability to sustain the trading system and the burial programs that depended upon it (Coe et al. 1986)."

"There is a 400-year hiatus in the development of major traditions in the Eastern Woodlands from 400 to 800 AD. Local traditions continued to develop, and some of them throve during this period… The Mississippian tradition was based on the introduction of new strains of maize from Mexico. Earlier strains of true plant domesticates may well have been limited to parts of the Eastern Woodlands that experienced growing seasons of 200 days or more. These included 12- and 14-row varieties of corn. The newer strains appear to have included some that could be adapted to growing seasons as short as 120 frost-free days. These included a hearty 8-row variety of corn. Furthermore, Mexican beans were added to the inventory of plant domesticates by at least 1000 AD. This addition provided an important protein supplement, releasing populations from density constraints based on the availability of wild animal protein sources. Meat was still a necessary part of the diet, but not the critical resource it had been previously. With beans the 'three sisters' trio of maize, beans and squash – the dominate domesticates of late prehistoric horticultural systems – was in place (Coe et al. 1986)."

"Indian bands, tribes and nations peripheral to the Mississippian florescence in many cases continued to practice mixed-hunting, gathering, fishing, and cultivating subsistence patterns. The older and simpler cultivating ecosystem continued to work well in the centuries following 800 AD. However, most communities that practiced any cultivation at all gradually adopted the more hardy strains of maize and beans. These could be cultivated as far north as the limit of 120 frost-free days. Beyond that limit even rudimentary cultivation was rarely practiced. Consequently the old cultivation pattern involving squash, gourds and an array of local plants such as goosefoot and sunflower was completely replaced by maize-beans-squash horticulture over the course of the last centuries of prehistory.

"However outside the limits of Mississippian expansion the new horticultural system did not involve intensive cultivation on alluvial floodplains. Instead, an extensive upland pattern emerged. The farmers of the Northeast, for example, practiced shifting cultivation, sometimes known as slash-and-burn or swidden agriculture. A single crop of corn was planted each year, inter-cropped with beans and squash. Old fields were abandoned when they became depleted or infested with pests, and new fields were cut from the forest every few years. Villages were moved to locations near new fields, often every 20 years or so. Wild foods, particularly deer, supplemented the crops. The large permanent cities of the more intensive Mississippian horticulturists were never possible in the context of this upland adaptation, but it was nonetheless successful until the arrival of European colonists (Coe et al. 1986)."

"The Iroquoians adopted an upland horticultural system based on improved strains of the three sisters (maize, beans and squash) by 1000 AD. The shift toward more dependence upon cultivation soon produced trends toward population nucleation on three levels. First, individual family households began joining together and living in multifamily dwellings. Women were almost certainly the traditional cultivators, and the new importance of horticulture reinforced their importance in subsistence and domestic affairs. The new multifamily houses were probably occupied by families connected by kinship through related women.

"The second level at which nucleation progressed was at the village level. Multifamily houses began to be built next to others of the same type in growing villages. To the extent that archeologists can be sure of such things, it is likely that each house came to hold a clan segment (lineage), and that villages developed councils of lineage heads to manage local affairs. Kinship relations, whether real or fictive, were extended to facilitate cooperative effort and the maintenance of order.

"At a third level the villages of Northern Iroquoians began to move toward each other in the course of regular relocations. By the time Early European explorers encountered them they had contracted from a very dispersed series of small scattered communities of 13 tribal areas. The 12 Iroquoian nations that resulted from this process of tribalization were the St Lawrence Iroquois, Mohawk, Oneida, Onondaga, Cayuga, Seneca, Susquehannock, Wenro, Erie, Neutral, Petun and Huron (Coe et al. 1986)."

"The changeover to agriculture was not an unmitigated blessing, if we are to judge by present archeological evidence. For one thing, its higher reliance on starchy foods meant an increase in dental problems. For another, corn was relied on too heavily in the diet, without adequate protein supplements, it could result in a population smaller in stature than with the ancestral hunters/gatherers and with shorter life spans. This happened with the Mississippian Mound Builders and may have been a factor in the decline of Cahokia. In the southeast, skeletal remains point to maize agriculture being physically more demanding than the hunting and gathering way of life (Dickason 1992)."

"By the proto-contact period, plants were, in general, and in spite of regional exceptions, the most important source of nourishment. When it came to medicines, however, the botanical world always had been, and remained, the major source (according to the Cherokee, animals brought diseases and plants provided the cures). Recollect missionary Chrestien Le Clercq (c. 1641 – after 1700) would report from Acadia, where he was from 1675 to 1686: 'Amerindians are all by nature physicians, apothecaries, and doctors, by virtue of the knowledge and experience they have of certain herbs, which they use successfully to cure ills that seem to us incurable.' That this knowledge had roots that went deep into the past is not questioned. The process by which the Amerindians acquired their herbal lore is not clearly understood, but there is no doubt about the results. More than 500 drugs in the medical pharmacopoeia today were originally used by Amerindians (Dickason 1992)."

***

For the rest of the discussion I'd like to focus in on a small portion of the Iroquoians and Algonquians of the Great Lakes region. For the Algonquians my focus will be on the Western Abenaki and the Algonquin.[1] When I use the term Iroquois, I am referring to the Five Nations of New York – Mohawk, Oneida, Onondaga, Cayuga, and Seneca. The Saint Lawrence Iroquois were displaced by their Algonquian enemies. Included in the Iroquois language group are the Huron (more properly the Wendot) who were allies of the Algonquians. Again, I will narrow the discussion primarily to the Wendot. Huronia consisted of five tribal areas – the Attignawantan, Ataronehronon, Arendaronon, Attigneenongnhac, and Tahontaenrat (Figure 7-3). The total pre-contact population of Huronia was estimated to be about 18,000 to 22,000. Post-contact (late 1630's and early 1640's) populations were reduced by about 50% due to epidemics of influenza and small pox to about 9,000.

Most of the ethnography available for the Wendot (Huron) comes from the annual reports, known as the Jesuit Relations, sent home to France by the Jesuit missionaries. The first Jesuit mission in the Americas was established in Acadia from 1611 to 1613. The missions in Canada along the St. Lawrence River and the Great Lakes began in 1625. The Jesuit Relations were discontinued after nearly 50 years in 1673. The Society of Jesus was dissolved by the Pope one hundred years later in 1773.

<?> Note that the Algonquin were a separate tribe within the larger Algonquian language family.

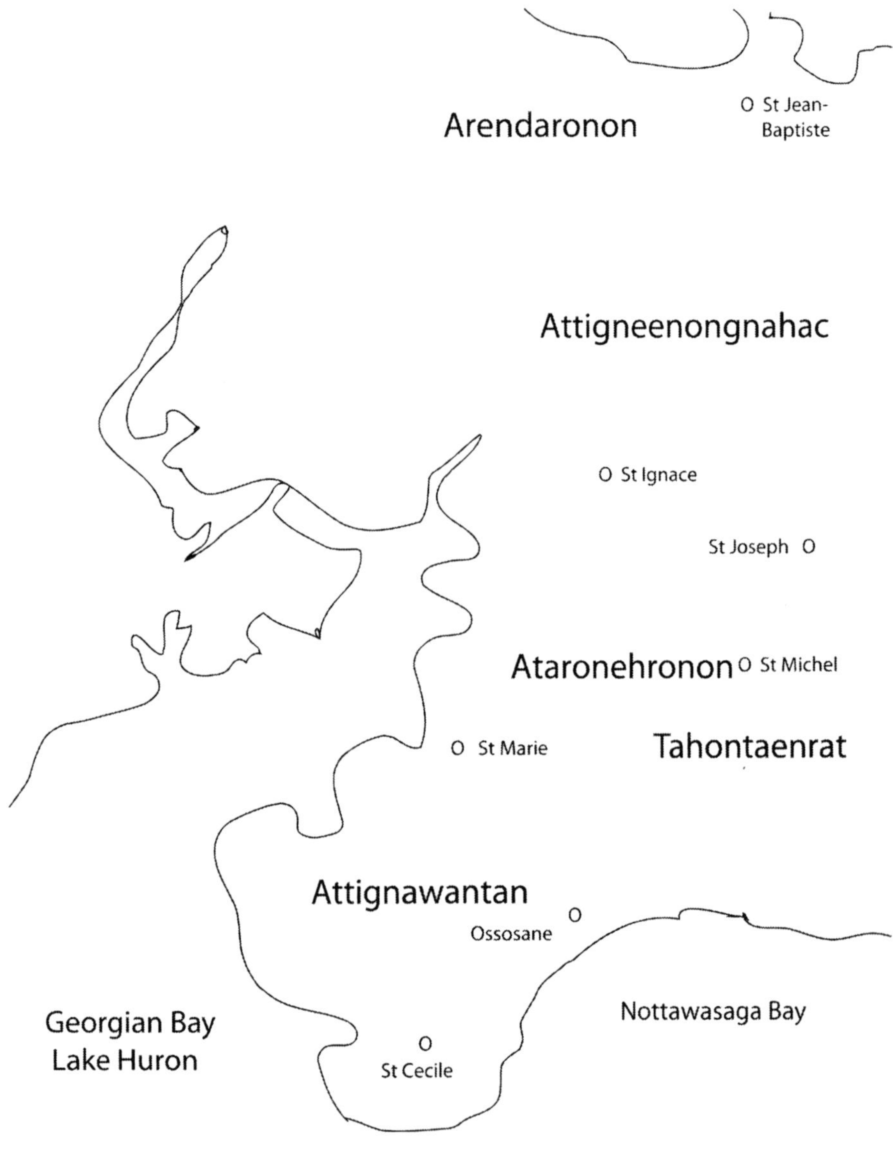

Figure 7-3. Tribal Areas of Huronia

From the Jesuit Relations we find that the lifestyle of the Wendot was one of agriculture supplemented by hunting, fishing, and gathering. The climate was suitable for growing the "three sisters" – corn, beans, and squash. They also grew tobacco. The lands of the Algonquin were too far north for agriculture and they, therefore, lived a strictly subsistence lifestyle. They traded extensively with the Wendot for agricultural products. It was said that Huronia was the granary of the Algonquin.

The Wendot practiced slash and burn agriculture. Men cleared the forest for the original fields. They felled the smaller trees and piled the trunk and branches around the stump and burned them. Larger trees were girdled and left in place to die. They were removed as they fell in time. The women maintained the fields after this initial clearance. They also planted the fields, weeded them, protected them, and harvested the crop.

They planted corn in mounds spaced about a pace apart. They planted about 10 corn seeds in each mound and hoed soil up around the stocks as the seedlings emerged. Once the corn was up they planted beans in the mounds. Squash and pumpkins were likely planted in separate fields. Tobacco was planted and tended by the men in small plots between the longhouses where it tended to be a bit warmer. The climate was not suitable for extensive fields of tobacco. Most of their tobacco came from trade with the Neutrals and Petun to the south. They planted enough corn to carry them through 2-3 years and still have a surplus to trade. Their fields were so extensive that it was said that it was easier to get lost in the corn fields than in the surrounding forests.

Fishing provided a major food supply to supplement the harvest. The spring and fall spawning runs was the major peak in fishing activity. The larger fish would be taken with a spear while smaller fish were taken in nets and weirs. The fish could be eaten fresh (got to love that sushi), smoked or dried. Whitefish were sometimes boiled and the oil skimmed off of the surface.

Gathering was not an important activity except in time of famine – which didn't occur very often. Abandoned fields provided an excellent habitat for the establishment of the fruits and berries they did collect like strawberries, blueberries, and raspberries. They also gathered hemp which was used to make twine and fishing nets. In the animal realm they gathered clams, crabs and turtles. They were also known to gather mice which were a continuous pest in their lodges.

As in the case of gathering, hunting was not a major source of food in their subsistence lifestyle. The primary game was deer. It was hunted singularly with the bow and arrow or in large communal hunts where the deer were driven to a triangular enclosure and killed with the bow and arrow or with spears. They were also driven into water where hunters in canoes waited for them with spears. The dog was domesticated and used for hunting and as a food source primarily in feasts and ceremonies. Other animals hunted occasionally were bear, moose, muskrat, beaver, woodchuck, fox, hare, turkey, ducks and other kinds of waterfowl. The only food taboo was crow.

"The early observers recognized the Huron followed a definite seasonal pattern. The cycle of activities began in early March when some of the men went hunting to places where deer had 'yarded' to escape the deep ice-encrusted snow of late winter. After returning from the hunt the men went fishing until mid-May to take advantage of the spring spawning runs of walleye, sucker, pike, and sturgeon. About this time the women gathered firewood and began preparing the fields by burning them over, a task the men helped with if new fields had to be cleared. Toward the end of May the women put the seed in the ground while the men departed on trading or war expeditions. A predetermined number of men would stay at home in case of enemy raids. These would help at odd jobs or go fishing and hunting near the village.

"Throughout the summer the villages were virtually deserted as the women and children stayed in the fields hoeing the corn and chasing away the pests. By the end of August and early September the corn was harvested, dried, and stored away. It was about this time that the men returned from trading and warfare and began to get ready for the large autumn deer hunts. Deer hunting was carried out a few days' journey to the south of Huronia in the oak (mast-producing) areas of Ontario. Here deer congregated for the rutting season, the last half of October. On returning from deer hunting the village populations concentrated on the fall spawning runs of white fish, lake trout, and cisco, which lasted from early November to the beginning of December. Most of the fish would be dried and stored away for the winter.

"During the late fall a number of Algonquian bands settled near the Huron villages to spend winter. The Nipissing usually settled among the Attignawantan in the west and various Ottawa Valley groups among the Arendaronon in eastern Huronia. On their way to Huronia the Algonquians would obtain fish, meat, and furs to trade for corn and other items. As soon as the ice broke in May, these groups would depart to the north.

"By early December everyone was back in the Villages marking the beginning of a variety of social activities to celebrate successful harvesting, hunting, fishing, and trading. These feasts were accompanied by gambling and gift giving. If captives had been taken, some were tortured at this time. Among the few activities performed during the winter were ice fishing and trade with Algonquian neighbors. The women would weave mats, manufacture fish nets, and prepare corn for next season's trade (Heidenreich 1978)."

Grandfather's father (known only as Ancetre Manitouabeouich)[2] was Algonquian and likely born into the Western Abenaki. As mentioned earlier he was adopted by the Wendot. Grandfather was born in Huronia – more specifically he was born between 1596 and 1600 into the Attignawantan (Nation of the Bear) in the village of Ossossane.

Jesuit missionary Jean de Brebeuf provides an excellent description of the Attignawantan people. "It is quite important, in my opinion, to note that they live assembled in villages, with sometimes as many as fifty, sixty, or a hundred cabins (that is to say, three hundred to four hundred households); that they cultivate the fields, from which they obtain sufficient food to maintain themselves year-round; and that they live together in peace and friendship. I certainly believe that there is not, perhaps, under heaven a nation more praiseworthy in this respect than the Nation of the Bear. Setting aside a few evil-minded persons, such as one meets almost everywhere, they have a gentleness and affability almost incredible for savages. They are not easily annoyed, and, moreover, if they have received wrong from anyone, they often conceal the resentment they feel; at least one finds here very few who allow themselves any display of anger or vengeance.

"They maintain such perfect harmony by visiting one another frequently, by helping one another in time of sickness, and by their feasts and their marriage alliances. When they are not busy with their fields, hunting, fishing, or trading, they spend less time in their own cabins than in those of their friends. If they fall sick or desire anything for their health, there is a rivalry as to who will show himself most obliging. If they have any unusually good delicacy, as I have already said, they offer a feast for their friends and hardly ever eat it alone. In their marriages there is this remarkable custom: They never marry anyone related to them in any way, either directly or collaterally; instead, they always make new marital alliances, which is very helpful in maintaining good-will. Moreover, by this habit of visiting one another – given that they possess, for the most part, rather good minds – they arouse and influence one another wonderfully.

"Consequently, almost none of them are incapable of conversing or of reasoning very well, and in good terms, on matters within their knowledge. The councils, too, held almost every day in the villages on almost all matters, improve their capacity for speaking. Anyone who wishes may be present and may express his opinion [not her- women were not allowed in council], though it is the elders who are in the ascendant and it is their judgment that decides issues. Let it be added also that propriety, courtesy, and civility, which are, as it were, the flower and charm of ordinary human conversation, are not lacking among these people. They call a polite person *Aiendawasti.* To be sure, you do not observe among them any of those hand kissings, compliments, and vain offers of service which do not pass beyond the lips. Yet they do render certain civilities to one another and preserve, through a sense of propriety, various ceremonies in their visits, dances, and feasts. Neglect of such courtesy and propriety would lead to immediate censure, and if anyone made such blunders repeatedly, he would soon become a byword in the village and would lose all his influence.

"When they meet, the only salutation they give is to call the other by name, or say 'my friend,' 'my comrade,' or 'my uncle,' if it is an elder. If an Indian finds himself in your cabin when you are eating, and if you present him a dish of food that you have scarcely touched, he will content himself with tasting it and will hand it back to you. But if you give him a dish for himself, he will not put his hand to it until he has shared it with his companions, and the latter would usually take only a spoonful. Admittedly, these are small things, but they nevertheless demonstrate that these people are not quite so rude and unpolished as one might suppose. (From the Jesuit Relations for 1636, in Greer 2000)."

Some of the principle villages were enclosed by strong wooden palisades made of pine trees from about 15 to 16 feet tall. They were often surrounded by a mote or natural depression. The palisades were normally in three rows and interlaced with one another. Large pieces of bark were placed between the palisades to about 8 to 9 feet high. The palisades were galleries or watch towers that could be manned in time of war. They were usually well stocked with stones to be hurled at the enemy and with water to put out fires which the enemy could lite from below. These villages would be moved about every 8 to 12 years as the soil in their fields became depleted or as they ran out of firewood within a short distance from the village. "They were moved a distance of 1, 2, or 3 leagues, but if forced to move as a result of enemy attacks, the distance was greater, as when Onondaga moved some 40 or 50 leagues (Tooker 1991)."

"Although the Huron were monogamous and formed nuclear families, the matrilineal extended family appears to have been the fundamental social and economic unit. Each nuclear family had its own place within the longhouse, but the members of the longhouse worked together as a closely cooperating unit. The daily running of the longhouse was done by the women, probably under the direction of the senior matron who exercised considerable authority over her daughters even when they were married. Major decisions were made by the male head of the household, probably in consultation with his wife and family members (Heidenreich 1978)." "Lalemont [a Jesuit missionary] described life in the longhouse as 'a miniature picture of hell,' in which dust and smoke blind the eyes, children and dogs give themselves free rein, and fleas and mice get at one's person and belongings (JR 17:13-15). The Huron of course saw things quite differently. To them the longhouse represented the physical manifestation of their social and economic system. It was at this level and in these surroundings that the Huron values of family solidarity, economic cooperation, and rule by the mutual agreement of adults found their basic expression. The values of the longhouse were projected to the village through kinship ties and ultimately to the tribal level (Heidenreich 1978)."

"It can be inferred from the ethnohistoric sources that every Huron was a member of one of eight clans. The best evidence for the Huron clan system comes from the historic Wyandot. These were fictive, apparently exogamous kinship groupings tracing descent to a common female ancestor. The historic Wyandot had eight clans: Turtle, Wolf, Bear, Beaver, Deer, Hawk, Porcupine, and Snake. Since the Huron confederacy was composed of four or five territorial tribes with a defined political hierarchy, the clan system with its chiefs must have been a social organization that cut across tribal divisions. Judging by the way the Huron political system operated, the clan system must have had mainly a ceremonial function rather than a political one.

"Localized clan segments were fundamental to political organization. The ethnohistoric sources make reference to 'large families' and 'important families' each of which had a 'distinct armorial bearing'. Large villages were made up of a number of clan segments each having its own civil and war chief who represented them in their respective councils; relatives of the chiefs acted as their councilors. Succession to the chieftainship was partly inherited in that the men who could be considered for the chieftainship were 'nephews' or 'grandsons' of an existing chief and partly elected in that the other family members had to approve the succession. Some lineages within each clan segment seem to have been more important than others. These furnished the chiefs who represented the larger clan segment within the village council and at higher levels (Heidenreich 1978)."

"Religion, mythology, feasts, dances, and other customs formed integrative institutions that were an important part of Huron life. It was that they called their *onderha*, or the 'foundation, prop, or maintenance' of the country. Huron ceremonial life was in the hands of the old men and transmitted by them (Heidenreich 1978)." "Feasts were an important means of maintaining friendship and gaining status. A feast giver was an honored and respected man because he shared his good fortune with others. Usually a great deal of food was consumed, and the majority of the feasts were probably prompted by abundance and a desire to share it. There were four main types of feasts: *enditeuhwa*, to express thanksgiving and gratitude; *awataerohi*, to cure someone of disease; *athataion*, a farewell feast from a dying man to an assembly of friends; and *atouront aochien*, a singing feast given prior to going to war, when a man wished to become renowned, or when a person received the name of a deceased chief. At some feasts, called *anondahoin*, only tobacco was smoked. At all feasts it was customary that the giver of the feast ate nothing or very little and spent his time entertaining his guests with song and talk.

"Games were played for pleasure, in response to dreams, and to cure illness. The most popular games were lacrosse, played by village teams or between entire villages, and a number of gambling games played by large or small numbers of individuals with heavy betting by players as well as bystanders. Like feasts and dances, games were important integrative institutions that brought people from different social segments and villages together. The gambling games in particular were also a mechanism for redistributing food surplus and other goods (Heidenreich 1978)."

The Wendot danced for one of four reasons: (1) to propitiate the spirits who they thought conferred benefits on them, (2) to welcome someone, (3) to rejoice for some victory, or (4) to prevent or cure disease. Tooker noted that the Wendot recognized three types of illness: (1) illnesses due to natural causes that were cured by natural remedies, (2) illnesses caused by the desires of the soul of the sick person that were cured by supplying these desires [called *ondinnonk*], and (3) illnesses caused by witchcraft that were cured by extracting the sorcerer's spell. "They have a faith in dreams which surpasses all belief. If Christians were to put into execution all their divine inspirations with as much care as our Indians carry out their dreams, no doubt they would very soon become great saints. They look upon their dreams as ordinances and irrevocable decrees; to delay the execution of them would be a crime (From the Jesuit Relations for 1636, in Greer 2000)."

"When they were to dance either naked or covered by a breechcloth, in accordance with the dream of a sick person or by order of the medicine man or the chiefs, a summons was given through all the streets of the village to notify and invite the young people. They were told the day and time of the feast, the reason for having the dance, and that they should come painted and wearing the finest things they had or else dressed as specifically directed. The people in the surrounding villages received the same notice and were also invited to be present; they came, if they wished so (Tooker 1991)."

"At the time of French contact early in the seventeenth century, the Huron were embroiled in a longstanding war with the five Iroquois tribes but especially with the Seneca. Within their memory, the Huron had also been at war with the Petun and possibly the Neutral. A peace had been concluded with the Petun shortly before Champlain's arrival in Huronia, while the Neutral tried to be on peaceful terms with both the Iroquois and Huron-Petun. The Neutral in turn were allied with the

Ottawa and probably the Petun against the Assistaeronnon of Michigan. Peaceful relations existed between the Huron and Algonquian groups of Ontario and Québec with who they were in contact. The Huron had an alliance with the Andaste (Susquehannock) who lived south of the Iroquois (Heidenreich 1978)." It was likely the alliance between the Huron and Algonquin that drove the St. Lawrence Iroquois south into New York after Jacques Cartier's first exploration in 1534.

"Prior to the late 1630's when both the motives and methods of Iroquoian warfare changed as a result of the fur trade, warfare was largely motivated by blood revenge, the gaining of personal prestige, and to some extent religious ideals. Wars for hunting rights or territorial gain were not fought prior to the 1640's. While it was easy to start a war, usually through murder and subsequent refusal to pay restitution, it was exceedingly difficult to arrange a peace. On several occasions when attempts were made to conclude a peace between members of the Huron confederacy and one or all of the Iroquoian tribes, either some groups on one side refused to join in the negotiations or the negotiations or peace collapsed because of a few murders. Murder had to be avenged, and the glory of a warrior depended on the perpetuation of the annual raids (Heidenreich 1978)."

"The fate of prisoners was variable. If too many were captured, or the retreat had to be conducted in a hurry, the prisoners were killed [and as you'll soon see they were the lucky ones], usually preceded by the women and children. If taken back to Huronia, the alternatives were torture and death, or particularly for the women and children, formal adoption by families who had lost someone through warfare (Heidenreich 1978)."

"Descriptions of torture and cannibalism among the Huron indicate that this ritual involved the transformation of normal group behavior into a universal outpouring of pent-up emotions. The person who was tortured became in fact a symbol of the tribe he represented, a hate object on whom the frustrations of life and past wrongs could be expended. It is significant that as much as the Jesuits abhorred the practice of torture and ritual cannibalism they were never able to put a stop to it (Heidenreich 1978)."

Jean de Brebeuf gives a rather lengthy and detailed description of torture as practiced by the Indians. "It all begins with an individual who raises a band of young men who gather, so it seems, for the purpose of avenging a private quarrel and the death of a friend, rather than for the honor and preservation of the nation. Then, if they succeed in capturing some of their enemies, they treat them with all the cruelty they can devise. Five or six days will sometimes pass in satiating their rage, burning the prisoners over a slow fire, and, not satisfied with seeing their skins entirely roasted, they cut open the legs, the thighs, the arms, and the most fleshy parts of the body and thrust into the wounds glowing brands or red-hot hatchets. Sometimes in the midst of these torments they compel them to sing – those who have the strength do so hurling forth a thousand imprecations against those who torment them.

"On the day of their death they must repeat the performance, if they have strength, and sometimes the kettle in which they are to be boiled will be on the fire, while these poor wretches are still singing as loudly as they can. This inhumanity is altogether intolerable; indeed, many people are unwilling to attend these fatal banquets. After having at last brained the victim, if he was a brave man, they tear out his heart, roast it on the coals, and distribute the pieces of it to the young men, as they think that this makes them courageous. Some make an incision in their own neck and cause some of his blood to run into it. They say that mingling the blood in this way has the power to ensure that they can never be surprised by the enemy, as they will always be aware of their approach, however secret it may be.

"They put him in the kettle piece by piece, and although at other feasts, the head – whether of a bear, a dog, a deer, or a large fish – is reserved for the captain, in this case the head is given to the lowest person in the company. Indeed, some people are horrified to taste this part, or any other portion of the body, whereas others eat it with pleasure. I have heard Indians in our cabin speak with gusto of the flesh of an Iroquois and praise its quality in the same terms as they would praise the flesh of a deer or a moose (From the Jesuit Relations for 1636, in Greer 2000)."

"The Huron did not fear death, believing that they went to an afterlife that was not too different from that of the living. A person who was dying was often prepared for death prior to the actual event. The dead person was placed on a mat in a flexed position and a chief announced the death to the village and friends in neighboring villages. While lying in state the body was attended by members of a clan or clan segment other than the dead person's own. If the deceased had been an important person, the chiefs from other villages would attend. Lamentations would begin as soon as a person died. At the same time various people who had known the deceased would give speeches extolling his virtues. On the third day after death, the chief in charge of the ceremonies would announce a feast at which the soul of the deceased took part. During the feast gifts were exchanged and heaped upon the deceased. Most of the gifts were given to the bereaved and to the person who directed the funeral. The body was placed on a scaffold 8 to 10 feet above the ground accompanied by some grave goods. No one spared himself in giving gifts; Brebeuf claimed that some funerals were so expensive that it seemed as if peopled worked and traded to acquire grave goods.

"Those killed in war or by drowning were buried below the ground. A shrine was built over the grave and encircled by a fence. Children less than one month of age were buried on a path or some other frequented place so that their spirit could rise and reenter the womb of a woman who passed that way.

"General mourning for the deceased lasted for about 10 days, but the wife or husband of the deceased was expected to mourn for one year. During that time they would not participate in any feasts or festivals, nor would they remarry.

"The burials described above were only temporary. About every 8 to 12 years the Huron would disinter all single burials and prepare them for the most important ceremony of all, the great Feast of the Dead. On the whole, the feast seems to have been a tribal affair in which the villages got together and buried their dead in a common grave. This was not only a religious ceremony but also an occasion to symbolize tribal union through common burial to renew friendship with the living and the dead. The Feast also acted as a general catharsis of personal grief (Heidenreich 1978)."

Jean de Brebeuf provides a rather lengthy and detailed description of the Feast of the Dead (in Greer 2000). I'll extract the most pertinent parts.

In each village they choose a fair day and proceed to the cemetery, where those called Aiheonde, who had charge of the funerals, take the bodies from the tombs in the presence of the relatives, as the latter renew their tears and feel afresh the grief they had on the day of the funeral...

The flesh of some is quite gone, and there is only a sort of parchment on their bones; other bodies look as if they had been dried and smoked, and show scarcely any signs of putrefaction; and still others are swarming with worms. When the relatives have gazed upon the bodies to their satisfaction, they cover them with handsome new beaver robes. Finally, after a time, they strip them of their flesh, taking off the skin and flesh that they throw into the fire along with the robes and mats in which the bodies were wrapped. They leave the bodies of the recently dead in the state in which they are, simply covering them with new robes...

The bones having been well cleaned, they put some of them into bags, some into fur robes, loaded them on their shoulders, and covered these packages with another beautiful hanging robe. As for the whole bodies, they put them on a sort of litter and carried them with all the others, each into his cabin, where each family made a feast of [made for, not of] its dead...

A day or two before setting out for the feast, they carried all these souls into one of the largest cabins of the village, where some of them were hung on the cabin poles and the rest were put on display inside...

The seven or eight days before the feast were spent in assembling the souls, as well as the foreigners who had been invited. Meanwhile, from morning until night, the living were continually making presents to the young people, in consideration of the dead. In one place, women were shooting with the bow for a prize, which was a quillwork belt, or a necklace or string of beads. In other parts of the village, the young men were throwing batons. The prize for victory in this contest was an ax, some knives, or even a beaver robe. Every day more souls arrived. It is heartening to see these processions, sometimes of two or three hundred persons, everyone carrying their souls – in other words, their bones – done up in parcels on their backs and under a handsome robe, as I indicated above. Some had arranged their parcels in the form of a man, ornamented with beaded necklaces and elegant garlands of long red fur. On setting out from the village, the whole band cried out haee, hae, and repeated this cry of the souls along the way. According to them, this cry greatly consoles them. Without it, this burden of souls would weigh very heavily on their backs and cause them a pain in the sides for the rest of their lives...

The bundled-up souls were then taken down [from the cabin poles], the relatives unwrapped them to say their last adieus, and then the tears flowed afresh. I admired the tenderness of one woman, the daughter of a captain who was a very important man in the country before he died at an advanced age, toward her dead father and children. She combed the hair of her father and handled his bones, one after the other, with as much affection as if she wished to restore him to life. She placed by his side his Atsatonewai, that is, his package of council sticks, which are all the books and papers of the country. As for her little children, she put beaded bracelets on their arms and bathed their bones with her tears. The others could scarcely tear her away, but they insisted, as it was time to depart. The man who bore the body of this old captain walked at the head; the men followed and then the women, walking in this order until they reached the pit.

The great pit was about 10 feet deep and five fathoms (30 feet) in diameter.

All around it was a scaffold and a well-constructed sort of theater, nine or ten fathoms in diameter [54 – 60 feet] and nine or ten feet high. Above this staging rose many poles, trimmed and well arranged, with cross poles to attach the bundles of souls. Since the whole bodies were to go in the bottom of the pit, they had been placed the preceding day at the edge of the hole, under the scaffold, stretched out upon bark or mats fastened to stakes at about the height of a man.

People arrived with their corpses at about one in the afternoon and divided themselves into their different cantons, and subdivided by family and village. They lay on the ground their parcels of souls, almost as earthen pots are displayed [in France] at a village fair. They unfolded also their bundles of robes and all presents they had brought, and placed them upon the poles, which extended from 500 to 600 fathoms [2,500 to 3,000 feet], so there were as many as twelve hundred presents. These remained thus on exhibition two full hours, in order to give foreigners time to view the wealth and magnificence of the country. I did not find the crowds as numerous as I had expected; probably there were no more than two thousand persons. At about three o'clock, everyone put away their various articles and folded up their robes.

Meanwhile, each captain took his turn to give the signal [to his people], and immediately they came running, all loaded with their bundles of souls, as if assaulting a town. They climbed to the theater on the ladders hung all around it, and hung their packages on the poles, each village having its designated section. That done all the ladders were taken away, but a few captains remained there and spent the rest of the afternoon until seven o'clock announcing the presents, which were made in the name of the dead to certain specified persons.

This, said they, is what such and such a dead man gives to such and such a relative. About five or six o'clock, they lined the bottom and sides of the pit with fine, large, new robes, each of ten beaver skins, in such a way that the robes extended more than a foot out of it. As they were preparing the robes that were to be employed for this purpose, some people went down to the bottom and brought up handfuls of sand. I asked what this ceremony meant and learned that they believe that this sand helps them win at games of chance. Of those twelve hundred presents that had been displayed, forty-eight robes served to line the bottom and sides of the pit; and each whole body, besides the robe in which it had been wrapped, had another one, and sometimes even two more, to cover it . . .

At seven o'clock, they took the whole bodies down into the pit. We had the greatest difficulty at getting near. Nothing has ever depicted for me so well the confusion there is among the damned. On all sides you could have seen them unloading half-decayed bodies, and on all sides was heard a horrible din of confused voices of people speaking without making themselves understood. Ten or twelve men were in the pit arranging the bodies side by side all around them. In the very middle of the pit, they put three large kettles, which could be of use only to souls: One had a hole through it, another had no handle, and a third was in no better condition. I saw very few beaded necklaces, though it is true that they put many on the bodies. This is all that was done on that day.

All the people passed the night on the spot. They lighted many fires to cook their meals. We withdrew for the night to the old village, intending to return the next morning at daybreak, when they were to throw the bones into the pit. But an accident occurred, and, as a result, we did not arrive in time, though we made great haste. One of the souls, which was not securely tied or was perhaps too heavy for the cord that fastened it, had fallen into the pit. The noise awakened the company, who immediately ran and crowded up on the scaffold, where they emptied indiscriminately all the bundles into the pit, though they kept the robes in which they had been wrapped. We had only set out from the village at that time, but the noise was so great that it seemed almost as if we were there. As we drew near, we saw nothing less than an image of Hell. The large space was quite full of fires and flames, and the air resounded in all directions with the confused voices of these barbarians. Eventually, the noise ceased and they began to sing, but in tones so sorrowful and lugubrious that it repre-

sented to us the horrible sadness and the abyss of despair into which these unhappy souls are forever plunged.

Almost everything had been thrown in by the time we arrived, for it was done almost in the blink of an eye; everyone had been in a great hurry, thinking there would not be room enough for all the souls. However, we saw enough of it to judge the rest. There were five or six men in the pit, using poles to arrange the bones. When the pit was within about two feet of being full, they turned back over the bones the robes that bordered the edge of the pit, and covered the remaining space with mats and bark. Then, in no special pattern, they heaped the pit with sand, poles, and wooden stakes. Some women brought some dishes of corn, and that day, and the following days, several cabins of the village provided baskets full of it, which were thrown upon the pit.

The whole morning passed in gift giving. Most of the robes in which the souls had been wrapped were cut into pieces and thrown from the height of the theater into the midst of the crowd, for anyone who could get them. It was very amusing when two or three people seized hold of a beaver skin, for then it had to be cut into as many pieces, and thus they found themselves almost empty-handed, for the fragment was scarcely worth the picking up. I admired the enterprising ingenuity of one Indian: Instead of running after these flying pelts, he would approach people disputing possession of a skin and offer them some tobacco, a precious commodity in short supply in the country this year, and thus he settled the matter to his own profit and advantage.

***

"The French had a unique approach to colonization. Partly because they came to the New World in comparatively small numbers and partly because they made their fortunes in Canada trading for furs with native hunters, they had come to terms with native cultures and interact exclusively with Indian peoples. The French fully occupied only a narrow territory in the St. Lawrence Valley centering on the forts (later towns) of Québec (1608), Three Rivers (1634), and Montreal

(1642). Trade with peoples living far to the west was facilitated by the waterways of the Great Lakes, and because natives generally traded only with friends, this commercial penetration led to the formation of military-diplomatic alliances that helped extend French influence over vast regions. A great inland empire took shape in the second half of the seventeenth century, but it was an empire not of domination and subjection, but of multistranded commercial/diplomatic links between France and the different Indian nations of the Canadian hinterland.

"Farther to the south, the English settled in larger numbers, and they came to depend heavily on agriculture for their food supply and export commodities. The New England and Chesapeake colonies developed a voracious appetite for Indian lands, an appetite that eventually led to war, extermination, and displacement. Whereas antagonism between natives and settler societies was a fundamental feature of colonization in English North America, in New France there were patterns of both cooperation and conflict. The French were not inherently kinder empire builders than the English, but their settlement of Canada did not require any substantial appropriation of Indian lands.

"Nor was New France built on the subjugation of Indians through military conquest, as the Spanish colonies of South and Central America were. In Spain's new world empire, Indians owed the colonizers labor service and other forms of tribute as both a symbol of their subjugation and a contribution to the Spanish economy. But in New France, the Indians were never really conquered. Violence and war were by no means absent from the colony's history, since military alliance naturally entailed conflict with nations outside the alliance, but the French were never in a position to impose their will by force of arms on the Indians in general. Allies, trading partners, and even, to some extent, defeated enemies, remained autonomous, though many of them came to recognize a certain French hegemony. Consequently, Jesuit missionaries had to work with natives who generally retained a high degree of independence, regardless of how entangled they became in the French network of alliance and trade (Greer 2000)."

In 1646 a young man stepped off the ship at Québec. Étienne de Lessart came to Canada to start a shipping business with his partner Martin Grouvel, a navigator from Québec. He had business relations with a number of merchants, among them Charles Audet, Mr. de la Chesnaie of Québec and David Aiaille, and Mr. St-Meur in Rochelle, France. Etienne was greeted at the dock by Olivier le Tardiff, a representative of Samuel de Champlain, and his scout Roch Manitouabeouich. Roch recalls a story his father told him about the first whites in Canada.

A great many years ago, when men with a white skin had never been seen in this land, some Indians who were out fishing, at a place where the sea widens, espied at a great distance something remarkably large floating on the water, and such as they had never seen before. These Indians immediately returning to the shore, apprised their countrymen of what they had observed, and pressed them to go out with them and discover what it might be. They hurried out together, and saw with astonishment the phenomenon which now appeared to their sight, but could not agree upon what it was; some believed it to be an uncommonly large fish or animal, while others were of the opinion that it must be a very big house floating on the sea. At length the spectators concluded that this wonderful object was moving towards the land, and that it must be an animal or something else that had life in it; it would therefore be proper to inform all the Indians on the inhabited islands of what they have seen, and put them on their guard. Accordingly they sent off a number of runners and watermen to carry the news to their scattered chiefs that they might send off in every direction for the warriors, with a message that they should come immediately. These arriving in numbers, and having themselves viewed the strange appearance, and observing that it was actually moving towards the entrance of the river or bay; concluded it to be a remarkably large house in which Mannitto (the Great or Supreme Being) himself was present, and that he probably was coming to visit them.

Of course he now knew that it wasn't the "Great or Supreme Being" but more likely Basque fishermen who were plying their trade on the east coast long before the first explorations of Jacques Cartier and Samuel de Champlain. What he may have sensed is that these Frenchmen now arriving, like Étienne, were coming to stay and they would significantly affect his and his peoples lifestyle indefinitely.

Étienne would receive from le Tardiff a concession at what would later become Beaupré. He would donate a piece of this land to build a chapel at Beaupré. This site would become the site of the Basilica of St. Anne de Beaupré. He would go on to farm a part of his concession while continuing his trading on the St. Lawrence River.

Grandfather would be baptized with Olivier le Tardiff as his godfather. He would be given the Christian name Roch after Saint Roch. Le Tardiff would adopt Roch's daughter and have her raised with the Ursuline nuns as a French women. She would be baptized Marie Sylvestré (Marie of the forest). Marie would marry Martin Prevost. This would be the first marriage of a Frenchman and a Native American woman.

It is with these two great-grandfathers, Roch Manitouabeouich and Étienne de Lessart that I began the search for my ancestry and that inspired this book.

Grandfather's Tale

The Origin of the Buffalo

Long ago, a tribe of Cheyenne hunters lived at the head of a rushing stream, which eventually emptied into a large cave.

Because of the great need for a new food supply for his people, the Chief called a council meeting.

"We should explore the large cave," he told his people. "How many brave hunters will offer to go on this venture? Of course, it may be very dangerous, but we have brave hunters." No one responded to the Chief's request.

Finally, one young brave painted himself for hunting and stepped forth, replying to the Chief, "I will go and sacrifice myself for our people."

He arrived at the cave, and to his surprise, First Brave found two other Cheyenne hunters near the opening, where the stream rushed underground.

"Are they here to taunt me," First Brave wondered? "Will they only pretend to jump when I do?"

But the other two braves assured him they would go.

"No, you are mistaken about us. We really do want to enter the cave with you," they said.

First Brave then joined hands with them and together they jumped into the huge opening of the cave. Because of the darkness, it took some time for their eyes to adjust. They then discovered what looked like a door. First Brave knocked, but there was no response. He knocked again, louder.

"What do you want, my brave ones?" asked an old Indian grandmother as she opened the door.

"Grandmother, we are searching for a new food supply for our tribe," First Brave replied. "Our people never seem to have enough food to eat."

"Are you hungry now?" she asked.

"Oh, yes, kind grandmother, we are very hungry," all three braves answered.

The old grandmother opened her door wide, inviting the young braves to enter.

"Look out there!" She pointed for them to look through her window. A beautiful wide prairie stretched before their eyes. Great herds of buffalo

were grazing contentedly. The young hunters could hardly believe what they saw!

The old grandmother brought each of them a stone pan full of buffalo meat. How good it tasted, as they ate and ate until they were filled. To their surprise, more buffalo meat remained in their stone pans!

"I want you to take your stone pans of buffalo meat back to your people at your camp," said the old grandmother. "Tell them that soon I will send some live buffalo."

"Thank you, thank you, thank you, kind grandmother," said the three young Cheyenne braves.

When the young hunters returned to their tribe with the gifts of buffalo meat, their people rejoiced over the new, good food. Their entire tribe ate heartily from the old grandmother's three magic pans, and were grateful.

When the Cheyenne's waked at dawn the next day, herds of buffalo had mysteriously appeared, surrounding their village! They were truly thankful to the old Indian grandmother and to the Sky Spirits for their good fortune.

Bibliography

Adovasio, J.M. and Jake Page. 2002. *The First Americans.* The Modern Library. New York.

Arsuaga, Juan Luis. 2001. *The Neanderthal's Necklace.* Four Walls Eight Windows. New York.

Bartlein, P.J., P.M. Anderson, M.E. Edwards, and P.F. McDowell. 1991. A Framework for Interpreting Paleoclimatic Variations in Eastern Beringia. Quat. Int. 10-12: 73-83.

Bell, Larry. 2011. *Climate of Corruption.* Green Leaf Book Group Press. Austin, TX.

Berger et al. 2010. *Australopithecus sediba*: A New Species of Homo-like *Australopiths* from South Africa. Science 328:195-204.

Blackmore, Susan. 1999. *The Meme Machine.* Oxford Univ. Press. Oxford.

Bonnichsen, Robson. 2005. An Introduction to Paleoamerican Origins. *In* Michael R. Waters and Ruth Gruhn, (eds.). *Paleoamerican Origin: Beyond Clovis.* Texas A & M University Press. College Station, Texas.

Boyd, Robert and Peter J. Richardson. 1996. Life in the Fast Lane: Rapid Cultural Change and the Human Evolutionary Process. *In* Jean-Pierre Changeux and Jean Chavaillon (eds.). *Origins of the Human Brain.* Clarendon Press. Oxford.

Burns, James A. (1996). Vertebrate Paleontology and the Alleged Ice-free Corridor: the Meat of the Matter. Quaternary International 32: 107-112.

Cavalli-Sforza, Luigi Luca. 2000. *Genes, People, and Languages*. University of California Press. Berkeley. Los Angles.

Cochran, Gregory and Henry Harpening. 2009. *The 10,000 Year Explosion: How Civilization Accelerated Human Evolution.* Perseus Book Group. New York.

Coe, Michael, Dean Snow, and Elizabeth Benson. 1987. *Atlas of Ancient America.* Facts on File. New York and Oxford.

Cohen, Mark Nathan. Undated. Origin of Agriculture. Answers.com.

Corballis, Michael C. 2011. *The Recursive Mind: The Origins of Human Language, Thought, and Civilization.* Princeton Univ. Press. Princeton and Oxford.

Darwin, Charles. *The Origin of the Species and The Decent of Man*. Random House, Toronto.

Dawkins, Richard. 1976. *The Selfish Gene*. Oxford University Press. Oxford.

Dawkins, Richard. 2004. The *Ancestor's Tale: A Pilgrimage to the Dawn of Evolution*. A Mariner Book, Houghton Mifflin Company, Boston and New York.

Dawkins, Richard. 2009. *The Greatest Show on Earth*. Free Press. New York.

Diamond, Jared. 2006. *The Third Chimpanzee: The Evolution and Future of the Human Animal.* Harper Perennial. New York, London, Toronto, Sidney.

Dickason, Olive Patricia. 1992. *Canada's First Nations*. University of Oklahoma Press. Norman.

Dillehay, Thomas D. 2000. *The Settlement of the Americas: A New Prehistory*. Basic Books. New York.

Dillehay, Tom D. 2002. Plant Food and its Implications for the Peopling of the New World: A View from South America. *In* Nina G. Jablonski (ed.). *The Fist Americans: The Pleistocene Colonization of the New World*. Memoirs of the California academy of Sciences Number 27. San Francisco, California.

Dixon, E. James. 1992. *Quest for the Origins of the First Americans*. The University of New Mexico Press. Albuquerque.

Dixon, E. James. 1999. *Bones, Boats, & Bison: Archeology and the First Colonization of Western America*. The University of New Mexico Press. Albuquerque.

Douglas, John and Linda Brown. 2002. The Migrations and Adaptations of the First Americans: Clovis and Pre-Clovis Viewed from South America. *In* Nina G. Jablonski (ed.). *The First Americans: The Pleistocene Colonization of the New World*. Memoirs of the California academy of Sciences Number 27. San Francisco, California.

Doidge, Norman, M.D. 2007. The Brain That Changes Itself. Penguin Books. London.

Duffy, Geoffrey G. http://www.climaterealists.org.nz/node/601

Dunbar, Robin. 2010. *Grooming, Gossip and the Evolution of Language*. Faber and Faber. London.

Edmonds, Margot and Ella E. Clark. 1989. *Voices of the Winds - Native American Legends*. Facts On File, Inc. New York.

Elias, Scott A. 2002. Setting the Stage: Environmental Conditions in Beringia as People Entered the New World. *In* Nina Jablonski (ed.). *The First Americans: The Pleistocene Colonization of the New World.* Memoirs of the California academy of Sciences Number 27. San Francisco, California.

Erlandson, Jon M. 2002. Anatomically Modern Humans, Maritime Voyaging, and the Pleistocene Colonization of the Americas. *In* Nina G. Jablonski (ed.). *The Fist Americans: The Pleistocene Colonization of the New World.* Memoirs of the California academy of Sciences Number 27. San Francisco, California.

Fagan, Brian. 1990. *The Journey from Eden.* Thames and Hudson Inc. New York.

Fagan, Brian. 2003. *The Great Journey.* university Press of Florida. Gainesville.

Fagan, Brian. 2010. *Cro-Magnon.* Bloomsbury Press. New York. Berlin. London.

Fiedel, Stuart J. 2005. Rapid Clovis Colonization of the America's: Chronological Evidence and Archaeological Analogues. *In* Michael R. Waters and Ruth Gruhn,(eds.). *Paleoamerican Origin: Beyond Clovis.* Texas A & M University Press. College Station, Texas.

Finlayson, Clive. 2009a. *The Humans That Went Extinct.* Oxford University Press.

Finlayson, Clive. 2009b. *Why Neanderthals Died Out and We Survived.* Oxford University Press.

Fischer, David Hackett. 2008. *Champlain's Dream.* Simon & Schuster. New York.

Gill, George W. 2002. Appearance of the "Mongoloid Skeletal Trait Complex" in the Northwestern Great Plains: Migration, Selection, or Both? *In* Michael R. Waters and Ruth Gruhn, (eds.). *Paleoamerican Origin: Beyond Clovis*. Texas A & M University Press. College Station, Texas.

Goddard, Ives. 1978. Central Algonquian Languages. *In* Bruce G. *Handbook of North American Indians. Volume 15. Northeast.* Smithsonian Institution. Washington.

Goddard, Ives. 1996. Introduction. *In* Ives Goddard (ed.) *Handbook of North American Indians. Volume 17. Languages.* Smithsonian Institution. Washington.

Goodyear, Albert C. 2005. Evidence of Pre-Clovis Sites in the Eastern United States. *In* Michael R. Waters and Ruth Gruhn, (eds.). *Paleoamerican Origin: Beyond Clovis*. Texas A & M University Press. College Station, Texas.

Green, Richard. 2010. A Draft Sequence of the Neandertal Genome. Science 328: 710-722.

Greenberg, Joseph H. 1987. *Languages in the Americas*. Stanford, California.

Haynes, Vance C., Jr. 2005. Clovis, Pre-Clovis, Climate Change and Extinctions. *In* Michael R. Waters and Ruth Gruhn, (eds.). *Paleoamerican Origin: Beyond Clovis*. Texas A & M University Press. College Station, Texas.

Houser, Marc. 2009. The Mind. Scientific American Sept 2009: 44-51.

Hurt, R. Douglas. 1987. *Indian Agriculture in America*. University Press of Kansas. Lawrence.

Johnson, Allen W. and Timothy Earle. 2000. *The Evolution of Human Societies: From Foraging to Agrarian State.* Stanford Univ. Press. Stanford.

Kenneally, Christine. 2007. *The First Word: Search for the Origin of Language.* Viking Penguin. London.

Kingdon, Jonathan. 2003. *Lowly Origins: Where, When and Why Our Ancestors First Stood Up.* Princeton University Press. Princeton and Oxford.

Klein, Richard G. and Blake Edgar. 2002). *The Dawn of Human Culture.* John Wiley & Sons, Inc. New York

Leakey, Richard. 1994. *The Origin of Humankind.* Basic Books, a division of HarperCollins Publishers, New York.

Lovejoy, Owen. 1981. The Origin of Man. Science 211: 341-350.

Lovejoy, C. Owen. 2009. Reexamining Human Origins in Light of *Ardipithecus ramidus.* Science 326: 74.

Lovejoy, C. Owen, Bruce Latmer, Gen Suwa, Berhane Asfaw, Tim D. White. 2009a. Combining Prehension and Propulsion: The Foot of *Ardipithecus ramidus.* Science 326: 72.

Lovejoy, C. Owen, Scott W. Simpson, Tim D. White, Berhane Asfaw, Gen Suwa. 2009b. Careful Climbers in the Miocene: The Forelimbs of *Ardipithecus ramidus* and Humans are Primitive. Science 326: 70.

Lovejoy, C. Owen, Gen Suwa, Scott W. Simpson, Jay H. Matterns, Tim D. White. 2009c. The Great Divides: *Ardipithecus ramidus* Reveals The Postcrania of Our Last Common Ancestors with African Apes. Science 326: 73.

Lynch, Gary and Richard Granger. 2008. *Big Brain: The Origins and Future of Human Intelligence.* Palgrave Macmillan. New York.

Marean, Curtis W. 2010. When the Sea Saved Humanity. Scientific American, August 2010. pp. 54-61.

Martin, Paul S. 2005. *Twilight of the Mammoths: Ice Age Extinction and the Rewilding of America.* University of California Press. Berekely. Los Angeles. London.

Meltzer, David J. 2002. What Do You Do When No One's Been There Before? Thoughts on the Exploration and Colonization of New Lands. *In* Nina G. Jablonski (ed.). *The First Americans: The Pleistocene Colonization of the New World.* Memoirs of the California academy of Sciences Number 27. San Francisco, California.

Merriwether, Andrew D. 2002. A Mitochondrial Perspective on the Peopling of the New World. *In* Nina G. Jablonski (ed.). *The Fist Americans: The Pleistocene Colonization of the New World.* Memoirs of the California academy of Sciences Number 27. San Francisco, California.

Morwood, Mike and Penny Van Oosterzee. 2007. *A new Human.* Left Coast Press. Walnut Creek, California.

Nichols, Johanna. 2002. The First American Languages. *In* Nina G. Jablonski (ed.). *The First Americans: The Pleistocene Colonization of the New World.* Memoirs of the California academy of Sciences Number 27. San Francisco, California.

Olson, Steve. 2002. *Mapping Human History: Genes, Race and Our Common Origins.* A Mariner Book. Houghton Mifflin Company. Boston. New York.

Page, Jake. 2003. *In the Hands of the Great Spirit.* Free Press. New York.

Pielou, E.C. 1991. *After the Ice Ages: The Return of life to Glaciated North America.* The University of Chicago Press. Chicago.

Pinker, Steven. 1996. Facts About Human Language Relevant to its Evolution. *In* Jean-Pierre Changeux and Jean Chavaillon (eds.). *Origins of the Human Brain.* Clarendon Press. Oxford.

Pinker, Steven. 1994. *The Language Instinct: How the Mind Creates Language.* Harper Perennial Modern Classics. New York, London, Toronto, Sydney.

Pinker, Steven. 2009. *How the Mind Works.* W.W. Norton and Company. New York, London.

Pringle, Heather, 1996. *In Search of Ancient America.* John Wiley & Sons, Inc. New York.

Renfrew, Colin. 2008. *Prehistory: The Making of the Human Mind.* Random House, Inc. New York

Ridley, Matt. 2003a. *The Red Queen.* Harper-Collins Publisher. New York.

Ridley, Matt. 2003b. *Nature Via Nurture.* Harper Collins Publishers Inc. New York.

Schurr, Theodore G. 2005. Tracking Genes through Time and Space: Changing Perspectives on New World Origins. *In* Michael R. Waters and Ruth Gruhn, (eds.). *Paleoamerican Origin: Beyond Clovis.* Texas A & M University Press. College Station, Texas.

Sheets-Johnstone, Maxine. 1990. *The Roots of Thinking.* Temple Univ. Press, Philadelphia.

Smith, Bruce D. 1989. Origins of Agriculture in Eastern North America. Science 246: 1556-1571.

Stanford, Dennis and Bruce Bradley. 2002. Ocean Trails and Prairie Paths? Thoughts about Clovis Origins. *In* Nina G. Jablonski (ed.). *The First Americans: The Pleistocene Colonization of the New World.* Memoirs of the California academy of Sciences Number 27. San Francisco, California.

Stanford, Dennis and Bruce A. Bradley. 2012. Across Atlantic Ice: The Origin of America's Clovis Culture. University of California Press. Berkeley.

Stanford, Dennis, Robson Bonnichsen, Betty Meggers, and D. Gentry Steele. 2005 Paleoamerican Origins: Models, Evidence, and Future Directions. *In* Michael R. Waters and Ruth Gruhn, (eds.). *Paleoamerican Origin: Beyond Clovis.* Texas A & M University Press. College Station, Texas..

Steele, Gentry D. and Joseph F. Powell. 2002. Facing the Past: A View of North American Human Fossil Record. *In* Nina G. Jablonski (ed.). *The First Americans: The Pleistocene Colonization of the New World.* Memoirs of the California academy of Sciences Number 27. San Francisco, California.

Stringer, Christopher and Robin McKie. 1996. *African Exodus: The Origins of Modern Humanity.* A John Macrae Book. Henry Holt and Company, New York.

Stringer, Chris. 2012. *Lone Survivors: How We Came to be the Only Humans on Earth.* Henry Holt and Company, LLC. New York

Swisher, Carl, Garniss Curtis, and Rodger Lewin. 2000. *Java Man.* Scribner. New York. London. Toronto. Sydney. Singapore.

Taylor, C.R. and V.J. Rowntree. 1973. Running on two or four legs: which consumes more energy? Science 179:186-187.

Tooker, Elizabeth. 1991. *The Ethnology of the Huron Indians 1615 - 1649.* Syracuse University Press. Syracuse.

Toth, Nicolas. 1987. The first technology. Scientific American, April 1987:112-121.

Turner, Christy G. II. 2002. Teeth, Needles, Dogs, and Siberia: Evidence for the Colonization of the New World. *In* Nina G. Jablonski (ed.). *The First Americans: The Pleistocene Colonization of the New World.* Memoirs of the California academy of Sciences Number 27. San Francisco, California.

Wade, Nicholas. 2006. *Before the Dawn*. The Penguin Press. New York.

Weiner, Jonathan. 1999. *Time, Love, Memory.* Alfred A. Knopf. New York.

Wells, Spenser. 2002. *The Journey of Man: A Genetic Odyssey.* Random House. Princeton, New Jersey.

Wells, Spencer. 2007. *Deep Ancestry.* National Geographic. Washington D.C.

White, Tim D., Stanley Ambrose, Gen Suwa, Denise F. Su, David De-Gusta, Raymond L Bernor, Jean-Renaud Boisserie, Michel Brunet, Eric Delson, Steven Frost, Nuria Garcia, Ioannis X. Giaourtsakis, Yohannes Haile-Selassie, F. Clark Howell, Thomas Lehmann, Andossa Likius, Cesur Pehlevan, Haruo Saegusa, Gina Semprebon, Maek Teaford, Elizabeth Vrba. 2009a. Macrovertebrate Paleontology and the Pliocene Habitat of *Ardipithecus ramidus*. Science 326:67.

White, Tim D., Berhane Asfaw, Yonas Beyene, Yohannes Haile-Selassie, C. Own Lovejoy, Gen Suwa, Giday WoldeGabriel. 2009b. *Ardipithecus ramidus* and the Paleobiology of Early Hominids. Science 326: 64.

Wilson, Peter J. 1980. *Man the Promising Primate: The Conditions of Human Evolution*. Yale University.

Wilson, P.J. 1988. *The Domestication of the Human Species*. Yale Univ. Press. New Haven.

Wong, Kate. 2009. Rethinking the Hobbit of Indonesia. Scientific American, Nov 2009: 66-73.

Wong, Kate. 2010a. Did Neandertals Think Like Us? Scientific American, June 2010:72 -75.

Wong, Kate. 2010b. Our Inner Neandertal. Scientific American, July 2010:18-20.

Wright, Robert. 1994. *The Moral Animal*. Vintage Books. New York.